CLYMER®

CHRYSLER

OUTBOARD SHOP MANUAL
3.5-140 HP • 1966-1984

The world's finest publisher of mechanical how-to manuals

CLYMER®

P.O. Box 12901, Overland Park, KS 66282-2901

Copyright ©1991 Penton Business Media, Inc.

FIRST EDITION
First Printing March, 1985
Second Printing November, 1985

SECOND EDITION
First Printing July, 1986
Second Printing November, 1986

THIRD EDITION
First Printing July, 1987

FOURTH EDITION
First Printing March, 1988
Second Printing August, 1988
Third Printing April, 1989
Fourth Printing January, 1990
Fifth Printing November, 1990

FIFTH EDITION
First Printing November, 1991
Second Printing May, 1993
Third Printing September, 1994
Fourth Printing August, 1996
Fifth Printing May, 1998
Sixth Printing August, 2000
Seventh Printing November, 2002
Eighth Printing May, 2004
Ninth Printing May, 2010

Printed in U.S.A.

CLYMER and colophon are registered trademarks of Penton Business Media, Inc.

ISBN-10: 0-89287-551-8

ISBN-13: 978-89287-551-1

Library of Congress: 91-57943

Special thanks to Marine Specialties, Sun Valley, California, and Ken's Boat Center, Burbank, California.

Chapter Two tools courtesy of Thorsen Tool, Dallas, Texas. Chapter Two test equipment courtesy of Dixson, Inc., Grand Junction, Colorado.

TECHNICAL ASSISTANCE: Tom Reha, Mason Motors, Pasadena, California.

TECHNICAL ILLUSTRATIONS: Mitzi McMarthy with additional illustrations courtesy of Chrysler Marine.

COVER: Photographed by Michael Brown Photographic Productions, Los Angeles, Calfornia. Assisted by Bill Masho.
- *Driver-Dubie.*
- *Model-Christine Parks.*
- *Photo boat courtesy of Cypress Gardens, Florida, and driven by Mike Monts do Oca.*
- *Facilities and boats arranged for by Marine Division, Crlysler Corporation.*

PRODUCTION: Shirley Renicker.

Publisher Ron Rogers

EDITORIAL

Editorial Director
James Grooms

Editor
Steven Thomas

Associate Editor
Rick Arens

Authors
Michael Morlan
George Parise
Ed Scott
Ron Wright

Technical Illustrators
Steve Amos
Errol McCarthy
Mitzi McCarthy
Bob Meyer

SALES

Sales Manager
Matt Tusken

CUSTOMER SERVICE

Customer Service Manager
Terri Cannon

Customer Service Representatives
Karen Barker
Dinah Bunnell
April LeBlond
Suzanne Myers
Sherry Rudkin

PRODUCTION

Group Production Manager
Dylan Goodwin

Production Manager
Greg Araujo

Production Editors
Holly McComas
Adriane Roberts

Associate Production Editor
Kendra Lueckert

Graphic Designer
Jason Hale

P.O. Box 12901, Overland Park, KS 66282-2901 • 800-262-1954 • 913-967-1719

More information available at *clymer.com*

Contents

QUICK REFERENCE DATA

RECOMMENDED LUBRICANTS AND SEALANTS

Chrysler lubricant	Use	Chrysler part No.
Marine Lubricant (Rykon No. 2EP)	All-purpose waterproof marine grease	T 2961
Marine Gear Lubricant	Extreme pressure E.P. 90 gearcase lubricant	597
Outboard Multi-Purpose Lubricant	Bearings and other moving parts	5H059
Anti-seize Compound	Propeller shaft splines	T 2987-1
LPSI Lubricant	Lubricating water displacer	T 8969
Loctite No. 75	—	T 8936-1
Loctite H (non-hardening)	Screw threads	T 2962-1
Loctite D (hardening)	Screw threads	T 2963-2
Locquic	Cleaner	T 8935
RTV sealant	Power head assembly	T 8983
EC-750 Industrial sealant	Main bearing seal grooves	T 8955
Gasoila	Screw threads	T 2960
Sealant primer	—	T 8935

SPARK PLUG RECOMMENDATIONS[1]

Model	Champion spark plug No.[2]
3.5 (1965-1969) and 3.6 hp	H8J
4.9 and Series 54/55 5 hp	L10
3.5 (1977-on), 4, 6 (1979-on) and 7.5 hp	L86
20 hp (Series 2000)	J4J
All Magnapower II models, 90, 1983 115, 125 and 140 hp Charger	UL18V
All other 2-cylinder models	L4J
All other 3-cylinder models	L20V

1. All conventional spark plugs are gapped to 0.030 in.
2. 50 hp factory recommendation is AC M42K plug.

SPARK PLUG CROSS-REFERENCE CHART[1]

Champion	NGK	AC	Autolite
H8J	—	—	2675
J4J[2]	B8S	M42K	353
L4J	B8S	M42FF	2635
L10	B5HS	45F	536
L20V	BUH	V40FFM	2852
L86	B6HS	—	414
UL18V	BUHX	VB40FFM	2892

1. Spark plugs produced by different manufacturers do not have identical heat range characteristics. If the spark plug recommended by the engine manufacturer is not available, use this chart to determine a suitable equivalent plug.
2. Champion J4J and J6J plugs are superceded by J6C; no cross-reference is available for J6C.

SELF-DISCHARGE RATE

Temperature	Approximate allowable self-discharge Per Day for first 10 days (specific gravity)
100° F (37.8° C)	0.0025 points
80° F (26.7° C)	0.0010 points
50° F (10.0° C)	0.0003 points

BATTERY CAPACITY (HOURS)

Accessory draw	Provides continuous power for:	Approximate recharge time
80 AMP-HOUR BATTERY		
5 amps	13.5 hours	16 hours
15 amps	3.5 hours	13 hours
25 amps	1.8 hours	12 hours
105 AMP-HOUR BATTERY		
5 amps	15.8 hours	16 hours
15 amps	4.2 hours	13 hours
25 amps	2.4 hours	12 hours

BATTERY CHARGE PERCENTAGE

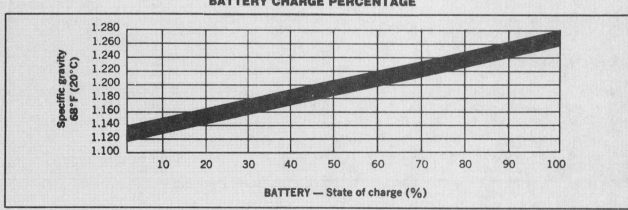

STANDARD TORQUE VALUES

Screw or nut size	in.-lb.
6-32	9
8-32	20
10-24	30
10-32	35
12-24	45
1/4-20	70
5/16-18	160
3/8-16	270

GEARCASE TIGHTENING TORQUES*

Fastener	in.-lb.
Bearing cage screws (4.4-8 hp)	70
Drive shaft nut	85 ft.-lb.
Gearcase cover screws	70
Lower gearcase-to-upper gearcase	260-270
Upper gearcase-to-motor leg	
3.5-4 hp	110
20-85 hp	
1-piece gearcase	160
2-piece gearcase	260-270
Standard torque values (screw or nut size)	
6-32	9
8-32	20
10-24	30
10-32	35
12-24	45
1/4-20	70
5/16-18	160
3/8-16	270

* Use standard torque values if specific fastener is not listed.

ELECTRICAL TIGHTENING TORQUES

Fastener	in.-lb.
Coil mounting screws	70
CD module mounting screws	
Delta	160
Motorola	90
Starter through-bolts	95-100
Standard torque values (screw or nut size)	
6-32	9
8-32	20
10-24	30
10-32	35
12-24	45
1/4-20	70
5/16-18	160
3/8-16	270

CYLINDER HEAD BOLT TORQUES*

Model	in.-lb.
3.5 hp	
1965-1969	NA
1977-on	130
3.6 hp	NA
4-4.5 hp	130
4.9-5 hp	80
6-8 hp	130
9.2 hp	65
9.9-15 hp	130
20 hp	
1965-1978	120
1979-on	190
25 hp	225
30 hp	
1965-1978	225
1979-on	190
35-55 hp	270
60-65 hp	225
70-140 hp	
3/8-16 bolts	265-275
5/16-18 bolts	220-230

* NA = Not available.

CHARGING SYSTEM SPECIFICATIONS (1980-ON)

Year/ model	Minimum amperage @ rpm	Cut-in rpm	Stator ohms	Maximum current draw
1980-on 9.9-15 hp	1.0 @ 2,000 3.5 @ 4,000	1,000	1.0-2.0	4
1980-on 20 and 30 hp; 1982-on 25 hp; 1983-on 35 hp	1.0 @ 2,000 2.0 @ 3,500	1,800	1.0-2.0	3
1980-1981 35 hp; 1980-on 45 and 50 hp	6.0 @ 3,500	1,100	0.1-0.4	6
1980-on 55 hp	0.5 @ 1,000 7.0 @ 4,000	800	1.0	6
1980-on 70-140 hp	7.0 @ 4,000	800	0.5-1.0	7

CLYMER®

CHRYSLER

OUTBOARD SHOP MANUAL
3.5-140 HP • 1966-1984

INTRODUCTION

This Clymer shop manual covers service and repair of all 1966-1984 Chrysler 3.5-140 hp outboards. Step-by-step instructions and hundreds of illustrations guide you through jobs ranging from simple maintenance to complete overhaul.

This manual can be used by anyone from a first time amateur to a professional mechanic. Easy to read type, detailed drawings and clear photographs give you all the information you need to do the work right.

Having a well-maintained engine will increase your enjoyment of your boat as well as assure your safety offshore. Keep this shop manual handy and use it often. It can save you hundreds of dollars in maintenance and repair bills and make yours a reliable, top-performing boat.

Chapter One

General Information

This detailed, comprehensive manual contains complete information covering maintenance, repair and overhaul. Hundreds of photos and drawings guide you throughout every procedure.

Troubleshooting, tune-up, maintenance and repair are not difficult if you know what tools and equipment to use and what to do. Anyone not afraid to get their hands dirty, of average intelligence and with some mechanical ability can perform most of the procedures in this manual. See Chapter Two for more information on tools and techniques.

A shop manual is a reference. You want to be able to find information quickly. Clymer books are designed with you in mind. All chapters are thumb tabbed and important items are indexed at the end of the manual. All procedures, tables, photos and instructions in this manual assume the reader may be working on the machine or using the manual for the first time.

Keep the manual in a handy place in your toolbox or boat. It will help you to better understand

how your boat runs, lower repair and maintenance costs and generally increase your enjoyment of your boat.

MANUAL ORGANIZATION

This chapter provides general information useful to boat owners and marine mechanics.

Chapter Two discusses the tools and techniques for preventative maintenance, troubleshooting and repair.

Chapter Three provides troubleshooting and testing procedures for all systems and individual components.

Following chapters describe specific systems, providing disassembly, inspection, assembly and adjustment procedures in simple step-by-step form. Specifications concerning a specific system are included at the end of the appropriate chapter.

NOTES, CAUTIONS AND WARNINGS

The terms NOTE, CAUTION and WARNING have specific meanings in this manual. A NOTE provides additional information to make a step or procedure easier or more clear. Disregarding a NOTE could cause inconvenience, but would not cause damage or personal injury.

A CAUTION emphasizes areas where equipment damage could cause permanent mechanical damage; however, personal injury is unlikely.

A WARNING emphasizes areas where personal injury or even death could result from negligence. Mechanical damage may also occur. WARNINGS *must* be taken seriously. In some cases, serious injury or death has resulted from disregarding similar warnings.

TORQUE SPECIFICATIONS

Torque specifications throughout this manual are given in foot-pounds (ft.-lb.), inch-pounds (in.-lb.) and newton meters (N•m.). Newton meters are being adopted in place of meter-kilograms (mkg) in accordance with the International Modernized Metric System. Existing torque wrenches calibrated in meter-kilograms can be used by performing a simple conversion: move the decimal point one place to the right. For example, 4.7 mkg = 47 N•m. This conversion is accurate enough for most mechanical operations even though the exact mathematical conversion is 3.5 mkg = 34.3 N•m.

ENGINE OPERATION

All marine engines, whether two or four-stroke, gasoline or diesel, operate on the Otto cycle of intake, compression, power and exhaust phases.

Two-Stroke Cycle

A two-stroke engine requires one crankshaft revolution (two strokes of the piston) to complete the Otto cycle. All engines covered in this manual are a two-stroke design. **Figure 1** shows gasoline two-stroke engine operation.

Four-Stroke Cycle

A four-stroke engine requires two crankshaft revolutions (four strokes of the piston) to complete the Otto cycle. **Figure 2** shows gasoline four-stroke engine operation.

FASTENERS

The material and design of the various fasteners used on marine equipment are carefully thought out and designed. Fastener design determines the type of tool required to work with the fastener. Fastener material is carefully selected to decrease the possibility of physical failure or corrosion. See *Galvanic Corrosion* in this chapter for information on marine materials.

Nuts, bolts and screws are manufactured in a wide range of thread patterns. To join a nut and bolt, the diameter of the bolt and the diameter of the hole in the nut must be the same. It is just as important that the threads are compatible.

The easiest way to determine if fastener threads are compatible is to turn the nut on the bolt, or bolt into its threaded opening, using fingers only. Be sure both pieces are clean. If much force is required, check the thread condition on each fastener. If the thread condition is good but the fasteners jam, the threads are not compatible.

Four important specifications describe the thread:
1. Diameter.
2. Threads per inch.
3. Thread pattern.

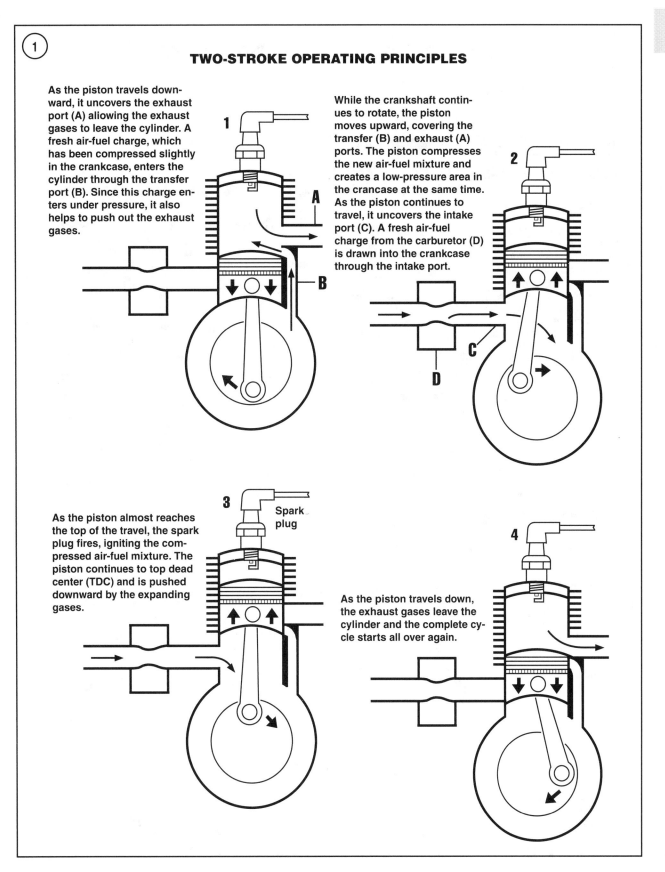

① **TWO-STROKE OPERATING PRINCIPLES**

As the piston travels downward, it uncovers the exhaust port (A) allowing the exhaust gases to leave the cylinder. A fresh air-fuel charge, which has been compressed slightly in the crankcase, enters the cylinder through the transfer port (B). Since this charge enters under pressure, it also helps to push out the exhaust gases.

While the crankshaft continues to rotate, the piston moves upward, covering the transfer (B) and exhaust (A) ports. The piston compresses the new air-fuel mixture and creates a low-pressure area in the crancase at the same time. As the piston continues to travel, it uncovers the intake port (C). A fresh air-fuel charge from the carburetor (D) is drawn into the crankcase through the intake port.

As the piston almost reaches the top of the travel, the spark plug fires, igniting the compressed air-fuel mixture. The piston continues to top dead center (TDC) and is pushed downward by the expanding gases.

As the piston travels down, the exhaust gases leave the cylinder and the complete cycle starts all over again.

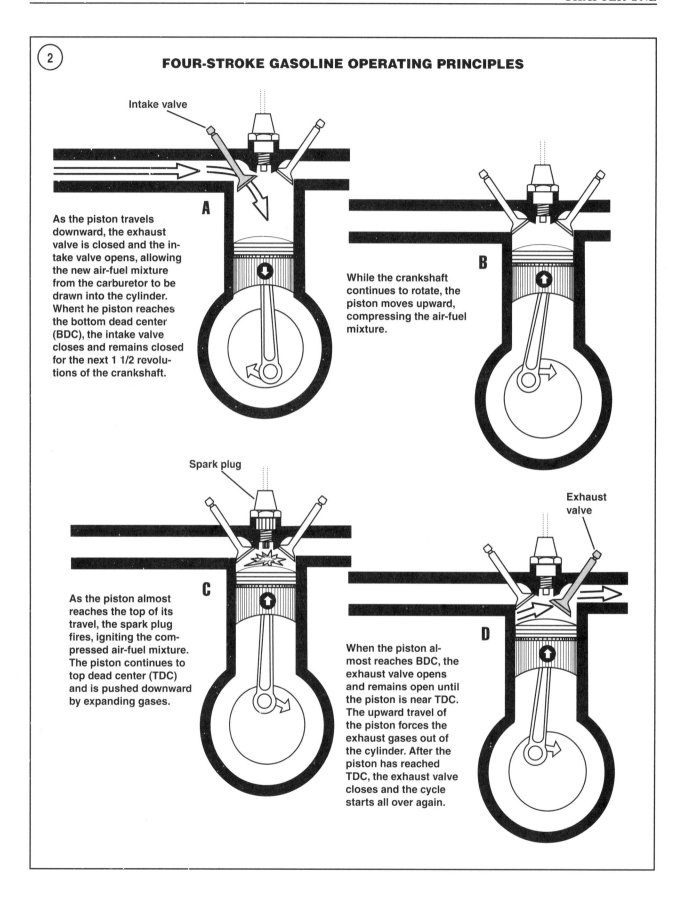

FOUR-STROKE GASOLINE OPERATING PRINCIPLES

Intake valve

A

As the piston travels downward, the exhaust valve is closed and the intake valve opens, allowing the new air-fuel mixture from the carburetor to be drawn into the cylinder. Whent he piston reaches the bottom dead center (BDC), the intake valve closes and remains closed for the next 1 1/2 revolutions of the crankshaft.

B

While the crankshaft continues to rotate, the piston moves upward, compressing the air-fuel mixture.

Spark plug

C

As the piston almost reaches the top of its travel, the spark plug fires, igniting the compressed air-fuel mixture. The piston continues to top dead center (TDC) and is pushed downward by expanding gases.

Exhaust valve

D

When the piston almost reaches BDC, the exhaust valve opens and remains open until the piston is near TDC. The upward travel of the piston forces the exhaust gases out of the cylinder. After the piston has reached TDC, the exhaust valve closes and the cycle starts all over again.

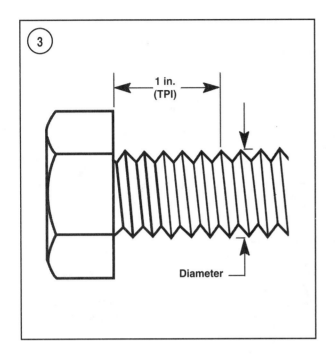

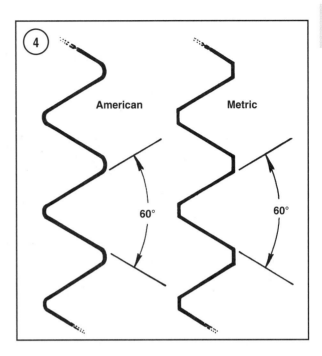

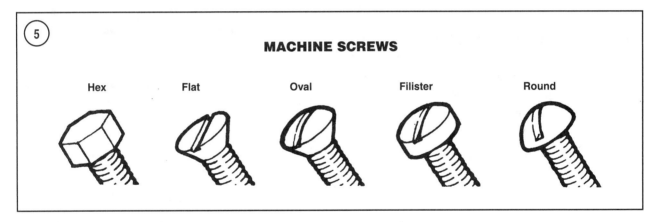

4. Thread direction

Figure 3 shows the first two specifications. Thread pattern is more subtle. Italian and British standards exist, but the most commonly used by marine equipment manufactures are American standard and metric standard. The root and top of the thread are cut differently as shown in **Figure 4**.

Most threads are cut so the fastener must be turned clockwise to tighten it. These are called right-hand threads. Some fasteners have left-hand threads; they must be turned counterclockwise to tighten. Left-hand threads are used in locations where normal rotation of the equipment would tend to loosen a right-hand threaded fastener. Assume all fasteners use right-hand threads unless the instructions specify otherwise.

Machine Screws

There are many different types of machine screws (**Figure 5**). Most are designed to protrude above the secured surface (rounded head) or be slightly recessed below the surface (flat head). In some applications the screw head is recessed well below the fastened sur-

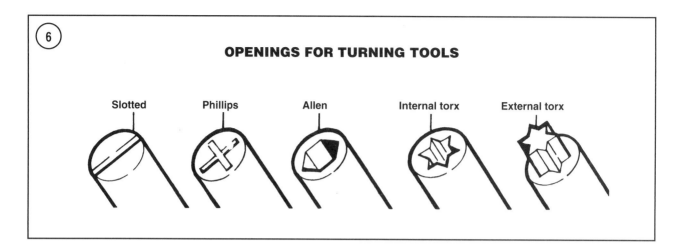

OPENINGS FOR TURNING TOOLS

Slotted   Phillips   Allen   Internal torx   External torx

face. **Figure 6** shows a number of screw heads requiring different types of turning tools.

Bolts

Commonly called bolts, the technical name for this fastener is cap screw. They are normally described by diameter, threads per inch and length. For example, 1/4-20 × 1 indicates a bolt 1/4 in. in diameter with 20 threads per inch, 1 in. long. The measurement across two flats of the bolt head indicates the proper wrench size required to turn the bolt.

Nuts

Nuts are manufactured in a variety of types and sizes. Most are hexagonal (six-sides) and fit on bolts, screws and studs with the same diameter and threads per inch.

Figure 7 shows several types of nuts. The common nut is usually used with some type of lockwasher. Self-locking nuts have a nylon insert that helps prevent the nut from loosening; no lockwasher is required. Wing nuts are designed for fast removal by hand. Wing nuts are used for convenience in non-critical locations.

To indicate the size of a nut, manufactures specify the diameter of the opening and the threads per inch. This is similar to a bolt specifi-

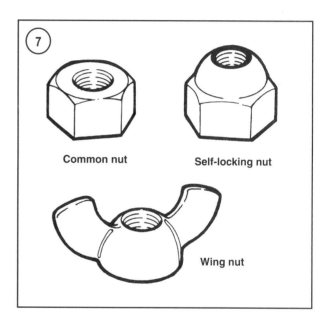

Common nut   Self-locking nut

Wing nut

cation, but without the length dimension. The measurement across two flats of the nut indicates the wrench size required to turn the nut.

Washers

There are two basic types of washers: flat washers and lockwashers. A flat washer is a simple disc with a hole that fits the screw or bolt. Lockwashers are designed to prevent a fastener from working loose due to vibration, expansion and contraction. **Figure 8** shows several types of lockwashers. Note that flat washers are often

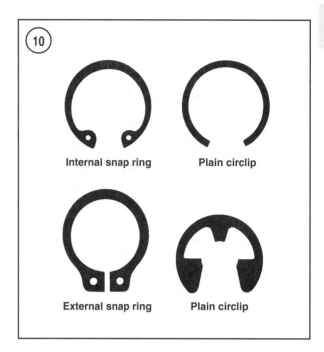

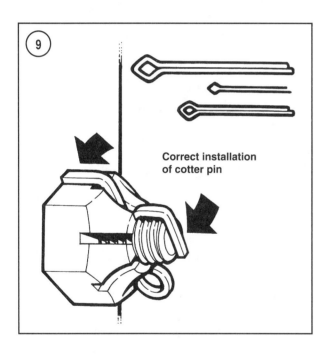

used between a lockwasher and a fastener to provide a smooth bearing surface. This allows the fastener to be turned easily with a tool.

Cotter Pins

In certain applications, a fastener must be secured so it cannot possibly loosen. The propeller nut on some marine drive systems is one such ap-

plication. For this purpose, a cotter pin (**Figure 9**) and slotted or castellated nut is often used. To use a cotter pin, first make sure the pin fits snugly, but not too tight. Then, align a slot in the fastener with the hole in the bolt or axle. Insert the cotter pin through the nut and bolt or propeller shaft and bend the ends over to secure the cotter pin tightly. If the holes do not align, tighten the nut just enough to obtain the proper alignment. Unless specifically instructed to do so, never loosen the fastener to align the slot and hole. Because the cotter pin is weakened after installation and removal, never reuse a cotter pin. Cotter pins are available in several styles, lengths and diameters. Measure cotter pin length from the bottom of its head to the tip of its shortest prong.

Snap Rings

Snap rings (**Figure 10**) can be an internal or external design. They are used to retain components on shafts (external type) or inside openings (internal type). Snap rings can be reused if they are not distorted during removal. In some applications, snap rings of varying thickness (selective fit) can be selected to position or control end play of parts assemblies.

LUBRICANTS

Periodic lubrication helps ensure long service life for any type of equipment. It is especially important with marine equipment because it is exposed to salt, brackish or polluted water and other harsh environments. The type of lubricant used is just as important as the lubrication service itself, although in an emergency, the wrong type of lubricant is better than none at all. The following paragraphs describe the types of lubricants most often used on marine equipment. Be sure to follow the equipment manufacture's recommendations for the lubricant types.

Generally, all liquid lubricants are called *oil*. They may be mineral-based (including petroleum bases), natural-based (vegetable and animal bases), synthetic-based or emulsions (mixtures). *Grease* is lubricating oil that has a thickening compound added. The resulting material then usually enhanced with anticorrosion, antioxidant and extreme pressure (EP) additives. Grease is often classified by the type of thickener added; lithium and calcium soap are the most commonly used.

Two-stroke Engine Oil

Lubrication for a two-stroke engine is provided by oil mixed with the incoming air/fuel mixture. Some of the oil mist settles out in the crankcase, lubricating the crankshaft, bearings and lower end of the connecting rod. The rest of the oil enters the combustion chamber to lubricate the piston, rings and the cylinder wall. This oil is then burned along with the air/fuel mixture during the combustion process.

Engine oil must have several special qualities to work well in a two-stroke engine. It must mix easily and stay in suspension in gasoline.

When burned, it cannot leave behind excessive deposits. It must also withstand the high operating temperature associated with two-stroke engines.

The National Marine Manufacturer's Association (NMMA) has set standards for oil used in two-stroke, water-cooled engines. This is the NMMA TC-W (two-cycle, water-cooled) grade. It indicates the oil's performance in the following areas:

1. Lubrication (preventing wear and scuffing).
2. Spark plug fouling.
3. Piston ring sticking.
4. Preignition.
5. Piston varnish.
6. General engine condition (including deposits).
7. Exhaust port blockage.
8. Rust prevention.
9. Mixing ability with gasoline.

In addition to oil grade, manufactures specify the ratio of gasoline and oil required during break-in and normal engine operation.

Gearcase Oil

Gearcase lubricants are assigned SAE viscosity numbers under the same system as four-stroke engine oil. Gearcase lubricant falls into the SAE 72-250 range. Some gearcase lubricants are multigrade. For example, SAE 80-90 is a common multigrade gear lubricant.

Three types of marine gearcase lubricants are generally available; SAE 90 hypoid gearcase lubricant is designed for older manual-shift units; type C gearcase lubricant contains additives designed for the electric shift mechanisms; high-viscosity gearcase lubricant is a heavier oil designed to withstand the shock loads of high performance engines or units subjected to severe duty use. Always use the gearcase lubricant specified by the manufacturer.

Grease

Greases are graded by the National Lubricating Grease Institute (NLGI). Greases are graded by number according to the consistency of the grease. These ratings range from No. 000 to No. 6, with No. 6 being the most solid. A typical multipurpose grease is NLGI No. 2. For specific applications, equipment manufactures may require grease with an additive such as molybdenum disulfide (MOS^2).

GASKET SEALANT

Gasket sealant is used instead of preformed gaskets on some applications, or as a gasket dressing on others. Three types of gasket sealant are commonly used: gasket sealing compound, room temperature vulcanizing (RTV) and anaerobic. Because these materials have different sealing properties, they cannot be used interchangeably.

Gasket Sealing Compound

This nonhardening liquid is used primarily as a gasket dressing. Gasket sealing compound is available in tubes or brush top containers. When exposed to air or heat it forms a rubber-like coating. The coating fills in small imperfections in gasket and sealing surfaces. Do not use gasket sealing compound that is old, has began to solidify or has darkened in color.

Applying Gasket Sealing Compound

Carefully scrape residual gasket material, corrosion deposits or paint from the mating surfaces. Use a blunt scraper and work carefully to avoid damaging the mating surfaces. Use quick drying solvent and a clean shop towel and wipe oil or other contaminants from the surfaces. Wipe or blow loose material or contaminants from the gasket. Brush a light coating on the mating surfaces and both sides of the gasket. Do not apply more compound than needed. Excess compound will be squeezed out as the surfaces mate and may contaminate other components. Do not allow compound into bolt or alignment pin holes

A hydraulic lock can occur as the bolt or pin compresses the compound, resulting in incorrect bolt torque.

RTV Sealant

This is a silicone gel supplied in tubes. Moisture in the air causes RTV to cure. Always place the cap on the tube as soon as possible if using RTV. RTV has a shelf life of approximately one year and will not cure properly after the shelf life expires. Check the expiration date on the tube and keep partially used tubes tightly sealed. RTV can generally fill gaps up to 1/4 in. (6.3 mm) and works well on slightly flexible surfaces.

Applying RTV Sealant

Carefully scrape all residual sealant and paint from the mating surfaces. Use a blunt scraper and work carefully to avoid damaging the mating surfaces. The mating surfaces must be absolutely free of gasket material, sealant, dirt, oil grease or other contamination. Lacquer thinner, acetone, isopropyl alcohol or similar solvents work well to clean the surfaces. Avoid using solvents with an oil, wax or petroleum base as they are not compatible with RTV compounds. Remove all sealant from bolt or alignment pin holes.

Apply RTV sealant in a continuous bead 0.08-0.12 in. (2-3 mm) thick. Circle all mounting bolt or alignment pin holes unless otherwise specified. Do not allow RTV sealant into bolt holes or other openings. A hydraulic lock can

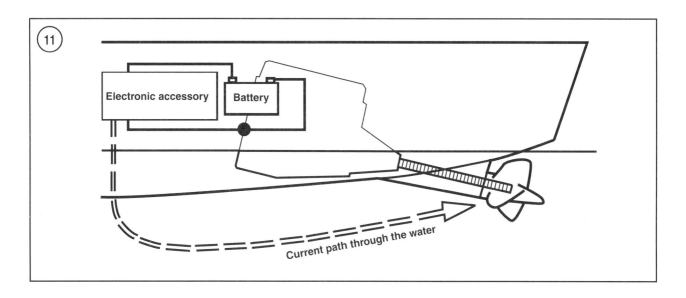

Current path through the water

occur as the bolt or pin compresses the sealant, resulting in incorrect bolt torque. Tighten the mounting fasteners within 10 minutes after application.

Anaerobic Sealant

This is a gel supplied in tubes. It cures only in the absence of air, as when squeezed tightly between two machined mating surfaces. For this reason, it will not spoil if the cap is left off the tube. Do not use anaerobic sealant if one of the surfaces is flexible. Anaerobic sealant is able to fill gaps up to 0.030 in. (0.8 mm) and generally works best on rigid, machined flanges or surfaces.

Applying Anaerobic Sealant

Carefully scrape all residual sealant from the mating surfaces. Use a blunt scraper and work carefully to avoid damaging the mating surfaces. The mating surfaces must be absolutely free of gasket material, sealant, dirt, oil grease or other contamination. Lacquer thinner, acetone, isopropyl alcohol or similar solvents work well to clean the surfaces. Avoid using solvents

with an oil, wax or petroleum base as they are not compatible with anaerobic compounds. Clean a sealant from the bolt or alignment pin holes. Apply anaerobic sealant in a 0.04 in. (1 mm) thick continuous bead onto one of the surfaces. Circle all bolt and alignment pin openings. Do not apply sealant into bolt holes or other openings. A hydraulic lock can occur as the bolt or pin compresses the sealant, resulting in incorrect bolt torque. Tighten the mounting fasteners within 10 minutes after application.

GALVANIC CORROSION

A chemical reaction occurs whenever two different types of metal are joined by an electrical conductor and immersed in an electrolytic solution such as water. Electrons transfer from one metal to the other through the electrolyte and return through the conductor.

The hardware on a boat is made of many different types of metal. The boat hull acts as a conductor between the metals. Even if the hull is wooden or fiberglass, the slightest film of water (electrolyte) on the hull provides conductivity. This combination creates a good environment for electron flow (**Figure 11**). Unfortunately, this electron flow results in galvanic corrosion of the metal involved, causing one of the metals to be corroded or eroded away. The amount of electron flow, and therefore the amount of corrosion, depends on several factors:

1. The types of metal involved.
2. The efficiency of the conductor.
3. The strength of the electrolyte.

Metals

The chemical composition of the metal used in marine equipment has a significant effect on the amount and speed of galvanic corrosion. Certain metals are more resistant to corrosion than others. These electrically negative metals are commonly called *noble*; they act as the cathode in any reaction. Metals that are more subject to corrosion are electrically positive; they act as the anode in a reaction. The more *noble* metals include titanium, 18-8 stainless steel and nickel. Less *noble* metals include zinc, aluminum and magnesium. Galvanic corrosion becomes more severe as the difference in electrical potential between the two metals increases.

In some cases, galvanic corrosion can occur within a single piece of metal. For example, brass is a mixture of zinc and copper, and, when immersed in an electrolyte, the zinc portion of the mixture will corrode away as a galvanic reaction occurs between the zinc and copper particles.

Conductors

The hull of the boat often acts as the conductor between different types of metal. Marine equipment, such as the drive unit can act as the conductor. Large masses of metal, firmly connected together, are more efficient conductors than water. Rubber mountings and vinyl-based paint can act as insulators between pieces of metal.

Electrolyte

The water in which a boat operates acts as the electrolyte for the corrosion process. The more efficient a conductor is, the more severe and rapid the corrosion will be.

Cold, clean freshwater is the poorest electrolyte. Pollutants increase conductivity; therefore, brackish or saltwater is an efficient electrolyte. This is one of the reasons that most manufacturers recommend a freshwater flush after operating in polluted, brackish or saltwater.

Protection From Galvanic Corrosion

Because of the environment in which marine equipment must operate, it is practically impossible to totally prevent galvanic corrosion. However, there are several ways in which the process can be slowed. After taking these precautions, the next step is to *fool* the process into occurring only where you want it to occur. This is the role of sacrificial anodes and impressed current systems.

Slowing Corrosion

Some simple precautions can help reduce the amount of corrosion taking place outside the hull. These precautions are not substitutes for the corrosion protection methods discussed under *Sacrificial Anodes* and *Impressed Current Systems* in this chapter, but they can help these methods reduce corrosion.

Use fasteners made of metal more noble than the parts they secure. If corrosion occurs, the parts they secure may suffer but the fasteners are protected. The larger secured parts are more able to withstand the loss of material. Also major problems could arise if the fasteners corrode to the point of failure.

Keep all painted surfaces in good condition. If paint is scraped off and bare metal exposed, corrosion rapidly increases. Use a vinyl- or plastic-based paint, which acts as an electrical insulator.

Be careful when applying metal-based antifouling paint to the boat. Do not apply antifouling paint to metal parts of the boat or the drive unit. If applied to metal surfaces, this type of paint reacts with the metal and results in corrosion between the metal and the layer of paint. Maintain a minimum 1 in. (25 mm) border between the painted surface and any metal parts. Organic-based paints are available for use on metal surfaces.

Where a corrosion protection device is used, remember that it must be immersed in the electrolyte along with the boat to provide any protection. If you raise the gearcase out of the water with the boat docked, any anodes on the gearcase may be removed from the corrosion process rendering them ineffective. Never paint or apply any coating to anodes or other protection devices. Paint or other coatings insulate them from the corrosion process.

Any change in the boat's equipment, such as the installation of a new stainless steel propeller, changes the electrical potential and may cause increased corrosion. Always consider this when adding equipment or changing exposed materials. Install additional anodes or other protection equipment as required ensuring the corrosion protection system is up to the task. The expense to repair corrosion damage usually far exceeds that of additional corrosion protection.

Sacrificial Anodes

Sacrificial anodes are specially designed to do nothing but corrode. Properly fastening such pieces to the boat causes them to act as the anode in any galvanic reaction that occurs; any other metal in the reaction acts as the cathode and is not damaged.

Anodes are usually made or zinc, a far from a noble material. Some anodes are manufactured of an aluminum and indium alloy. This alloy is less noble than the aluminum alloy in drive system components, providing the desired sacrificial properties. The aluminum and indium alloy is more resistant to oxide coating than zinc anodes. Oxide coating occurs as the anode material reacts with oxygen in the water. An oxide coating will insulate the anode, dramatically reducing corrosion protection.

Anodes must be used properly to be effective. Simply fastening anodes to the boat in random locations will not do the job.

First determine how much anode surface is required to adequately protect the equipment's surface area. A good starting point is provided by the Military Specification MIL-A-818001, which states that one square inch of new anode protects either:
1. 800 square inches of freshly painted steel.
2. 250 square inches of bare steel or bare aluminum alloy.
3. 100 square inches of copper or copper alloy.

This rule is valid for a boat at rest. If underway, additional anode area is required to protect the same surface area.

The anode must be in good electrical contact with the metal that it protects. If possible, attach an anode to all metal surfaces requiring protection.

Good quality anodes have inserts around the fastener holes that are made of a more noble material. Otherwise, the anode could erode away around the fastener hole, allowing the anode to loosen or possibly fall off, thereby loosing needed protection.

Impressed Current System

An impressed current system can be added to any boat. The system generally consists of the anode, controller and reference electrode. The anode in this system is coated with a very noble

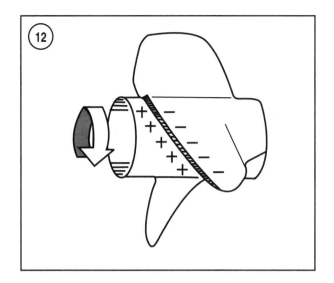

metal, such as platinum, so that it is almost corrosion-free and can last almost indefinitely. The reference electrode, under the boat's waterline, allows the control module to monitor the potential for corrosion. If the module senses that corrosion is occurring, it applies positive battery voltage to the anode. Current then flows from the anode to all other metal component, regardless of how noble or non-noble these components may be. Essentially, the electrical current from the battery counteracts the galvanic reaction to dramatically reduce corrosion damage.

Only a small amount of current is needed to counteract corrosion. Using input from the sensor, the control module provides only the amount of current needed to suppress galvanic corrosion. Most systems consume a maximum of 0.2 Ah at full demand. Under normal conditions, these systems can provide protection for 8-12 weeks without recharging the battery. Remember that this system must have constant connection to the battery. Often the battery supply to the system is connected to a battery switching device causing the operator to inadvertently shut off the system while docked.

An impressed current system is more expensive to install than sacrificial anodes but, considering its low maintenance requirements and the superior protection it provides, the long term cost may be lower.

PROPELLERS

The propeller is the final link between the boat's drive system and the water. A perfectly maintained engine and hull are useless if the propeller is the wrong type, is damaged or is deteriorated. Although propeller selection for a specific application is beyond the scope of this manual, the following provides the basic information needed to make an informed decision. The professional at a reputable marine dealership is the best source for a propeller recommendation.

How a Propeller Works

As the curved blades of a propeller rotate through the water, a high-pressure area forms on one side of the blade and a low-pressure area forms on the other side of the blade (**Figure 12**). The propeller moves toward the low-pressure area, carrying the boat with it.

Propeller Parts

Although a propeller is usually a one-piece unit, it is made of several different parts (**Figure 13**). Variations in the design of these parts make different propellers suitable for different applications.

The blade tip is the point of the blade furthest from the center of the propeller hub or propeller shaft bore. The blade tip separates the leading edge from the trailing edge.

The leading edge is the edge of the blade nearest the boat. During forward operation, this is the area of the blade that first cuts through the water.

The trailing edge is the surface of the blade furthest from the boat. During reverse operation,

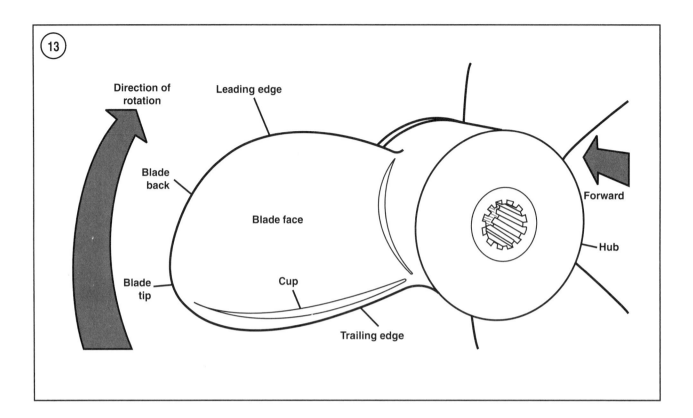

(13)

Direction of rotation

Leading edge

Blade back

Blade face

Forward

Hub

Blade tip

Cup

Trailing edge

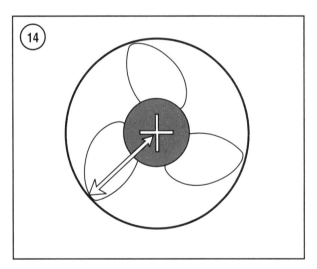

(14)

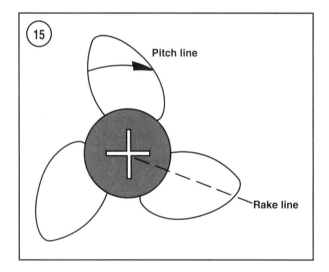

(15)

Pitch line

Rake line

this is the area of the blade that first cuts through the water.

The blade face is the surface of the blade that faces away from the boat. During forward operation, high-pressure forms on this side of the blade.

The blade back is the surface of the blade that faces toward the boat. During forward gear operation, low-pressure forms on this side of the blade.

The cup is a small curve or lip on the trailing edge of the blade. Cupped propeller blades generally perform better than non-cupped propeller blades.

The hub is the center portion of the propeller. It connects the blades to the propeller shaft. On most drive systems, engine exhaust is routed through the hub; in this case, the hub is made up of an outer and inner portion, connected by ribs.

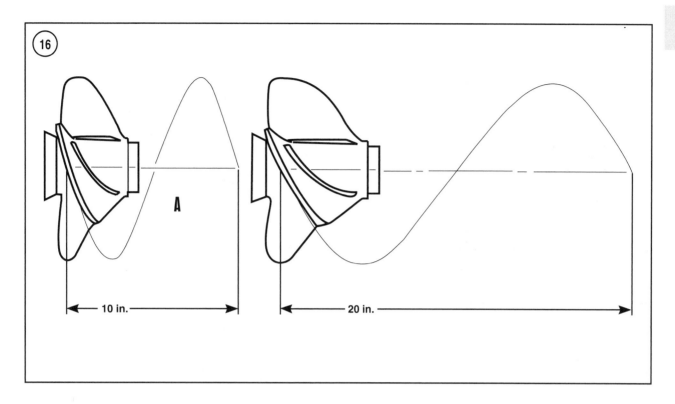

The diffuser ring is used on though- hub exhaust models to prevent exhaust gasses from entering the blade area.

Propeller Design

Changes in length, angle, thickness and material of propeller parts make different propellers suitable for different applications.

Diameter

Propeller diameter is the distance from the center of the hub to the blade tip, multiplied by two. Essentially it is the diameter of the circle formed by the blade tips during propeller rotation (**Figure 14**).

Pitch and rake

Propeller pitch and rake describe the placement of the blades in relation to the hub (**Figure 15**).

Pitch describes the theoretical distance the propeller would travel in one revolution. In A, **Figure 16**, the propeller would travel 10 inches in one revolution. In B, **Figure 16**, the propeller would travel 20 inches in one revolution. This distance is only theoretical; during operation, the propeller achieves only 75-85% of its pitch. Slip rate describes the difference in actual travel relative to the pitch. Lighter, faster boats typically achieve a lower slip rate than heavier, slower boats.

Propeller blades can be constructed with constant pitch (**Figure 17**) or progressive pitch (**Figure 18**). On a progressive propeller, the pitch starts low at the leading edge and increases toward the trailing edge. The propeller pitch specification is the average of the pitch across the entire blade. Propellers with progressive pitch usually provide better overall performance than constant pitch propellers.

Blade rake is specified in degrees and is measured along a line from the center of the hub to the blade tip. A blade that is perpendicular to the

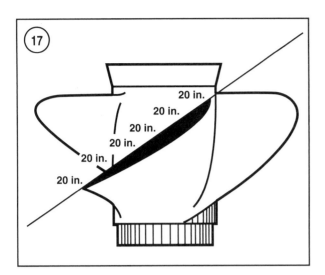

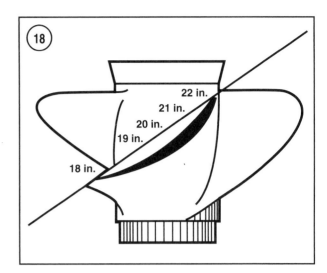

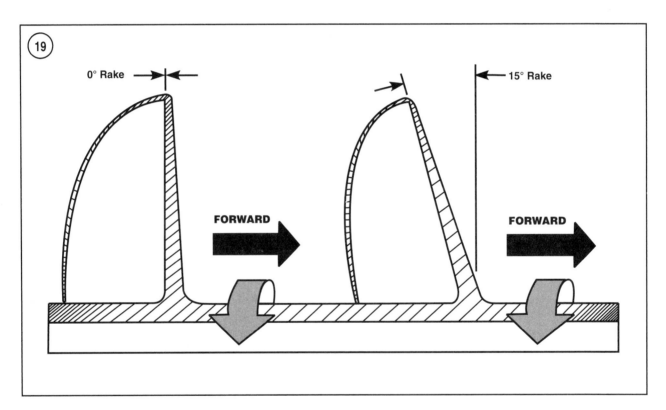

hub (**Figure 19**) has 0° rake. A blade that is angled from perpendicular (**Figure 19**) has a rake expressed by its difference from perpendicular. Most propellers have rakes ranging from 0-20°. Lighter faster boats generally perform better with propeller with a greater amount of rake. Heavier, slower boats generally perform better using a propeller with less rake.

Blade thickness

Blade thickness in not uniform at all points along the blade. For efficiency, blades are as thin a possible at all points while retaining enough strength to move the boat. Blades are thicker where they meet the hub and thinner at the blade tips (**Figure 20**). This is necessary to support the

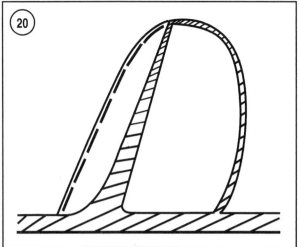

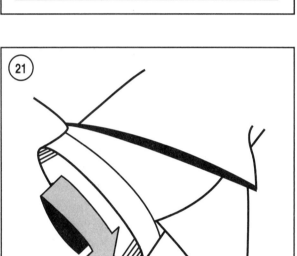

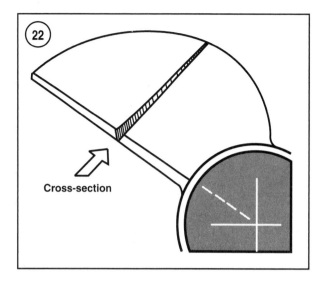

Cross-section

heavier loads at the hub section of the blade. Overall blade thickness is dependent on the strength of the material used.

When cut along a line from the leading edge to the trailing edge in the central portion of the blade (**Figure 21**), the propeller blade resembles and airplane wing. The blade face, where high-pressure exists during forward rotation, is almost flat. The blade back, where low-pressure exists during forward rotation, is curved, with the thinnest portions at the edges and the thickest portion at the center.

Propellers that run only partially submerged, as in racing applications, may have a wedge shaped cross-section (**Figure 22**). The leading edge is very thin and the blade thickness increases toward the trailing edge, where it is thickest. If a propeller such as this is run totally submerged, it is very inefficient.

Number of blades

The number of blades used on a propeller is a compromise between efficiency and vibration. A one-bladed propeller would the most efficient, but it would create an unacceptable amount of vibration. As blades are added, efficiency decreases, but so does vibration. Most propellers have three or four blades, representing the most practical trade-off between efficiency and vibration.

Material

Propeller materials are chosen for strength, corrosion resistance and economy. Stainless steel, aluminum, plastic and bronze are the most commonly used materials. Bronze is quite strong but rather expensive. Stainless steel is more common than bronze because of its combination of strength and lower cost. Aluminum alloy and plastic materials are the least expensive

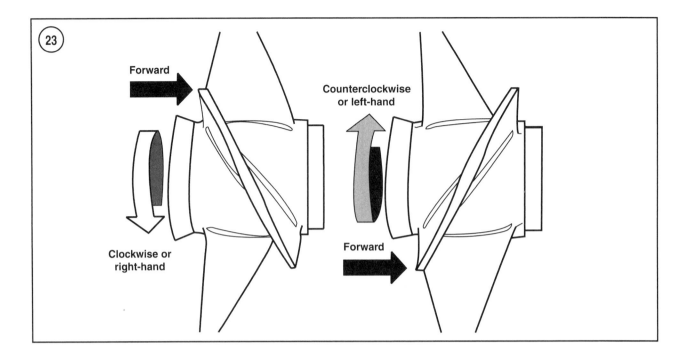

Forward

Counterclockwise
or left-hand

Clockwise or
right-hand

Forward

but usually lack the strength of stainless steel. Plastic propellers are more suited for lower horsepower applications.

Direction of rotation

Propellers are made for both right-hand and left hand rotations although right-hand is the most commonly used. As viewed from the rear of the boat while in forward gear, a right-hand propeller turns clockwise and a left-hand propeller turns counterclockwise. Off the boat, the direction of rotation is determined by observing the angle of the blades (**Figure 23**). A right-hand propeller's blade slant from the upper left to the lower right; a left-hand propeller's blades are opposite.

Cavitation and Ventilation

Cavitation and ventilation are *not* interchangeable terms; they refer to two distinct problems encountered during propeller operation.

To help understand cavitation, consider the relationship between pressure and the boiling point of water. At sea level, water boils at 212° F (100° C). As pressure increases, such as within an engine cooling system, the boiling point of the water increases—it boils at a temperature higher than 212° F (100° C). The opposite is also true. As pressure decreases, water boils at a temperature lower than 212° F (100° C). It the pressure drops low enough, water will boil at normal room temperature.

During normal propeller operation, low pressure forms on the blade back. Normally the pressure does not drop low enough for boiling to occur. However, poor propeller design, damaged blades or using the wrong propeller can cause unusually low pressure on the blade surface (**Figure 24**). If the pressure drops low enough, boiling occurs and bubbles form on the blade surfaces. As the boiling water moves to a higher pressure area of the blade, the boiling ceases and the bubbles collapse. The collapsing bubbles release energy that erodes the surface of the propeller blade.

Corroded surfaces, physical damage or even marine growth combined with high-speed operation can cause low pressure and cavitation on gearcase surfaces. In such cases, low pressure

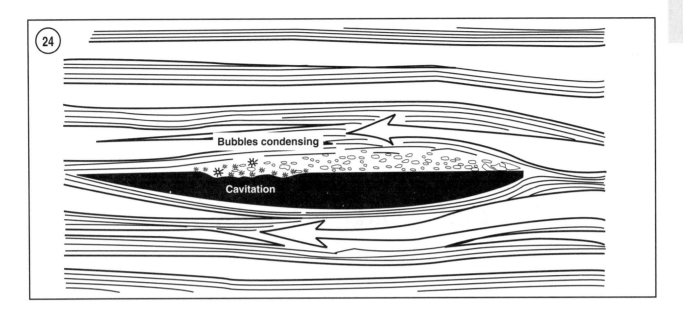

Bubbles condensing

Cavitation

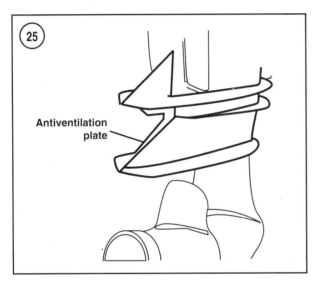

Antiventilation
plate

forms as water flows over a protrusion or rough surface. The boiling water forms bubbles that collapse as they move to a higher pressure area toward the rear of the surface imperfection.

This entire process of pressure drop, boiling and bubble collapse is called *cavitation*. The ensuing damage is called *cavitation burn*. Cavitation is caused by a decrease in pressure, not an increase in temperature.

Ventilation is not as complex a process as cavitation. Ventilation refers to air entering the blade area, either from above the water surface or from a though-hub exhaust system. As the blades meet the air, the propeller momentarily looses it bite with the water and subsequently loses most of its thrust. An added complication is that the propeller and engine over-rev, causing very low pressure on the blade back and massive cavitation.

Most marine drive systems have a plate (**Figure 25**) above the propeller designed to prevent surface air from entering the blade area. This plate is correctly called an *anti-ventilation plate*, although it is often incorrectly called an *anticavitation plate*.

Most propellers have a flared section at the rear of the propeller called a diffuser ring. This feature forms a barrier, and extends the exhaust passage far enough aft to prevent the exhaust gases from ventilating the propeller.

A close fit of the propeller to the gearcase is necessary to keep exhaust gasses from exiting and ventilating the propeller. Using the wrong propeller attaching hardware can position the propeller too far aft, preventing a close fit. The wrong hardware can also allow the propeller to rub heavily against the gearcase, causing rapid wear to both components. Wear or damage to these surfaces will allow the propeller to ventilate.

Chapter Two

Tools and Techniques

This chapter describes the common tools required for marine engine repair and troubleshooting. Techniques that make the work easier and more effective are also described. Some of the procedures in this book require special skills or expertise; it some cases it is better to entrust the job to a specialist or qualified dealership.

SAFETY FIRST

Professional mechanics can work for years and never suffer a serious injury. Avoiding injury is as simple as following a few rules and using common sense. Ignoring the rules can of often does lead to physical injury and/or damaged equipment.

1. Never use gasoline as a cleaning solvent.
2. Never smoke or use a torch near flammable liquids, such as cleaning solvent. Dirty or solvent soaked shop towels are extremely flamma-ble. If working in a garage, remember that most home gas appliances have pilot lights.
3. Never smoke or use a torch in an area where a battery is being charged. Highly explosive hydrogen gas is formed during the charging process.
4. Use the proper size wrench to avoid damaged fasteners and bodily injury.
5. If loosening a tight or stuck fastener, consider what could happen if the wrench slips. Protect yourself accordingly.
6. Keep the work area clean, uncluttered and well lighted.
7. Wear safety goggles while using any type of tool. This is especially important when drilling, grinding or using a cold chisel.
8. Never use worn or damaged tools.
9. Keep a Coast Guard approved fire extinguisher handy. Ensure it is rated for gasoline (Class B) and electrical (Class C) fires.

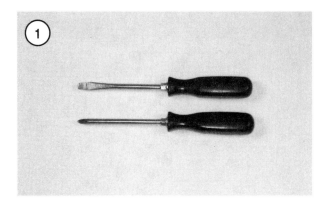

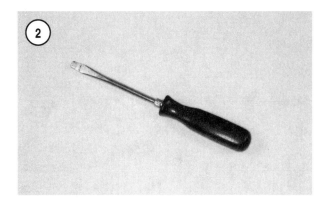

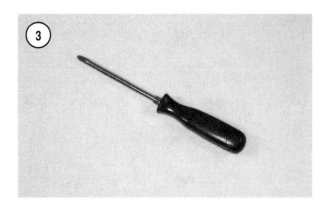

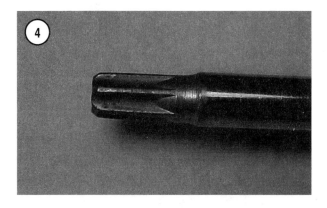

BASIC HAND TOOLS

A number of tools are required to maintain and repair a marine engine. Most of these tools are also used for home and automobile repair. Some tools are made especially for working on marine engines; these tools can be purchased from a marine dealership. Having the required tools always makes the job easier and more effective.

Keep the tools clean and in a suitable box. Keep them organized with related tools stored together. After using a tool, wipe it clean using a shop towel.

The following tools are required to perform virtually any repair job. Each tool is described and the recommended size given for starting a tool collection. Additional tools and some duplication may be added as you become more familiar with the equipment. You may need all U.S. standard tools, all metric size tools or a mixture of both.

Screwdrivers

A screwdriver (**Figure 1**) is a very basic tool, but if used improperly can do more damage than good. The slot on a screw has a definite dimension and shape. Always select a screwdriver that conforms to the shape of the screw. Use a small screwdriver for small screws and a large one for large screws or the screw head are damaged.

Three types of screwdrivers are commonly required: a slotted (flat-blade) screwdriver (**Figure 2**), Phillips screwdriver (**Figure 3**) and Torx screwdriver (**Figure 4**).

Screwdrivers are available in sets, which often include an assortment of slotted Phillips and Torx blades. If you buy them individually, buy at least the following:

   a. Slotted screwdriver—5/16 × 6 in. blade.
   b. Slotted screwdriver—3/8 × 12 in. blade.
   c. Phillips screwdriver—No. 2 tip, 6 in. blade.

d. Phillips screwdriver—No. 3 tip, 6 in. blade.

e. Torx screwdriver—T15 tip, 6 in. blade.

f. Torx screwdriver—T20 tip, 6 in. blade.

g. Torx screwdriver—T25 tip, 6 in. blade.

Use screwdrivers only for driving screws. Never use a screwdriver for prying or chiseling. Do not attempt to remove a Phillips, Torx or Allen head screw with a slotted screwdriver; you can damage the screw head so that even the proper tool is unable to remove it.

Keep the tip of a slotted screwdriver in good condition. Carefully grind the tip to the proper size and taper if it is worn or damaged. The sides of the blade must be parallel and the blade tip must be flat. Replace a Phillips or Torx screwdriver if its tip is worn or damaged.

Pliers

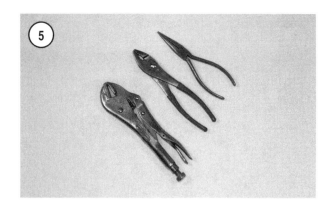

Pliers come in a wide range of types and sizes. Pliers are useful for cutting, gripping, bending and crimping. Never use pliers to cut hardened objects or turn bolts or nuts. **Figure 5** shows several types of pliers.

Each type of pliers has a specialized function. General-purpose pliers are mainly used for gripping and bending. Locking pliers are used for gripping objects very tightly, like a vise. Use needlenose pliers to grip or bend small objects. Adjustable or slip-joint pliers (**Figure 6**) can be adjusted to grip various sized objects; the jaws remain parallel for gripping objects such as pipe or tubing. There are many more types of pliers. The ones described here are the most common.

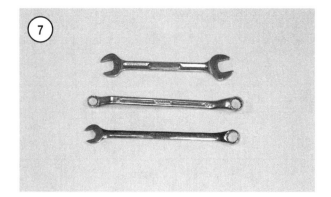

Box-end and Open-end Wrenches

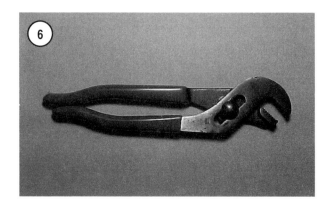

Box-end and open-end wrenches (**Figure 7**) are available in sets in a variety of sizes. The number stamped near the end of the wrench refers to the distance between two parallel flats on the hex head bolt or nut.

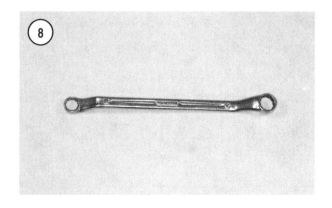

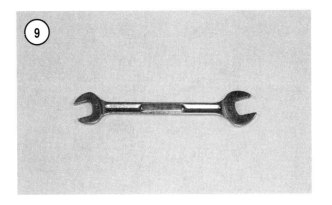

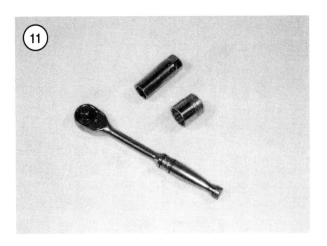

superior holding power; the 12-point allow a shorter swing if working in tight quarters.

Use an open-end wrench if a box-end wrench cannot be positioned over the nut or bolt. To prevent damage to the fastener, avoid using and open-end wrench if a large amount of tightening or loosening toque is required.

A combination wrench has both a box-end and open-end. Both ends are the same size.

Adjustable Wrenches

An adjustable wrench (**Figure 10**) can be adjusted to fit virtually any nut or bolt head. However, it can loosen and slip from the nut or bolt, causing damage to the nut and possible physical injury. Use an adjustable wrench only if a proper size open-end or box-end wrench in not available. Avoid using an adjustable wrench if a large amount of tightening or loosening torque is required.

Adjustable wrenches come in sized ranging from 4-18 in. overall length. A 6 or 8 in. size is recommended as an all-purpose wrench.

Socket Wrenches

A socket wrench (**Figure 11**) is generally faster, safer and more convenient to use than a common wrench. Sockets, which attach to a suitable handle, are available with six-point or 12-point openings and use 1/4, 3/8, and 1/2 in. drive sizes. The drive size corresponds to the square hole that mates with the ratchet or flex handle.

Torque Wrench

A torque wrench (**Figure 12**) is used with a socket to measure how tight a nut or bolt is installed. They come in a wide price range and in 1/4, 3/8, and 1/2 in. drive sizes. The drive size

Box-end wrenches (**Figure 8**) provide a better grip on the nut and are stronger than open end wrenches. An open-end wrench (**Figure 9**) grips the nut on only two flats. Unless it fits well, it may slip and round off the points on the nut. A box-end wrench grips all six flats. Box-end wrenches are available with six-point or 12 point openings. The six-point opening provides

corresponds to the square hole that mates with the socket.

A typical 1/4 in. drive torque wrench measures in in.-lb. increments, and has a range of 20-150 in.-lb. (2.2-17 N•m). A typical 3/8 or 1/2 in. torque measures in ft.-lb. increments, and has a range of 10-150 ft.-lb. (14-203 N•m).

Impact Driver

An impact driver (**Figure 13**) makes removal of tight fasteners easy and reduces damage to bolts and screws. Interchangeable bits allow use on a variety of fasteners.

Snap Ring Pliers

Snap ring pliers are required to remove snap rings. Snap ring pliers (**Figure 14**) usually come with different size tips; many designs can be switched to handle internal or external type snap rings.

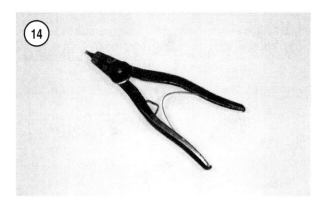

Hammers

Various types of hammers (**Figure 15**) are available to accommodate a number of applications. Use a ball-peen hammer to strike another tool, such as a punch or chisel. Use a soft-face hammer to strike a metal object without damaging it.

Never use a metal-faced hammer on engine and drive system components as severe damage will occur. You can always produce the same amount of force with a soft-faced hammer.

Always wear eye protection when using hammers. Make sure the hammer is in good condition and that the handle is not cracked. Select the correct hammer for the job and always strike the object squarely. Do not use the handle or the side of the hammer head to stroke an object.

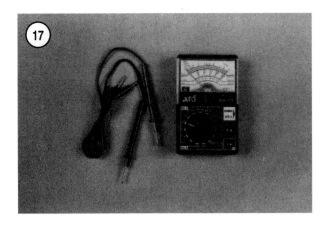

Feeler Gauges

This tool has either flat or wire measuring gauges (**Figure 16**). Use wire gauges to measure spark plug gap; use flat gauges for other measurements. A nonmagnetic (brass) gauge may be specified if working around magnetized components.

Other Special Tools

Many of the maintenance and repair procedures require special tools. Most of the necessary tools are available from a marine dealership or from tool suppliers. Instructions for their use and the manufacture's part number are included in the appropriate chapter.

Purchase the required tools from a local marine dealership or tool supplier. A qualified machinist, often at a lower price, can make some tools locally. Many marine dealerships and rental outlets will rent some of the required tools. Avoid using makeshift tools. Their use may result in damaged parts that cost far more than the recommended tool.

TEST EQUIPMENT

This section describes equipment used to perform testing, adjustments and measurements on marine engines. Most of these tools are available from a local marine dealership or automotive parts store.

Multimeter

This instrument is invaluable for electrical troubleshooting and service. It combines a voltmeter, ohmmeter and an ammeter in one unit. It is often called a VOM.

Two types of mutimeter are available, analog and digital. Analog meters (**Figure 17**) have a moving needle with marked bands on the meter face indicating the volt, ohm and amperage scales. An analog meter must be calibrated each time the scale is changed.

A digital meter (**Figure 18**) is ideally suited for electrical troubleshooting because it is easy to read and more accurate than an analog meter. Most models are auto-ranging, have automatic polarity compensation and internal overload protection circuits.

Either type of meter is suitable for most electrical testing described in this manual. An analog meter is better suited for testing pulsing voltage signals such as those produced by the ignition system. A digital meter is better suited for testing very low resistance or voltage reading (less than 1 volt or 1 ohm). The test procedure will indicate if a specific type of meter is required.

The ignition system produces electrical pulses that are too short in duration for accurate measurement with a using a conventional multimeter. Use a meter with peak-volt reading ca pability to test the ignition system. This type of meter captures the peak voltage reached during an electrical pulse.

Scale selection, meter specifications and test connections vary by the manufacturer and model of the meter. Thoroughly read the instructions supplied with the meter before performing any test. The meter and certain electrical components on the engine can be damaged if tested incorrectly. Have the test performed by a qualified professional if you are unfamiliar with the testing or general meter usage. The expense to replace damaged equipment can far exceed the cost of having the test performed by a professional.

Strobe Timing Light

This instrument is necessary for dynamic tuning (setting ignition timing while the engine is running). By flashing a light at the precise instant the spark plug fires, the position of the timing mark can be seen. The flashing light makes a moving mark appear to stand still next to a stationary mark.

Timing lights (**Figure 19**) range from inexpensive models with a neon bulb to expensive models with a xenon bulb, built in tachometer and timing advance compensator. A built in tachometer is very useful as most ignition timing

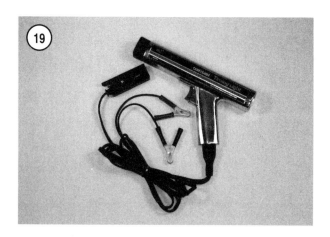

specifications are based on a specific engine speed.

A timing advance compensator delays the strobe enough to bring the timing mark to a certain place on the scale. Although useful for troubleshooting purposes, this feature should not be used to check or adjust the base ignition timing.

Tachometer/Dwell Meter

A portable tachometer (**Figure 20**) is needed to tune and test most marine engines. Ignition timing and carburetor adjustments must be performed at a specified engine speed. Tachometers are available with either an analog or digital display.

The fuel/air mixture must be adjusted with the engine running at idle speed. If using an analog

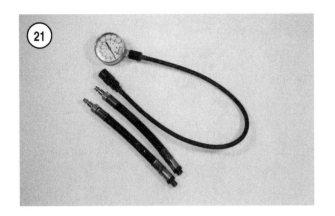

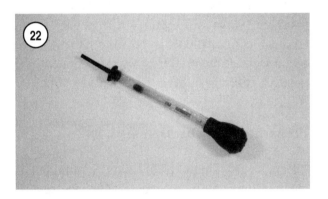

vide accurate measurement at all speeds without the need to change the range or scale. Many of these use an inductive pickup to receive the signal from the ignition system.

A dwell meter is often incorporated into the tachometer to allow testing and/or adjustments to engines with a breaker point ignition system.

Compression Gauge

This tool (**Figure 21**) measures the amount of pressure created in the combustion chamber during the compression stroke. Compression indicates the general engine condition making it one of the most useful troubleshooting tools.

The easiest type to use has screw-in adapters that fit the spark plug holes. Rubber tipped, press-in type gauges are also available. This type must be held firmly in the spark plug hole to prevent leakage and inaccurate test results..

Hydrometer

Use a hydrometer to measure specific gravity in the battery. Specific gravity is the density of the battery electrolyte as compared to pure water and indicates the battery's state of charge. Choose a hydrometer (**Figure 22**) with automatic temperature compensation; otherwise the electrolyte temperature must be measured during charging to determine the actual specific gravity.

Precision Measuring Tools

Various tools are required to make precision measurements. A dial indicator (**Figure 23**), for example, is used to determine piston position in the cylinder, runout and end play of shafts and assemblies. It is also used to measure free movement between the gear teeth (backlash) in the drive unit.

tachometer, choose one with a low range of 0-1000 rpm or 0-2000 rpm range and a high range of 0-6000 rpm. The high range setting is needed for testing purposes but lacks the accuracy needed at lower speeds. At lower speeds the meter must be capable of detecting changes of 25 rpm or less.

Digital tachometers are generally easier to use than most analog type tachometers. They pro-

Venier calipers (**Figure 24**), micrometers (**Figure 25**) and other precision tools are used to measure the size of parts, such as the piston.

Precision measuring equipment must be stored, handled and used carefully or it will not remain accurate.

SERVICE HINTS

Most of the service procedures in this manual are straightforward and can be performed by anyone reasonably handy with tools. It is suggested, however, that you consider your skills and available tools and equipment before attempting a repair involving major disassembly of the engine or drive unit.

Some operations, for example, require the use of a press. Other operations require precision measurement. Have the procedure or measurements performed by a professional if you do not have access to the correct equipment or are unfamiliar with its use.

Special Battery Precautions

Disconnecting or connecting the battery can create a spike or surge of current throughout the electrical system. This spike or surge can damage certain components of the charging system. Always verify the ignition switch is in the OFF position before connecting or disconnecting the battery or changing the selection on a battery switch.

Always disconnect both battery cables and remove the battery from the boat for charging. If the battery cables are connected, the charger may induce a damaging spike or surge of current into the electrical system. During charging, batteries produce explosive and corrosive gasses. These gases can cause corrosion in the battery compartment and creates an extremely hazardous condition.

Disconnect the cables from the battery prior to testing, adjusting or repairing many of the systems or components on the engine. This is nec-

essary for safety, to prevent damage to test equipment and to ensure accurate testing or adjustment. Always disconnect the negative battery cable first, then the positive cable. When reconnecting the battery, always connect the positive cable first, then the negative cable.

Preparation for Disassembly

Repairs go much faster if the equipment is clean before you begin work. There are special cleaners such as Gunk or Bel-Ray Degreaser, for cleaning the engine and related components. Just spray or brush on the cleaning solution, let it stand, then rinse with a garden hose.

Use pressurized water to remove marine growth and corrosion or mineral deposits from external components such as the gearcase, drive shaft housing and clamp brackets. Avoid directing pressurized water directly as seals or gaskets; pressurized water can flow past seal and gasket surfaces and contaminate lubricating fluids.

> *WARNING*
> *Never use gasoline as a cleaning agent. It presents an extreme fire hazard. Always work in a well-ventilated area if using cleaning solvent. Keep a Coast Guard approved fire extinguisher, rated for gasoline fires, readily accessible in the work area.*

Much of the labor charged for a job performed at a dealership is usually for removal and disas-

sembly of other parts to access defective parts or assemblies. It is frequently possible to perform most of the disassembly then take the defective part or assembly to the dealership for repair.

If you decide to perform the job yourself, read the appropriate section in this manual, in its entirety. Study the illustrations and text until you fully understand what is involved to complete the job. Make arrangements to purchase or rent all required special tools and equipment before starting.

Disassembly Precautions

During disassembly, keep a few general precautions in mind. Force is rarely needed to get things apart. If parts fit tightly, such as a bearing on a shaft, there is usually a tool designed to separate them. Never use a screwdriver to separate parts with a machined mating surface, such as the cylinder head or manifold. The surfaces will be damaged and leak.

Make diagrams or take instant photographs wherever similar-appearing parts are found. Often, disassembled parts are left for several days or longer before resuming work. You may not remember where everything came from, or carefully arranged parts may become disturbed.

Cover all openings after removing parts to keep contamination or other parts from entering.

Tag all similar internal parts for location and mounting direction. Reinstall all internal components in the same location and mounting direction as removed. Record the thickness and mounting location of any shims as they are removed. Place small bolts and parts in plastic sandwich bags. Seal and label the bags with masking tape.

Tag all wires, hoses and connections and make a sketch of the routing. Never rely on memory alone; it may be several days or longer before you resume work.

Protect all painted surfaces from physical damage. Never allow gasoline or cleaning solvent on these surfaces.

Assembly Precautions

No parts, except those assembled with a press fit, require unusual force during assembly. If a part is hard to remove or install, find out why before proceeding.

When assembling parts, start all fasteners, then tighten evenly in an alternating or crossing pattern unless a specific tightening sequence or procedure is given.

When assembling parts, be sure all shims, spacers and washers are installed in the same position and location as removed.

Whenever a rotating part butts against a stationary part, look for a shim or washer. Use new gaskets, seals and O-rings if there is any doubt about the conditions of the used ones. Unless otherwise specified, a thin coating of oil on gaskets may help them seal more effectively. Use heavy grease to hold small parts in place if they tend to fall out during assembly.

Use emery cloth and oil to remove high spots from piston surfaces. Use a dull screwdriver to remove carbon deposits from the cylinder head, ports and piston crown. *Do not* scratch or gouge these surfaces. Wipe the surfaces clean with a *clean* shop towel when finished.

If the carburetor must be repaired, completely disassemble it and soak all metal parts in a commercial carburetor cleaner. Never soak gaskets and rubber or plastic parts in these cleaners.

Clean rubber or plastic parts in warm soapy water. Never use a wire to clean jets and small passages because they are easily damaged. Use compressed air to blow debris from all passages in the carburetor body.

Take your time and do the job right. Break-in procedure for a newly rebuilt engine or drive is the same as for a new one. Use the recommended break-in oil and follow the instructions provided in the appropriate chapter.

SPECIAL TIPS

Because of the extreme demands placed on marine equipment, several points must be kept in mind when performing service and repair. The following are general suggestions that may improve the overall life of the machine and help avoid costly failure.

1. Unless otherwise specified, apply a threadlocking compound, such as Loctite Threadlocker, to all bolts and nuts, even if secured with a lockwasher. Use only the specified grade of threadlocking compound. A screw or bolt lost from an engine cover or bearing retainer could easily cause serious and expensive damage before the loss is noticed. When applying threadlocking compound, use only enough to lightly coat the threads. If too much is used, it can work its way down the threads and contaminate seals or bearings.

2. If self-locking fasteners are used, replace them with new ones. Do not install standard fasteners in place of self-locking ones.

3. Use caution when using air tools to remove stainless steel nuts or bolts. The heat generated during rapid spinning easily damages the threads of stainless steel fasteners. To prevent thread damage, apply penetrating oil as a cooling agent and loosen or tighten them slowly.

4. Use a wide chisel to straighten the tab of a fold-over type lockwasher. Such a tool provides a better contact surface than a screwdriver or pry bar, making straightening easier. During installa-

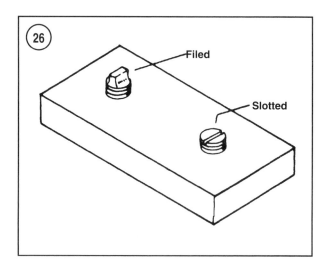

tion, use a new fold-over type lockwasher. If a new lockwasher is not available, fold over a tab on the washer that has not been previously used. Reusing the same tab may cause the washer to break, resulting in a loss of locking ability and a loose piece of metal adrift in the engine. When folding the tab into position, carefully pry it toward the flat on the bolt or nut. Use a pair or plies to bend the tab against the fastener. Do not use a punch and hammer to drive the tab into position. The resulting fold may be too sharp, weakening the washer and increasing its chance of failure.

5. Use only the specified replacement parts if replacing a missing or damaged bolt, screw or nut. Many fasteners are specially hardened for the application.

6. Install only the specified gaskets. Unless specified otherwise, install them without sealant. Many gaskets are made with a material that swells when it contacts oil. Gasket sealer prevents them from swelling as intended and can result in oil leakage. Most gaskets must be a specific thickness. Installing a gasket that is too thin or too thick in a critical area could cause expensive damage.

7. Make sure all shims and washers are reinstalled in the same location and position. Whenever a rotating part contacts a stationary part, look for a shim or washer.

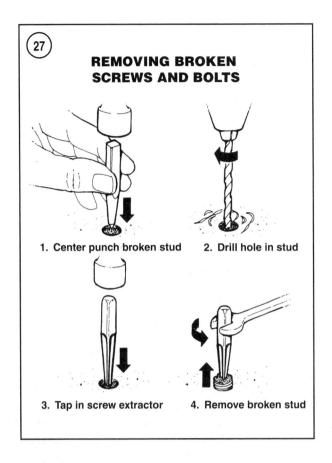

REMOVING BROKEN SCREWS AND BOLTS

1. Center punch broken stud

2. Drill hole in stud

3. Tap in screw extractor

4. Remove broken stud

MECHANICS TECHNIQUES

Marine engines are subjected to conditions very different from most engines. They are repeatedly subjected to a corrosive environment followed by periods of non-use for weeks or longer. Such use invites corrosion damage to fasteners, causing difficulty or breakage during removal. This section provides information that is useful for removing stuck or broken fasteners and repairing damaged threads.

Removing Stuck Fasteners

When a nut or bolt corrodes and cannot be removed, several methods may be used to loosen it. First, apply penetrating oil, such as Liquid Wrench or WD-40. Apply it liberally to the threads and allow it to penetrate for 10-15 minutes. Tap the fastener several times with a small

hammer; however, do not hit it hard enough to cause damage. Reapply the penetrating oil if necessary.

For stuck screws, apply penetrating oil as described, then insert a screwdriver in the slot. Tap the top of the screwdriver with a hammer. This looses the corrosion in the threads allowing it to turn. If the screw head is too damaged to use a screwdriver, grip the head with locking pliers and twist the screw from the assembly.

A Phillips, Allen or Torx screwdriver may start to slip in the screw during removal. If slippage occurs, stop immediately and apply a dab of course valve lapping compound onto the tip of the screwdriver. Valve lapping compound or a special screw removal compound is available from most hardware and automotive parts stores. Insert the driver into the screw and apply downward pressure while turning. The gritty material in the compound improves the grip on the screw, allowing more rotational force before slippage occurs. Keep the compound away from any other engine components. It is very abrasive and can cause rapid wear if applied onto moving or sliding surfaces.

Avoid applying heat unless specifically instructed because it may melt, warp or remove the temper from parts.

Removing Broken Bolts or Screws

The head of bolt or screw may unexpectedly twist off during removal. Several methods are available for removing the remaining portion of the bolt or screw.

If a large portion of the bolt or screw projects out, try gripping it with locking pliers. If the projecting portion is too small, file it to fit a wrench or cut a slot in it to fit a screwdriver (**Figure 26**). If the head breaks off flush or cannot be turned with a screwdriver or wrench, use a screw extractor (**Figure 27**). To do this, center punch the remaining portion of the screw or bolt. Select the proper size of extractor for the size of the fastener. Using the drill size specified on the extractor, drill a hole into the fastener. Do not drill deeper than the remaining fastener. Carefully tap the extractor into the hole and back the remnant out using a wrench on the extractor.

Remedying Stripped Threads

Occasionally, threads are stripped through carelessness or impact damage. Often the threads can be repaired by running a tap (for internal threads on nuts) or die (for external threads on bolts) through threads (**Figure 28**).

To clean or repair spark plug threads, use a spark plug tap. If an internal thread is damaged, it may be necessary to install a Helicoil or some other type of thread insert. Follow the manufacturer's instructions when installing their insert.

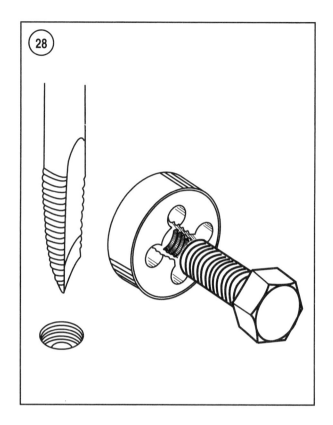

Chapter Three

Troubleshooting

Troubleshooting is a relatively simple matter when it is done logically. The first step in any troubleshooting procedure is to define the symptoms as closely as possible and then localize the problem. Subsequent steps involve testing and analyzing those areas which could cause the symptoms. A haphazard approach may eventually solve the problem, but it can be very costly in terms of wasted time and unnecessary parts replacement.

Proper lubrication, maintenance and periodic tune-ups as described in Chapter Four and Chapter Five will reduce the necessity for troubleshooting. Even with the best of care, however, an outboard motor is prone to problems which will require troubleshooting. This chapter contains brief descriptions of each operating system and troubleshooting procedures to be used. **Tables 1-4** present typical starting, ignition and fuel system problems with their probable causes and solutions. **Tables 1-7** are at the end of the chapter.

OPERATING REQUIREMENTS

Every outboard motor requires 3 basic things to run properly: an uninterrupted supply of fuel and air in the correct proportions, proper ignition at the right time and adequate compression. If any of these are lacking, the motor will not run.

The electrical system is the weakest link in the chain. More problems result from electrical malfunctions than from any other source. Keep this in mind before you blame the fuel system and start making unnecessary carburetor adjustments.

If a motor has been sitting for any length of time and refuses to start, check the condition of the battery first to make sure it has an adequate charge, then look to the fuel delivery system. This includes the gas tank, fuel pump, fuel lines and carburetor(s). Rust may have formed in the tank, obstructing fuel flow. Gasoline deposits may have gummed up carburetor jets and air passages. Gasoline tends to lose its potency after standing for long periods. Condensation may contaminate it with water. Drain the old gas and try starting with a fresh tankful.

STARTING SYSTEM

Description

Chrysler 9.2 hp and larger outboard motors may be equipped with an electric starter motor (**Figure 1**). The motor is mounted vertically on the engine. When battery current is supplied to the starter motor, its pinion gear is thrust upward to engage the teeth on the engine flywheel. Once the engine starts, the pinion gear disengages from the flywheel. This is similar to the method used in cranking an automotive engine.

Autolectric models use a starter-generator assembly bracket-mounted above the flywheel. The armature shaft of the starter-generator connects directly to the crankshaft. When battery current is supplied to the starter-generator, the armature shaft rotates the crankshaft to start the engine and continues rotating with the crankshaft to produce electrical current.

The electric starting system requires a fully charged battery to provide the large amount of current required to operate the starter motor or starter-generator. On models with an automotive-type starter, the battery may be charged externally or by a lighting coil on the alternator stator plate which keeps the battery charged while the engine is running. On Autolectric models, the starter-generator produces sufficient current to maintain a full battery charge.

The starting circuit on all outboards equipped with an electric starting system consists of the battery, an ignition or starter switch, an interlock switch, the starter motor or starter-generator and connecting wiring. On smaller displacement engines, electrical current is transmitted from the battery to the starter motor through a heavy-duty starter switch. See **Figure 2**. Depressing the starter switch completes the circuit between the battery and starter motor.

Larger displacement engines use a solenoid or starter relay to carry the heavy electrical

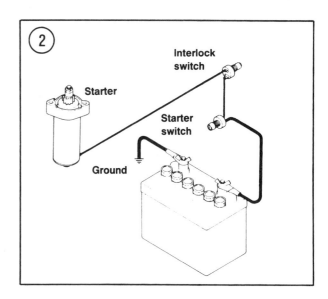

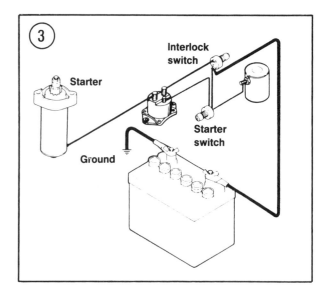

current to the motor. See **Figure 3**. Depressing the starter switch or turning the ignition switch to the START position allows current to flow through the solenoid coil. The solenoid contacts close and allow current to flow from the battery through the solenoid to the starter motor or starter-generator.

An interlock switch in all starting circuits prevents current flow to the starter motor or starter-generator if the shift mechanism is not in NEUTRAL.

The starting circuit on some models may also include one or more of the following:

    a. Stop or ignition "kill" switch.
    b. Circuit breaker.
    c. Choke solenoid.

Figure 4 (35-55 hp) and **Figure 5** (20 hp Autolectric) are simplified schematics of typical starting systems.

CAUTION
Do not operate an electric starter motor or starter-generator continuously for more than 15 seconds. Allow the motor to cool for at least 3 minutes between attempts to start the engine.

Troubleshooting Procedures

Refer to **Table 1** at the end of the chapter. Before troubleshooting the starting circuit, make sure that:

    a. The battery is fully charged.
    b. The shift mechanism is in NEUTRAL.

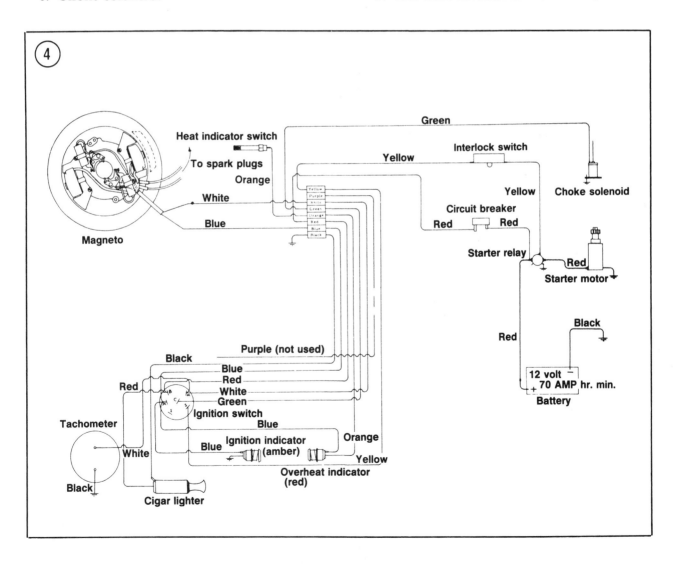

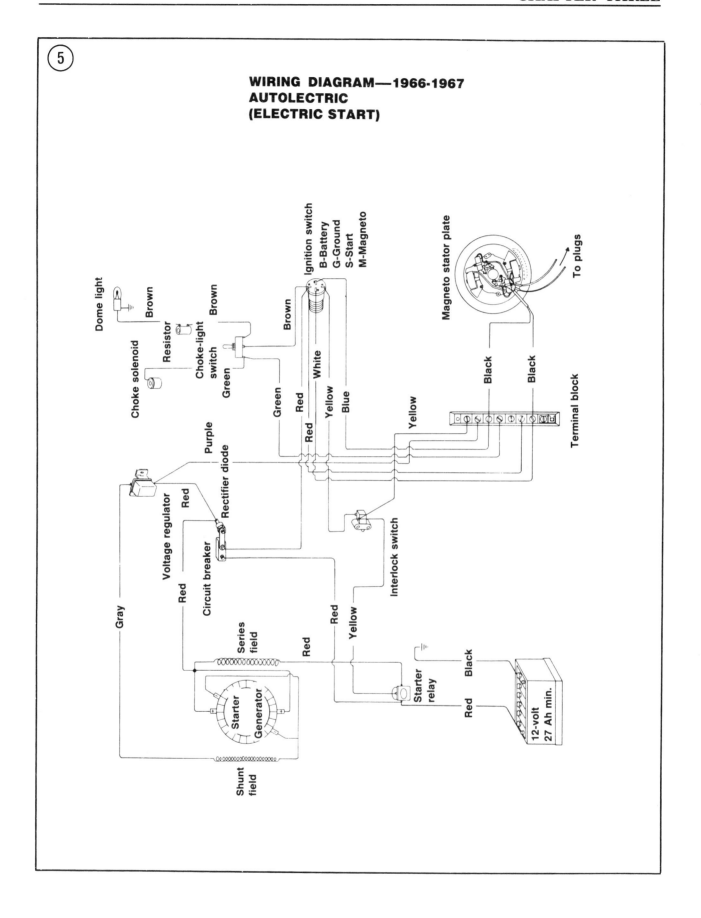

⑤

**WIRING DIAGRAM—1966-1967
AUTOLECTRIC
(ELECTRIC START)**

c. All electrical connections are clean and tight.

d. The wiring harness is in good condition, with no worn or frayed insulation.

e. Battery cables are the proper size and length. Replace cables that are undersize or relocate the battery to shorten the distance between battery and starter solenoid.

f. The fuel system is filled with an adequate supply of fresh gasoline that has been properly mixed with Chrysler Outboard Oil. See Chapter Four.

A 12-volt test light is a very simple and accurate device for troubleshooting the starting system when used with a wiring diagram for the engine being tested. One lead of the test light must be connected to ground at all times, unless otherwise specified. When the other lead is connected to a hot circuit, the light will come on.

The troubleshooting procedures in this chapter are intended only to isolate a malfunction to a certain component. If further bench testing is then necessary, remove the suspected component and have it tested by an authorized service center. It is often less expensive to just replace a faulty component than to have it tested and repaired. Refer to Chapter Seven for component removal and installation procedures.

Autolectric Models

Refer to **Figure 6** (9.2 hp), **Figure 7** (9.9 and 12.9 hp), **Figure 5** (1966-1967 20 hp) or **Figure 8** (1968-on 20 hp) for this procedure.

1. Check test light by connecting its leads to the positive and negative battery terminals. The light should come on. If it does not, check the test leads for loose terminals or replace the test light bulb.

2. Rotate engine twist grip to STOP position to close interlock switch.

3. Connect one test lead to a good engine ground.

4. Connect the other test lead to the input side of the starter relay. If the lamp does not light, the battery cable terminal connections are loose or there is an open in the cable between the battery and the starter relay. Tighten terminal connections or replace the battery cable as required.

5. Connect the test lead to the output side of the starter relay and turn the ignition switch to the START position. The light should come on and the starter should turn the engine over. If the light comes on but the engine does not turn over or turns over slowly, remove the starter and have it bench tested.

6. If the light does not come on in Step 5, connect the test light lead to the small terminal (yellow lead) of the starter relay and turn the ignition switch to the START position. The light should not come on. If it does, the starter relay is defective.

7. Connect the test lead to the outer terminal of the interlock switch. The light should not come on. If it does, check the yellow wire between the interlock switch and starter relay for an open circuit or loose connections.

8. Connect the test lead to the inner terminal of the interlock switch. If the light comes on, replace the interlock switch.

9. Connect the test lead to the start (S) terminal of the ignition switch (yellow lead). See **Figure 9**. Turn the ignition switch to the START position. The light should not come on. If it does, repair or replace the wire between the ignition and interlock switches as required.

10. Connect the test lead to the battery (B) terminal of the ignition switch (red lead). See **Figure 10**. Turn the ignition switch to the START position. The light should not come on. If it does, replace the ignition switch.

3

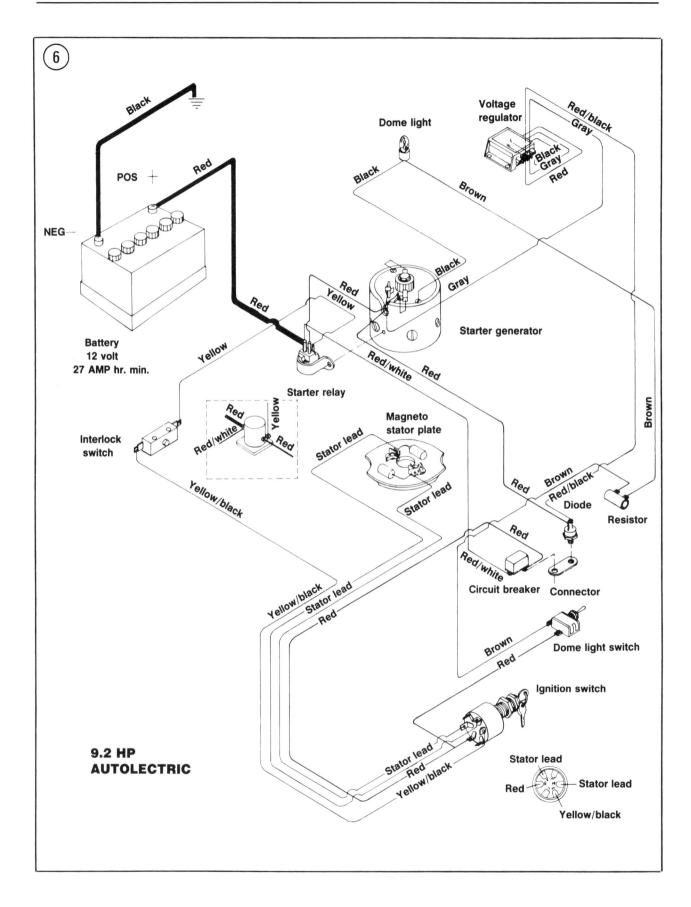

(6)

Black

POS +   Red

NEG—

Battery
12 volt
27 AMP hr. min.

Yellow

Red
Yellow

Red

Interlock
switch

Red
Red/white   Yellow
Red

Starter relay

Yellow/black

Dome light

Black

Brown

Black
Gray

Starter generator

Red/white   Red

Voltage
regulator

Red/black
Gray

Black
Gray
Red

Brown

Magneto
stator plate

Stator lead

Stator lead

Red   Brown
Red/black

Diode

Resistor

Red

Red/white

Circuit breaker   Connector

Dome light switch

Brown
Red

Ignition switch

Yellow/black
Stator lead
Red

Stator lead
Red
Yellow/black

Stator lead
Red
Yellow/black

Stator lead

Red   Stator lead

Yellow/black

9.2 HP
AUTOLECTRIC

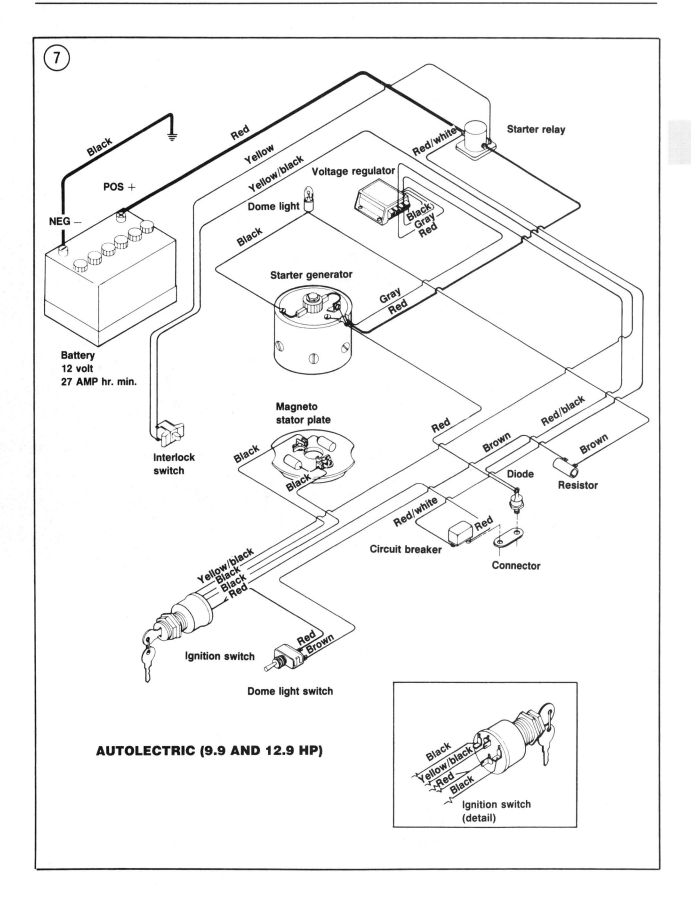

AUTOLECTRIC (9.9 AND 12.9 HP)

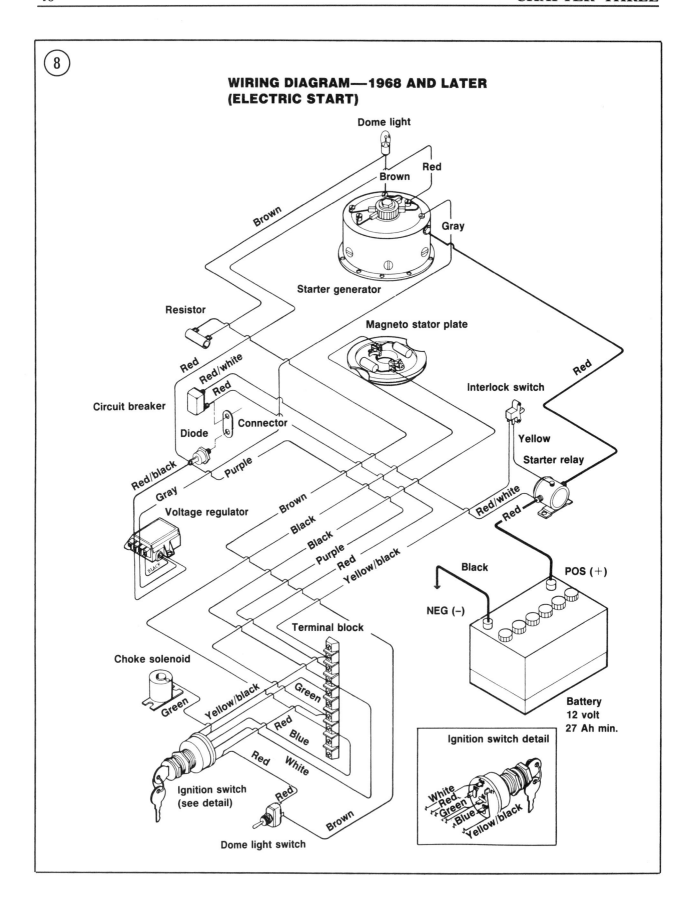

⑧

**WIRING DIAGRAM—1968 AND LATER
(ELECTRIC START)**

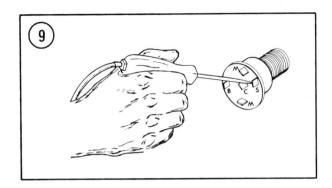

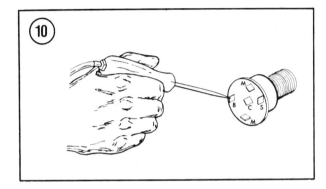

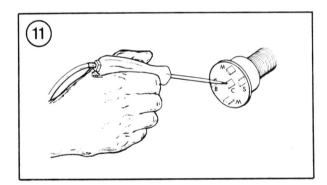

11. Connect the test lead to the long terminal on the circuit breaker. The light should not come on. If it does, repair or replace the wire between the ignition switch and circuit breaker as required.

12. Connect the test lead to the short terminal on the circuit breaker. The light should not come on. If it does, replace the circuit breaker.

13. If the light does not come on in Step 12 and the engine will not start, check for a defect in the wire between the circuit breaker and starter relay.

NOTE
Steps 14-16 apply only to 20 hp models.

14. Connect the test lead to the choke terminal on the choke solenoid. Turn the ignition switch to the ON position and depress ignition key to activate the solenoid. If the light comes on, replace the choke solenoid.

15. Connect the test lead to the choke (C) terminal on the ignition switch (green lead). See **Figure 11**. The light should not come on. If it does, repair or replace the lead between the ignition switch and choke solenoid.

16. If the light does not come on in Step 15 and the engine will not start, replace the ignition switch.

CAUTION
The engine must be provided with an adequate supply of water while performing Step 17 and Step 18. Install a flushing device, place the engine in a test tank or perform the steps with the boat in the water.

17. Connect a tachometer according to manufacturer's instructions. Connect the test lead to the gray lead of the voltage regulator. Start the engine and run until the test light comes on. Note the tachometer reading. The light should come on below 3,700 rpm. If the light does not come on until engine speed exceeds 3,700 rpm, shut the engine off and proceed with the next step.

18. Disconnect the blocking diode red lead from the voltage regulator ignition terminal. Disconnect the starter frame assembly gray lead from the voltage regulator field terminal. Connect the red and gray leads together with the test light lead. Start the engine and run at full throttle. If the light comes on, replace the voltage regulator. If the light does not come on, the problem is in the starter-generator.

19. Disconnect the spark plug leads.

3

20. Connect a volt-ammeter to the starter-generator according to manufacturer's instructions.

21. Turn the ignition key to the START position and note the starter current draw when the meter needle stabilizes. If the reading is 55-75 amps (9.2-15 hp) or 65-70 amps (20 hp), the problem is not in the starting circuit. If the reading is outside the specified range, continue testing.

22. Remove the spark plugs. Shift engine into NEUTRAL.

23. Connect a tachometer according to manufacturer's instructions.

24. Repeat Step 21. The engine should crank at 700-900 rpm with a current draw of approximately 40 amps. If it does not, refer to **Table 2**.

All Others
(Without Starter Solenoid or Relay)

Refer to **Figure 12** for this procedure.

1. Place the engine in NEUTRAL and connect one test light lead to a good engine ground.

2. Connect the other lead to the starter switch input terminal (1, **Figure 12**). The test light should come on. If the lamp does not light or is very dim, the battery cable terminal connections are loose or there is an open in the cable between the battery and the ignition switch. Tighten terminal connections or replace the battery cable as required.

3. Connect the test lead to the starter switch output terminal (2, **Figure 12**) and depress the switch button. If the light does not come on, replace the starter switch.

4. Connect the test lead to the interlock switch input terminal (3, **Figure 12**) and depress the switch button. If the light does not come on, repair or replace the wire between the interlock and ignition switches.

5. Connect the test lead to the interlock switch output terminal (4, **Figure 12**) and depress the switch button. If the light does not come on, check the gear shift linkage adjustment. See Chapter Nine. If adjustment is correct, replace the interlock switch.

6. Connect the test lead to the starter motor terminal (5, **Figure 12**) and depress the starter switch button. The test lamp should not light. If it does, replace the starter.

7. If the test lamp does not light in Step 6 and the engine still does not start, repair or replace the starter to interlock switch lead as required.

All Others
(With Starter Solenoid or Relay)

Refer to **Figure 13** (typical) for this procedure.

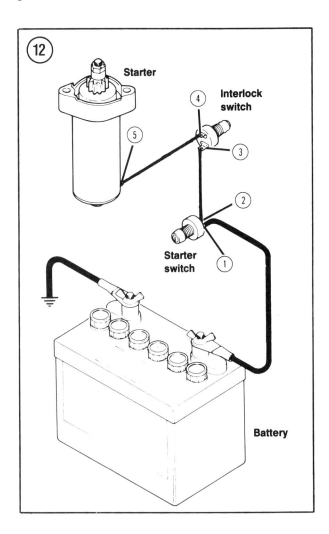

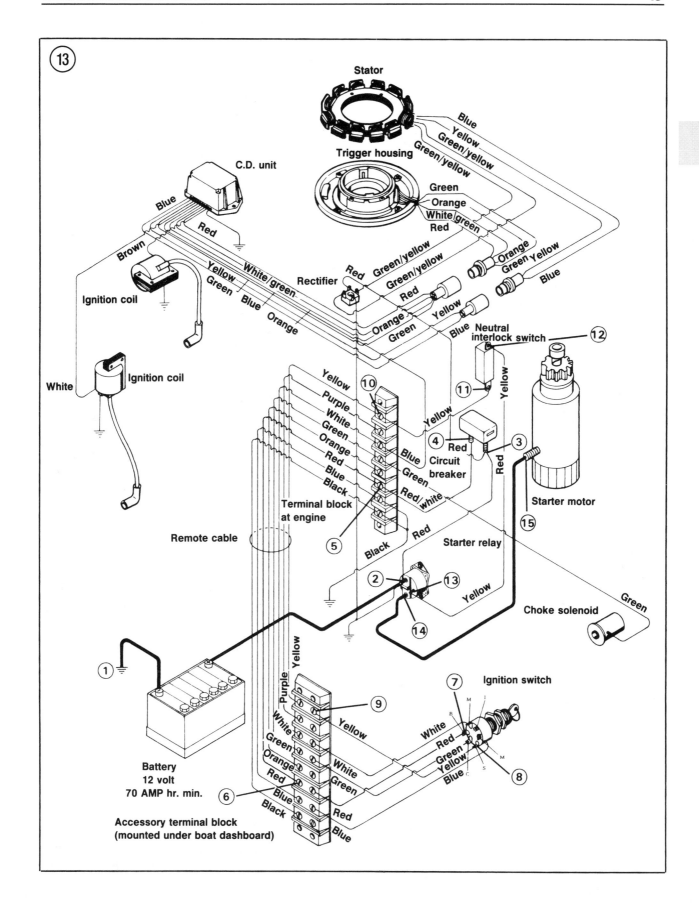

1. Place the engine in NEUTRAL and connect one test light lead to a good engine ground.

2. Connect the other lead to the starter relay input terminal (2, **Figure 13**). The test light should come on. If the lamp does not light or is very dim, the battery cable terminal connections are loose or there is an open in the cable between the battery and the starter relay. Tighten terminal connections or replace the battery cable as required.

3. Connect the test lead to the circuit breaker input terminal (3, **Figure 13**). If the light does not come on, repair or replace the wire between the starter relay and circuit breaker as required.

NOTE
If the circuit breaker reset button pops out when depressed in Step 4, the circuit breaker is not defective. There is a short in the starting circuit that must be located and corrected.

4. Connect the test lead to the circuit breaker output terminal (4, **Figure 13**). If the light does not come on, depress the reset button. If the light still does not come on, replace the circuit breaker.

5. Connect the test lead to the red wire terminal on the terminal block (5, **Figure 13**), if so equipped. If the light does not come on, repair or replace the wire between the circuit breaker and terminal block as required.

6. Connect the test lead to the red wire terminal on the accessory block (6, **Figure 13**), if so equipped. If the light does not come on, repair or replace the red lead in the remote cable as required.

7. Connect the test lead to the battery (B) terminal on the ignition switch (7, **Figure 13**). If the light does not come on, repair or replace the red wire between the ignition switch and accessory block (if so equipped) or ignition switch and circuit breaker as required.

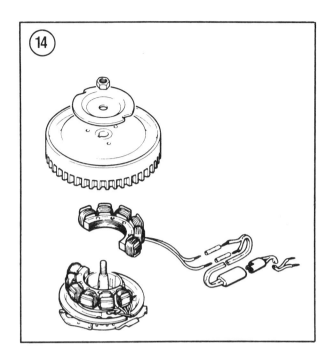

8. Connect the test lead to the start (S) terminal on the ignition switch (8, **Figure 13**). If the light does not come on, check for an open or short in the yellow wire from the ignition switch to the starter relay. If the wiring is good, replace the ignition switch.

NOTE
The ignition switch must be in the START position during Steps 9-15.

9. Connect the test lead to the yellow wire terminal of the accessory block (9, **Figure 13**), if so equipped. If the light does not come on, repair or replace the yellow wire between the ignition switch and accessory block.

10. Connect the test lead to the yellow wire terminal of the terminal block (10, **Figure 13**), if so equipped. If the light does not come on, repair or replace the yellow wire in the remote cable as required.

11. Connect the test lead to the interlock switch terminal (11, **Figure 13**). If the light does not come on, repair or replace the yellow lead between the interlock and ignition switches (models with a terminal block) or the red lead between the interlock

switch and battery (models without a terminal block).

12. Connect the test lead to the interlock switch terminal (12, **Figure 13**). If the light does not come on, manually depress the interlock switch plunger. If the light still does not come on, replace the switch.

13. Connect the test lead to the starter relay yellow lead terminal (models with a terminal block) or red lead terminal (models without a terminal block). If the light does not come on, repair or replace the lead between the starter relay and interlock switch as required.

14. Connect the test lead to the starter relay-to-starter terminal (14, **Figure 13**). If the light does not come on, replace the starter relay.

15. Connect the test lead to the starter motor positive terminal (15, **Figure 13**). The light should not come on. If it does, replace the starter.

NOTE
The ignition switch must be depressed and held in the CHOKE position for the remaining steps.

16. Connect the test lead to the choke (C) terminal on the ignition switch (green wire). If the light does not come on, replace the ignition switch.

17. Connect the test lead to the green wire terminal on the accessory block, if so equipped. If the light does not come on, repair or replace the green wire between the accessory block and ignition switch.

18. Connect the test lead to the green wire terminal on the terminal block, if so equipped. If the light does not come on, repair or replace the green wire in the remote cable as required.

19. Connect the test lead to the choke solenoid terminal. If the light comes on, disconnect the negative battery cable and connect an ohmmeter between the solenoid

terminal and case. If the meter does not read 0.5-1.35 ohms, replace the solenoid.

CHARGING SYSTEM

Four types of charging systems are used on Chrysler outboards. An AC lighting system or DC battery charging system may be fitted to 3.5-7.5 hp models. A more sophisticated alternator charging system is used on 9.9 hp and larger engines. Autolectric models use a starter-generator.

AC LIGHTING SYSTEM

The AC lighting system consists of an alternator stator and flywheel (**Figure 14**). The system provides alternating current to operate accessories such as boat lights only when the engine is running. Poor connections, defective wiring insulation or the use of too many accessories are major causes of AC lighting system problems.

Troubleshooting

1. Disconnect the male and female power supply plugs at the engine.

2. Carefully inspect all wiring for worn or cracked insulation and corroded or loose connections. Replace wiring or clean and tighten connections as required.

NOTE
Do not connect test lamp in series with any lights or accessories as they will drain the test lamp batteries.

3. Use a self-powered test lamp to check each length of wiring in the circuit between connection points. If the test lamp does not light when connected into the circuit, replace that section of wiring.

4. When the entire circuit has been checked, start the engine and check operation of the lights and accessories. If they still do not work, shut the engine off and connect the test

3

lamp leads to the female power supply plug at the engine. The test lamp should light.

5. If the test lamp does not light in Step 4, disconnect the stator lead splices and remove the female plug at the engine. Connect the test lamp leads between the stator leads. If the test lamp lights, one or both splices were defective. If the test lamp does not light, connect an ohmmeter between the stator leads. If the meter does not read aproximately one ohm, the stator leads are open or grounded.

6. If the ohmmeter reads one ohm in Step 5 but the lighting system still does not function properly, connect a tachometer according to manufacturer's instructions.

CAUTION
The engine must be provided with an adequate supply of water while performing Step 7. Install a flushing device, place the engine in a test tank or perform the step with the boat in the water.

7. Connect an AC ammeter between the stator leads. Start the engine and note the ammeter reading as engine speed is gradually increased to approximately 4,500 rpm. The stator should start to function at a cut-in speed of approximately 1,000 rpm, delivering a minimum output of 1 amp at 2,000 rpm and 3.5 amps at 4,000 rpm. If it does not meet these specifications, replace the stator.

DC BATTERY CHARGING SYSTEM

Models equipped with an AC lighting system may be converted to a DC battery charging system by installing Battery Charging Kit part No. 5H091 (3.5-4 hp) or part No. 5H105 (6-7.5 hp). This incorporates a rectifier to convert the AC lighting current to DC battery charging current. On 3.5-4 hp models, the rectifier and bracket assembly is installed on the fuel tank leg and grounded to the cylinder head. On 6-7.5 hp models, the rectifier and bracket assembly is installed at the top of the cylinder head and grounded to the exhaust cover as shown in **Figure 15**.

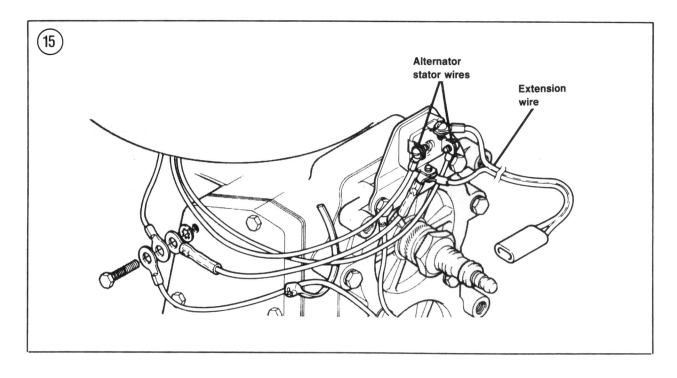

Troubleshooting

> *CAUTION*
> *The engine must be provided with an adequate supply of water while it is running during this procedure. Install a flushing device, place the engine in a test tank or perform the procedure with the boat in the water.*

Refer to **Figure 16** for this procedure.

1. Connect a tachometer according to manufacturer's instructions.

2. Start the engine and run at 2,500 rpm.

3. Connect one lead of a 12-volt test lamp to a good engine ground. Probe rectifier terminals A and B with the other test lead. The light should come on at each of the 3 terminals, but its intensity should be greater at terminal B than at either of the A terminals.

4. If the light does not come on or if the intensity is not as specified in Step 3, shut the engine off and disconnect the stator leads at the rectifier A terminals.

5. Connect an ohmmeter between the stator leads. If the meter does not read approximately one ohm, the stator leads are open or grounded. If the meter reads one ohm, continue testing.

6. Connect an AC ammeter between the stator leads. Start the engine and note the ammeter reading as engine speed is gradually increased to approximately 4,500 rpm. The stator should start to function at 1,000 rpm, delivering a minimum output of 1 amp at 2,000 rpm and 3.5 amps at 4,000 rpm. If it does not meet these specifications, replace the stator.

> *NOTE*
> *Some self-powered test lamps may have reverse polarity which will affect the results in Steps 7-10. If the results are the opposite of those stated, repeat the steps with the test leads reversed.*

7. If the stator is good, shut the engine off and connect the red lead of a self-powered test lamp to ground. Connect the black test lamp lead to one rectifier A terminal, then to the other A terminal. The light should glow brightly at each terminal.

8. Move the black test lamp lead to rectifier terminal B. The light should glow dimly.

9. Move the red test lead to terminal B. Connect the black test lead to one rectifier A terminal, then to the other A terminal. The light should not come on.

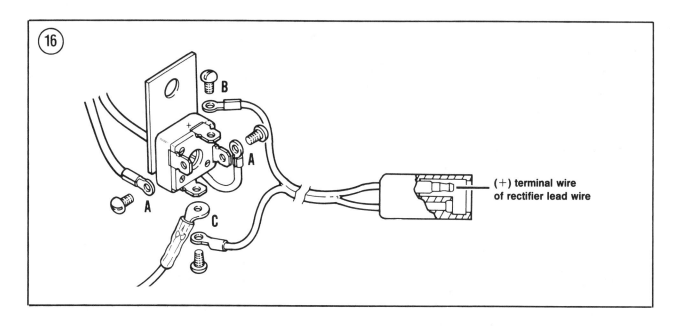

(+) **terminal wire of rectifier lead wire**

10. Repeat Step 9 with the test light leads reversed. The light should come on.

11. Replace the rectifier if it does not perform as indicated in Steps 7-10.

ALTERNATOR CHARGING SYSTEM

The charging system consists of the alternator stator coils, permanent magnets located within the flywheel, a rectifier (Prestolite) or regulator-rectifier (Motorola) to change alternating current to direct current, the circuit breaker, battery and connecting wiring. See **Figure 17**.

A malfunction in the alternator charging system generally causes the battery to remain undercharged. Since the alternator stator is protected by its location underneath the flywheel, it is more likely that the battery, rectifier, circuit breaker or connecting wiring will cause problems. The following conditions will cause rectifier damage:

   a. Battery leads reversed.
   b. Running the engine with the battery leads disconnected.
   c. A broken wire or loose connection resulting in an open circuit.

Preliminary Checks

Before troubleshooting the alternator charging system, visually check the following.

1. Make sure the red cable is connected to the positive battery terminal. If polarity is reversed, check for a damaged rectifier.

NOTE
A damaged rectifier will generally be discolored or have a burned appearance.

2. Check for corroded or loose connections. Clean, tighten and insulate with liquid neoprene as required.

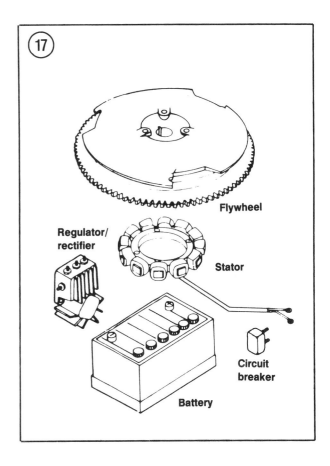

3. Check battery condition. Clean and recharge as required.

4. Check wiring harness between the stator and battery for damaged or deteriorated insulation and corroded, loose or disconnected connections. Repair, tighten or replace as required.

Troubleshooting (1966-1979 35-55 hp)

CAUTION
The engine must be provided with an adequate supply of water while running during this procedure. Install a flushing device, place the engine in a test tank or perform the procedure with the boat in the water.

1. Connect a tachometer according to manufacturer's instructions.

2. Start the engine and run at 1,000-1,100 rpm.

3. Briefly disconnect one of the battery cables at the battery. If the alternator is producing sufficient current to operate the ignition, the engine will continue running. Reconnect the battery cable.

4. Connect a 12-volt test lamp between ground and each of the 3 rectifier input terminals marked with a yellow dot. If the lamp does not light at all 3 terminals with the engine running, replace the stator.

5. Shut the engine off. Leave the ignition switch in the ON position. Connect the test lamp between ground and the rectifier output terminal (red wire). If the lamp does not light, repair or replace the red wire between the rectifier and the circuit breaker.

6. Repeat Step 4. If the lamp lights at any of the 3 terminals, replace the rectifier.

Troubleshooting
(1966-1979 70-140 hp)

CAUTION
The engine must be provided with an adequate supply of water while running during this procedure. Install a flushing device, place the engine in a test tank or perform the procedure with the boat in the water.

1. Connect a tachometer according to manufacturer's instructions.

2. Start the engine. Connect a 10-amp ammeter between the positive battery cable and positive battery terminal post.

3. Connect a voltmeter between the circuit breaker and ground.

4. Note the ammeter and voltmeter readings as engine speed is gradually increased to approximately 4,500 rpm:

   a. The stator should start to function at a cut-in speed of approximately 800 rpm, delivering a minimum output of 7 amps at 4,000 rpm.

   b. The voltage regulator should maintain an operational voltage of 14-15 volts, with a maximum of 17.5 volts.

5. If the stator cut-in speed or output is not as specified in Step 4, continue testing. Replace the voltage regulator if it does not maintain the operational voltage specified in Step 4.

6. Shut the engine off and disconnect the stator leads at the rectifier terminals. Connect a self-powered test lamp between the stator leads. If the lamp does not light, replace the stator.

7. Mark the remaining red and purple leads and the terminals to which they are connected for correct reinstallation and disconnect the leads.

8. Connect the black test lamp lead to the red lead terminal and probe each of the remaining terminals with the red test lead. The lamp should light at each terminal.

9. Repeat Step 8 with the test leads reversed. The lamp should not light at any terminal.

10. Move the red test lead to the rectifier ground terminal or mounting stud. Probe each of the rectifier stator terminals with the black test lead. The lamp should light at each terminal.

11. Replace the rectifier if it does not perform as specified in Steps 7-10. If the rectifier is good, reconnect all leads to their proper terminals.

Troubleshooting
(1980-on 20-140 hp)

1. Disconnect the negative battery cable.

2. Carefully inspect all wiring for worn or cracked insulation and corroded or loose connections. Replace wiring or clean and tighten connections as required.

NOTE
Do not connect test lamp in series with any lights or accessories as they will drain the test lamp batteries.

3. Use a self-powered test lamp to check each length of wiring in the circuit between

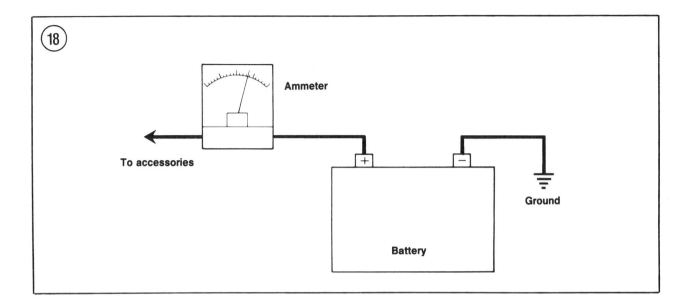

(18)

Ammeter

To accessories

+   −

Ground

Battery

connection points. If the test lamp does not light when connected into the circuit, replace that section of wiring.

4. Connect a tachometer according to manufacturer's instructions.

5. Connect an ammeter between the positive battery terminal and the accessory feed wire. See **Figure 18**. Reconnect the negative battery cable.

> *CAUTION*
> *The engine must be provided with an adequate supply of water while performing Step 6 and Step 7. Install a flushing device, place the engine in a test tank or perform the steps with the boat in the water.*

6. Start the motor and turn on all accessories. Note the ammeter reading and compare to the maximum current draw provided in **Table 5**. If current draw exceeds the specified amperage, reduce the number of accessories used with the charging system.

7. With the engine running, connect the ammeter between the positive battery cable and terminal. See **Figure 19**. Note the ammeter and tachometer readings as engine speed is gradually increased to approximately 4,500 rpm. Compare the readings to **Table 5**.

Replace the stator if it does not meet the specifications.

8. Shut the engine off. Disconnect the green/yellow stator leads at the rectifier. Connect an ohmmeter between the stator leads and compare the reading to **Table 5**. Replace the stator if the reading is not within specifications.

9. Perform the rectifier test as described in this chapter. If the rectifier is satisfactory, perform the circuit breaker test as described in this chapter.

Rectifier Test (Square Design)

1. Disconnect all rectifier leads.

2. Connect a self-powered test lamp between the negative terminal and one AC terminal as shown in **Figure 20**, then move the black test lead to the other AC terminal. The light should glow brightly at each terminal.

3. Move the black test lead to the positive (+) terminal. The light should glow dimly.

4. Move the red test lead to the positive (+) terminal. Connect the black test lead to one rectifier AC terminal, then to the other AC terminal. The light should not come on.

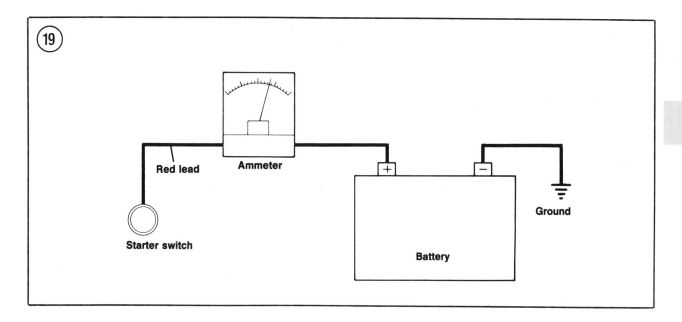

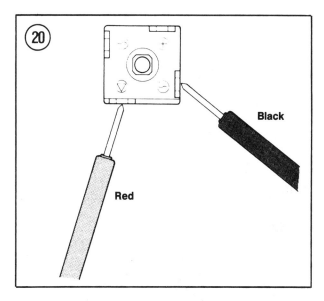

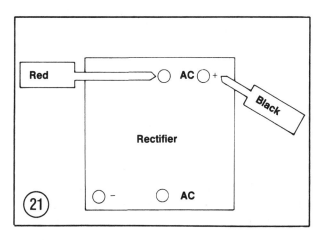

5. Connect the black test lead to the negative (–) terminal. Connect the red test lead to one rectifier AC terminal, then to the other AC terminal. The light should not come on at either terminal.

6. Replace the rectifier if it does not perform as indicated in Steps 2-5.

**Rectifier Test
(Rectangular Design)**

1. Disconnect all rectifier leads.

2. Connect a self-powered test lamp between the positive terminal and the top AC terminal as shown in **Figure 21**, then move the red test lead to the bottom AC terminal. The light should glow brightly at each terminal.

3. Move the red test lead to the negative (–) terminal. The light should glow dimly.

4. Move the red test lead to the positive (+) terminal. Connect the black test lead to the top rectifier AC terminal, then to the bottom AC terminal. The light should not come on at either terminal.

5. Replace the rectifier if it does not perform as indicated in Steps 2-4.

Regulator/Rectifier Test

CAUTION
The engine must be provided with an
adequate supply of water while running
during this procedure. Install a flushing
device, place the engine in a test tank or
perform the procedure with the boat in
the water.

1. Connect a tachometer according to manufacturer's instructions.
2. Disconnect the red/white lead at the battery side of the starter solenoid or relay.
3. Connect the disconnected wire and terminal with a jumper lead to start the engine.
4. Start the engine and run at idle. Remove the jumper lead installed in Step 3.

NOTE
Chrysler service replacement capacitor
part No. 404030 may be used in Step 5.

5. Connect a voltmeter and 500 microfarad capacitor to the disconnected red/white wire and ground as shown in **Figure 22**.
6. Increase engine speed to approximately 4,000 rpm and note the voltmeter reading. If less than 13.5 volts or greater than 17.5 volts, replace the regulator/rectifier.

Circuit Breaker Test

1. Depress the circuit breaker reset button.
2. Connect a test lamp between the circuit breaker terminals. Replace the circuit breaker if the lamp does not light.

AUTOLECTRIC MODELS

A combined starter-generator is bracket-mounted above the flywheel on Autolectric 9.2-20 hp engines. The armature shaft of the starter-generator is in constant engagement with the engine crankshaft. As the two shafts rotate, the generator provides electrical

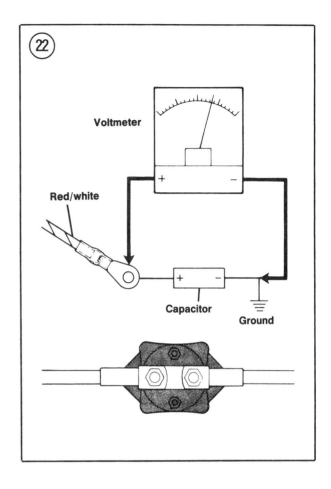

current to operate the engine and charge the battery.

Troubleshooting

CAUTION
The engine must be provided with an
adequate supply of water while
performing this procedure. Install a
flushing device, place the engine in a
test tank or perform the procedure with
the boat in the water.

Refer to **Figure 6** (9.2 hp), **Figure 7** (9.9 and 12.9 hp), **Figure 5** (1966-1967 20 hp) or **Figure 8** (1968-on 20 hp) for this procedure.
1. Connect an ammeter and tachometer according to manufacturer's instructions.
2. Start the engine and slowly accelerate until the ammeter reads one ampere. Note the tachometer reading. The generator should cut in at a speed below 3,700 rpm.

3. Run the engine at wide-open throttle. The ammeter should read a minimum of 6 amps output.

4. Shut the engine off and connect one lead of a voltmeter to a good engine ground. Connect the other lead to the long terminal on the circuit breaker.

5. Start the engine and run at approximately 4,000 rpm. Disconnect the positive battery terminal. If the voltmeter reads more than 16 volts, replace the voltage regulator.

6. If there is no reading in Step 5, disconnect the wiring harness red lead from the ignition side of the voltage regulator and the starter frame assembly gray lead from the field side of the voltage regulator.

7. Connect the disconnected red and gray leads to the positive voltmeter lead. Start the engine and slowly increase its speed until the voltmeter reads approximately 14.5 volts. If the meter reading reaches 14.5 volts, replace the voltage regulator. If there is no meter reading, remove the starter-generator assembly for bench testing.

8. If the system passes all of the steps in this procedure but still does not keep the battery recharged, the problem is caused by improper engine operation, prolonged trolling or operation below the generator cut-in speed.

IGNITION SYSTEM

The wiring harness used between the ignition switch and engine is adequate to handle the electrical needs of the outboard. It *will not* handle the electrical needs of accessories. Whenever an accessory is added, run new wiring between the battery and accessory, installing a separate fuse panel on the instrument panel.

If the ignition switch requires replacement, *never* install an automotive-type switch. A marine-type switch must always be used.

Description

Variations of three different ignition systems have been used on Chrysler outboards since 1966. See Chapter Seven for a full description. For the purposes of troubleshooting, the ignition systems can be divided into 4 basic types:

1. A breaker point magneto ignition.
2. A breaker point battery ignition.
3. A breaker point CD (capacitor discharge) ignition.
4. A breakerless CD (capacitor discharge) ignition.

As a rule of thumb, magneto ignitions are generally used on manual start models, while electric start models use some form of CD or battery ignition. General troubleshooting procedures are provided in **Table 3**.

Basic Troubleshooting Precautions

Several precautions should be strictly observed to avoid damage to the ignition system.

1. Do not reverse the battery connections. This reverses polarity and can damage the rectifier(s) or CD unit.

2. Do not "spark" the battery terminals with the battery cable connections to check polarity.

3. Do not disconnect the battery cables with the engine running unless specified in a test procedure.

4. Do not crank the engine if the CD unit is not properly grounded.

5. Do not touch or disconnect any ignition components when the engine is running, while the ignition switch is ON or while the battery cables are connected unless specified in a test procedure.

Troubleshooting Preparation
(All Ignition Systems)

NOTE
To test the wiring harness for poor solder connections in Step 1, bend the molded rubber connector while checking each wire for resistance.

1. Check the wiring harness and all plug-in connections to make sure that all terminals are free of corrosion, all connectors are tight and the wiring insulation is in good condition.
2. Check all electrical components that are grounded to the engine for a good ground connection.
3. Make sure that all ground wires are properly connected and that the connections are clean and tight.
4. Check remainder of the wiring for disconnected wires and short or open circuits.
5. Make sure there is an adequate supply of fresh and properly mixed fuel available to the engine.
6. If the engine is to be operated during the test procedure, make sure that sufficient water is provided for proper cooling to avoid power head or gearcase damage. Use a flush device, place the engine in a test tank or perform the procedure with the boat in the water.
7. Check the battery condition. Clean terminals and recharge battery, if necessary.
8. Check spark plug cable routing. Make sure the cables are properly connected to their respective spark plugs.
9. Keep all spark plugs in the order of their removal. Check the condition of each plug. See Chapter Four.

MAGNETO IGNITION SYSTEM

A magneto ignition system generally contains 7 major components:

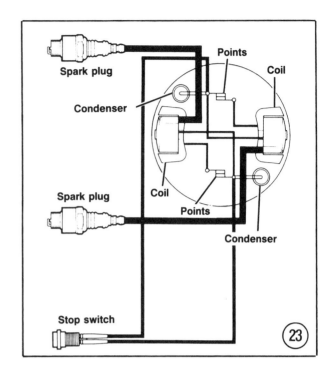

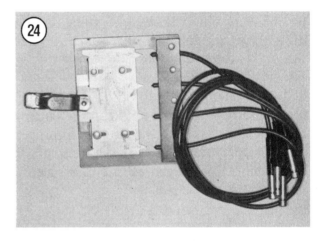

a. Flywheel.
b. Stator plate.
c. Throttle cam.
d. Breaker point set(s).
e. Condenser(s).
f. Coil(s).
g. Spark plug(s).

A typical magneto ignition system is shown in **Figure 23**.

Coils and condensers can only be tested with a suitable ignition analyzer. Chrysler recommends the use of a Merc-O-Tronic

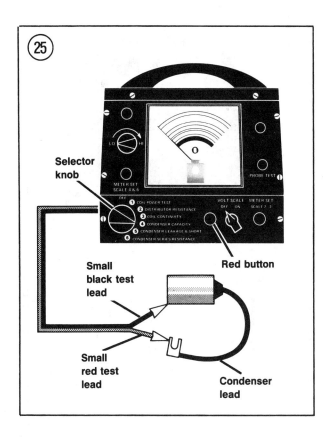

analyzer. This test equipment is available through any marine dealer and comes complete with instructions and test specifications.

Troubleshooting

1. Install a spark tester (**Figure 24**) between the plug wire and a good ground to check for spark at each cylinder. If a spark tester is not available, remove each spark plug and reconnect the proper plug cable to one plug. Lay the plug against the cylinder head so its base makes a good connection and turn the engine over. There should be a bright blue spark at each cylinder.

2. If there is no spark or only a weak yellowish or red spark in Step 1, check for loose connections at the coil and battery and correct as required.

3. Repeat Step 1 with each remaining plug if a spark tester is not used. If the connections are good, continue testing.

4. If a good spark is not produced at each cylinder in Steps 1-3, disconnect the leads from the ignition stop or "kill" switch and repeat Step 1. If a good spark is produced with the switch out of the circuit, replace the switch.

5. If the spark is still not satisfactory, remove the flywheel. See Chapter Eight. Check the breaker point condition and gap. Replace or adjust the points as required. See Chapter Four.

6. Remove the condenser(s) from the stator plate. Remove the coil(s) from the stator plate or power head. See Chapter Seven.

7. Test the condenser(s) and coil(s) as described in this chapter.

Condenser Capacity and Leakage Test

1. Plug the Merc-O-Tronic analyzer into an AC outlet and set the selector knob on position 4. Clip the analyzer test leads together, depress the red button and rotate the meter set knob to align the meter needle with the SET line on the meter scale. Release the red button and unclip the test leads.

2. Connect the small red test lead to the condenser lead and the small black test lead to the condenser body. See **Figure 25**.

3. Depress the red button and note the reading on the CONDENSER CAPACITY scale. If the reading is not within specifications (**Table 6**), replace the condenser. If the reading is within specifications, turn the selector knob OFF and continue testing.

4. With the test leads connected as shown in **Figure 25**, turn the selector knob to position 5. Depress the red button for 15 seconds and note the meter needle movement. It should move all the way to the right on the CONDENSER LEAKAGE AND SHORT scale, then return all the way to the left. Replace the condenser if the needle does not

return. Turn the selector knob OFF, disconnect the test leads and continue testing.

6. Set the selector knob to position 6 and repeat Step 1 to align the meter needle.

7. Repeat Step 2, then note the meter needle movement. Replace the condenser if the needle moves outside the green area at the right of the CONDENSER SERIES RESISTANCE scale.

8. Disconnect the test leads and unplug the analyzer.

Coil Test Sequence

WARNING
Perform this test on a wooden or insulated workbench to prevent leakage or shock hazards.

1. Plug the Merc-O-Tronic analyzer into an AC outlet. Connect the test leads as shown in **Figure 26** or **Figure 27**.

2. With the current control knob on LO, turn the selector knob to position 1.

3. Slowly rotate the current control knob clockwise until the meter needle reads 1.0 (external coil) or 1.5 (stator-mounted coil) on the COIL POWER TEST scale. If the spark gap is firing uniformly at this point, the coil is good. Rotate the current control knob back to LO and turn the selector knob OFF.

4. With the analyzer test leads connected as shown in **Figure 26** or **Figure 27**, turn the selector knob to position 1.

CAUTION
Complete Step 5 as quickly as possible to prevent tester or coil damage.

5. Rotate the current control knob clockwise until the meter needle reaches full scale. If the spark gap is firing uniformly at this point, the coil is good. Rotate the current control knob back to LO and turn the selector knob OFF.

6. Disconnect the large red test lead from the coil and plug the test probe into the analyzer jack.

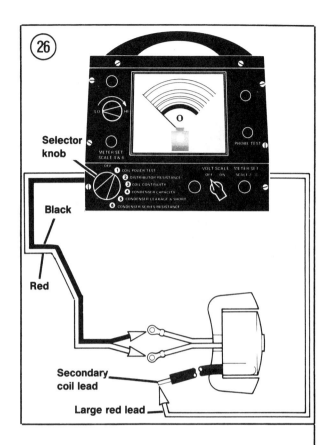

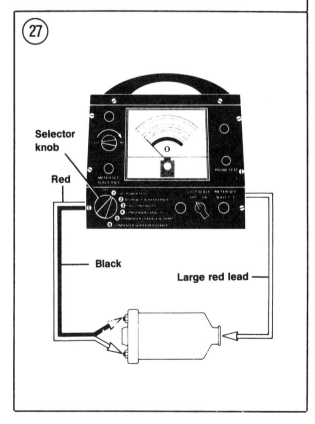

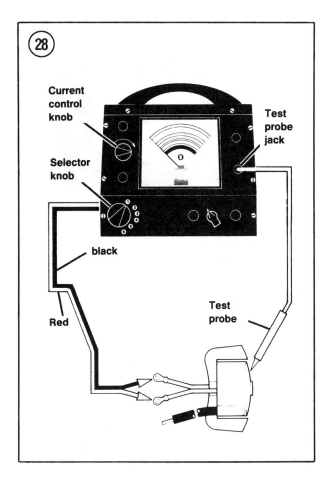

7. Turn the selector knob to position 1 and rotate the current control knob clockwise until the meter needle reaches full scale. Quickly move the test probe over the insulated surfaces of the coil and spark plug lead. See **Figure 28**. A faint spark at the probe is acceptable; strong sparks indicate coil leakage. Rotate the current control knob back to LO and turn the selector knob OFF. Remove the test leads.

8. Turn the selector knob to position 2. Holding the test leads apart, rotate the meter set knob to align the needle with the SET line on the meter scale.

9. Connect the test leads as shown in **Figure 29** or **Figure 30**. Note the meter reading on the DISTRIBUTOR RESISTANCE scale and compare to the primary resistance specifications (**Table 7**). Replace the coil if the reading is not within specifications. Turn the selector knob OFF and disconnect the test leads.

10. Turn the selector knob to position 3. Clip the analyzer test leads together, rotate the

3

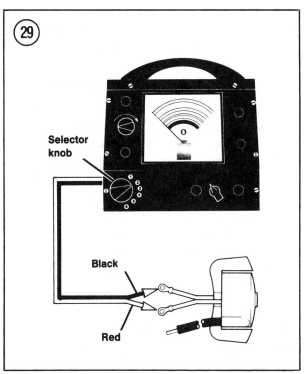

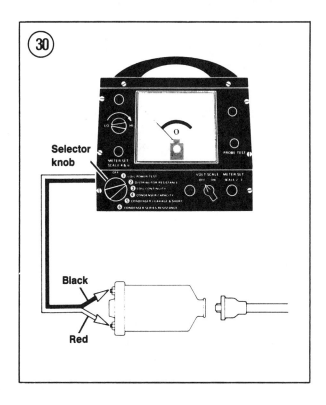

meter set knob to align the meter needle with the SET line on the meter scale and unclip the test leads.

11. Connect the red test lead to the spark plug lead. Connect the black test lead to either primary terminal. See **Figure 31** or **Figure 32**. Note the meter reading on the COIL CONTINUITY scale and compare to the secondary resistance specifications (**Table 7**). Replace the coil if the reading is not within specifications. Turn the selector knob OFF and disconnect the test leads.

BATTERY IGNITION SYSTEM

A battery ignition system may use a magneto breaker plate and cam located under the flywheel or a distributor mounted on the power head and belt-driven by the crankshaft. In addition, the system contains the following components:
a. Breaker points.
b. Condensers.
c. Coils.
d. Spark plugs.

Coils and condensers can only be tested with a suitable ignition analyzer. Chrysler recommends the use of a Merc-O-Tronic analyzer. This test equipment is available through any marine dealer and comes complete with instructions and test specifications.

Troubleshooting

The following procedures are performed with a 12-volt test lamp and require a fully charged battery with properly connected cables that are in good condition. It is assumed that all engine wiring and remote electric harness wiring is connected and correct.

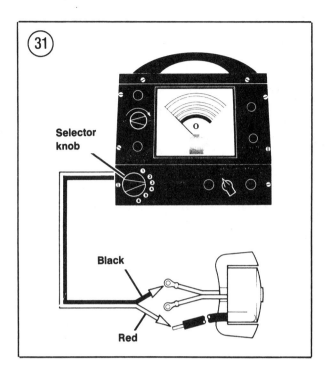

Selector knob

Black

Red

Battery to coil test

The ignition switch must be in the ON position with the engine not running (avoid leaving the switch in the ON position for prolonged periods). If the 12-volt test lamp does not light during one step, proceed to the next one. A dim light at any point in the procedure indicates a poor or shorted connection.

1. Connect one lead of the test lamp to a good engine ground. Probe the positive (+) terminal of each coil with the other test lead. If the test lamp lights at each coil positive terminal, the ignition circuit is satisfactory between the battery and coils.

2. If the test lamp does not light at one or more coils in Step 1, move the test lead probe to the terminal block blue connector. If the lamp lights, repair or replace the blue wire between the terminal block and coils as required.

3. Move the test lead probe to the ignition switch I terminal. If the lamp lights, repair or replace the blue wire in the remote cable harness.

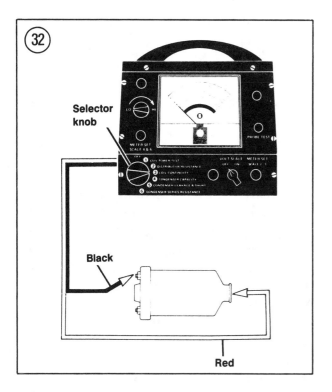

4. Move the test lead probe to the ignition switch B terminal. If the lamp lights, replace the ignition switch. If the lamp does not light, look for an open circuit in the wiring between the ignition switch and battery.

Coil and condenser to breaker point test

The ignition switch must be in the ON position with the engine not running (avoid leaving the switch in the ON position for prolonged periods). If the 12-volt test lamp does not light during one step, proceed to the next one. A dim light at any point in the procedure indicates a poor or shorted connection.

1. Remove the spark plugs to prevent the engine from starting. See Chapter Four.
2. Connect one lead of the test light to a good engine ground. Probe the negative (–) terminal of the coil with the other test lead. If the breaker points are closed, the test light will not come on. If the breaker points are open, the light will come on.

3. Rotate the flywheel 360 degrees and watch the test light. It should come on for 90 degrees and remain off for 270 degrees.
4. If the test light remains on during the 360 degree rotation, there is an open circuit caused by:
  a. Breaker points insulated or not closing.
  b. A defective lead wire between the coil and breaker points.
5. If the test light remains off during the 360 degree rotation, there is a closed or shorted circuit. Proceed as follows:
  a. Rotate the flywheel until one set of breaker points is open.
  b. Disconnect the breaker point lead at the negative (–) terminal of the coil. If the test light now shows current at the negative (–) terminal of the coil, either the lead wire or the breaker point set is shorted.
  c. If the test lamp does not light, disconnect the condenser lead wire. If the test lamp lights, the condenser is shorted.
6. Repeat Step 5 to test the other breaker point set and coil.
7. No. 1 cylinder only—Disconnect the tachometer lead wire (white) from the No. 1 coil negative (–) terminal. If the test light comes on when the terminal is probed, there is a short in the tachometer or tachometer circuit.
8. Reinstall the spark plugs. See Chapter Four.

Ignition circuit test (engine running)

> *CAUTION*
> *The engine must be provided with an adequate supply of water while performing this procedure. Install a flushing device, place the engine in a test tank or perform the procedure with the boat in the water.*

This test is useful in locating a misfiring or dead cylinder.

1. Connect a tachometer according to manufacturer's instructions.

2. Start the engine and run at 800-1,000 rpm in NEUTRAL.

3. Connect one test light probe to a good engine ground. Connect the other probe to the negative (–) terminal of the coil. The test light should flicker on and off as the breaker points open and close. If the test light remains on or off, shut the engine off and perform the *Coil and Condenser to Breaker Point Test* in this chapter.

4. If the test light flickers in Step 3 as it should and there is no loss of engine rpm, the primary side of the ignition circuit is satisfactory. Continue testing to locate the problem in the secondary circuit.

5. Shut the engine off. Remove the spark plugs (Chapter Four), reconnect the plug wires to the plugs and ground the plugs to the cylinder head.

6. Turn the ignition switch ON and crank the engine. If the plugs do not fire with a bright blue spark, check the spark plug wire and its connection to the coil and spark plug. If the wire and its connections are good, test the coil as described in this chapter.

Condenser Capacity and Leakage Test

Test the condenser(s) as described for magneto ignition systems in this chapter.

Coil Test Sequence

Test the coils as described for magneto ignition systems in this chapter.

CD IGNITION SYSTEM

A wide variety of CD (capacitor discharge) ignition systems have been used on Chrysler engines since the early Seventies. Through the 1979 model year, they carried various trade names—Magnapower or Magnapower I, Magnapower Electronic, Magnapower II, Magnapower III, etc.—according to system design and model usage. From 1980-on, all are referred to under the collective heading of "CD ignition."

While all Magnapower I ignitions (except the Magnapower Electronic) function essentially the same, there are several design variations because Delta, Prestolite and Motorola each produced the basic Magnapower or CD ignition used by Chrysler Marine. For example, 70-135 hp engines may be equipped with either a Delta, Motorola, Motorola 2, Motorola Electronic Magnapower or CD ignition without regard for year or model. Therefore, component shape, size, location and, in some cases, wiring and/or function differ according to the manufacturer.

Although outboards fitted with a Magnapower ignition were shipped from the factory with a Magnapower decal on the power head, the decal did not always identify the ignition manufacturer and may have become damaged or lost.

To help in determining which Magnapower I system is used on a particular 70-135 hp engine, compare the part No. on the CD unit with the listing below:

a. Delta—part Nos. A 321301, A 321301-1 or A 321301-2.

b. Motorola—part Nos. 404301 or 404301-1.

c. Motorola 2 or Electronic—part Nos. 404301-2, 404301-3, 523301 or 523301-1.

Since 1979, one CD unit (part No. A 523301-1) has been used as a replacement for all Magnapower I ignitions. If the original CD unit has been replaced, it probably carries this part number and will not help you to identify the system.

For this reason, the wiring diagrams at the end of the book should be consulted before using the troubleshooting procedures provided in this chapter if you are uncertain of which Magnapower ignition system is used on your motor. Select the appropriate wiring diagram according to engine size and component design and use it to locate exact test points.

All Magnapower and CD ignitions require the use of a Chrysler CD Tester T8953. In addition, some also require the use of Chrysler Ignition Tester T8996 and Plug Adapter T11201 or T11237.

DELTA AND MOTOROLA MAGNAPOWER I (1974-1979) AND CD IGNITION (1980-ON)

Circuit Test

This procedure will troubleshoot the Delta, Motorola, Motorola 2, Motorola Electronic and CD ignitions on 70-135 hp engines.

1. Turn the ignition switch ON and connect a 12-volt test lamp between the terminal block blue wire terminal and a good engine ground. If the test lamp does not light, the problem is between the terminal block and the ignition switch.
2. Move the test lamp probe to the starter relay input terminal. If the lamp does not light, repair or replace the wiring between the battery and starter relay.
3. Move the test lamp to the circuit breaker input terminal. If the lamp does not light, repair or replace the wiring between the circuit breaker and starter relay.
4. Move the test lamp to the circuit breaker output terminal. If the lamp does not light, depress the reset button. If the lamp still does not light, replace the circuit breaker.
5. Move the test probe to the terminal block red wire terminal. If the lamp does not light,

repair or replace the wiring between the terminal block and ignition switch.
6. Move the test probe to the ignition switch B terminal. If the lamp lights, replace the switch.
7. Move the test probe to the ignition switch I terminal. If the lamp does not light, replace the switch.
8. Move the test probe to the terminal block blue wire terminal. If the lamp does not light, repair or replace the wiring between the terminal block and ignition switch.
9. Move the test probe to the distributor blue terminal. The test lamp should not light. If it does, perform Step 10 and then move to Step 14.
10. Disconnect the blue wire from the terminal block to prevent the engine from starting.

CAUTION
Make sure the tester CD IN lead does not contact any metal, as it is live once the 2 leads are connected in Step 11.

11. Connect the BATT N lead of tester T8953 to a good engine ground. Connect the tester BATT P lead to the starter relay. See **Figure 33**.
12. Rotate the tester voltage knob clockwise as far as possible. If the tester does not show at least 12 volts, recharge or replace the battery as required.
13. Crank the engine and note the tester reading. If less than 9 volts, test the starter circuit as described in this chapter before proceeding with Step 5.
14. Connect the tester leads as shown in **Figure 34** (Delta), **Figure 35** (Motorola) or **Figure 36** (Motorola 2, CD and Electronic).
15. Rotate the tester voltage knob counterclockwise until the meter reads 12 volts. Turn the indicator knob clockwise as far as possible. The mark on the indicator knob should align with the "100" mark on the tester scale.

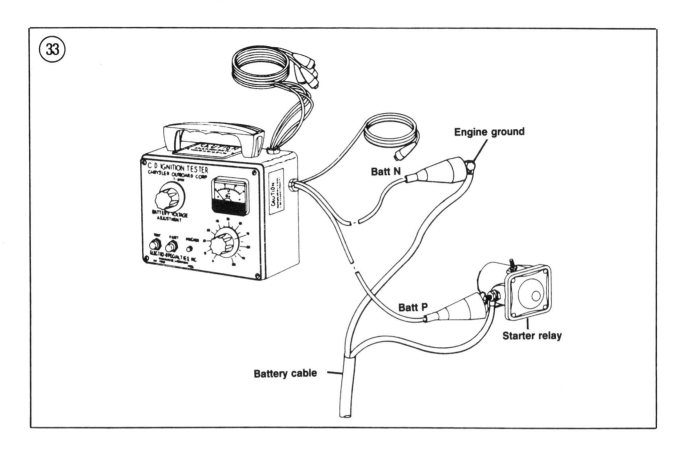

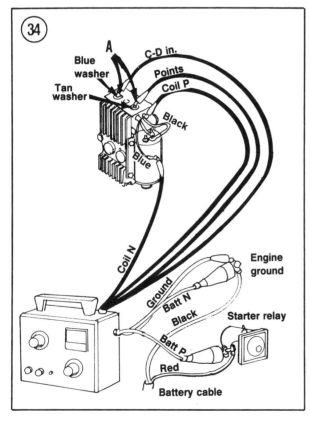

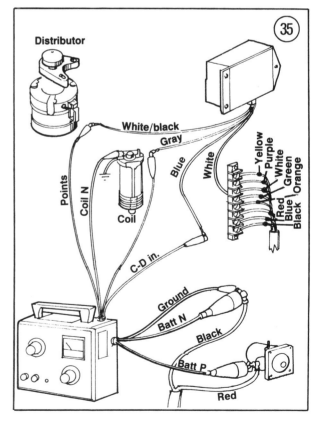

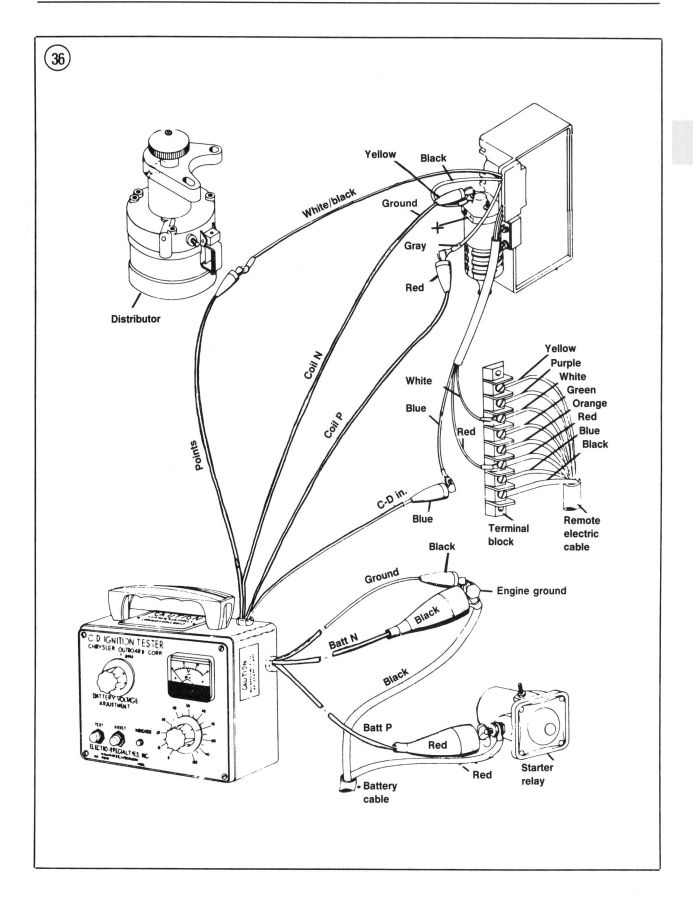

③⑥

Distributor

Yellow
Black
White/black
Ground
Gray
Red

Coil N
Coil P
Points

White
Blue
Red

Yellow
Purple
White
Green
Orange
Red
Blue
Black

C-D in.
Blue

Terminal
block

Remote
electric
cable

Black
Ground
Engine ground
Batt N
Black

C.D. IGNITION TESTER
CHRYSLER OUTBOARD CORP.

BATTERY VOLTAGE
ADJUSTMENT

TEST   RESET   INDICATOR

ELECTRO-SPECIALTIES, INC.

Black
Batt P
Red
Red

Battery
cable

Starter
relay

3

16. While cycling the tester TEST button 2-3 times per second to simulate breaker point action, slowly turn the indicator knob counterclockwise until the tester indicator lamp lights. This should occur at a scale setting of approximately "80" at 12 volts.

17. Depress tester RESET button. If the indicator lamp does not go out, replace the CD unit.

18. Cycle the TEST button at least 5 times. If the indicator lamp does not light for each cycle, replace the CD unit.

19. Turn the indicator knob clockwise to align the knob mark with the "100" mark on the tester scale and adjust tester voltage knob until the meter reads 12 volts.

20. Repeat Steps 16-18. The results should be the same with the indicator knob reading within 2 points of the setting in Step 16.

21. If the indicator knob reading is less than specified in Step 20, scrape sufficient paint from the CD unit to provide a good ground and connect the tester ground lead directly to the CD unit. Repeat Steps 16-20. If the indicator knob reading does not change, replace the CD unit.

22. If the reading does change in Step 21, the CD unit is not grounded properly. Correct the poor ground condition, return the tester ground lead to its engine ground and repeat Steps 16-20 to make sure ground is satisfactorily.

23. If the CD unit has checked satisfactorily to this point, adjust the tester voltage knob until the meter reads 9 volts and align the indicator knob mark with the "100" mark on the tester scale.

24. Repeat Steps 16-20. The results should be the same with the indicator knob reading a minimum of 65 (Motorola) or 73 (Delta). If the readings are below the minimum, replace the CD unit.

25A. Motorola Electronic and CD—If the circuit and CD unit are satisfactory to this point, test the distributor as described in this chapter.

25B. All others—If the circuit and CD unit are satisfactory to this point, check the distributor breaker point condition and gap. Replace or adjust the points as required. See Chapter Four.

Motorola Electronic Distributor Test

1. Rotate the flywheel to position the TDC or 0 degree mark with the timing pointer.

2. Slip the distributor drive belt off the distributor pulley and turn the ignition switch ON.

3. Disconnect the coil secondary wire at the coil tower. Install a 12 inch length of secondary wire in the coil tower.

4. Grasp the end of the wire installed in Step 3 with a pair of insulated pliers and hold the end of it about 1/2 inch from a ground.

5. Spin the distributor pulley. If a spark jumps the gap between the wire and ground, the problem is in the secondary circuit.

6. If there is no spark in Step 5, turn the ignition switch OFF, disconnect the battery cables and remove the distributor cap. Rotate the distributor pulley to expose the preamplifier.

7. Disconnect the white/black wire at the distributor and connect the negative lead of a voltmeter to the terminal. Connect the positive voltmeter lead to the preamplifier blue wire and reconnect the battery cables.

8. Turn ignition switch to ON or RUN position (do not start engine).

9. Insert a credit card, match book cover or similar opaque item between the preamplifier and distributor contact to activate the preamplifier make-break circuit. The voltmeter should react as follows:

   a. Red preamplifier—tripping the circuit will cause the voltmeter needle to move upscale and then return to zero.

b. Black preamplifer—tripping the circuit will cause the voltmeter to read battery voltage until the opaque item is removed.

10. If the voltmeter reacts as described in Step 9, the preamplifier is presumed to be good although this is not necessarily true. There is no test to isolate a preamplifier that breaks down under load or heat conditions. If the engine gradually dies or slows down as if running out of fuel, restarts immediately but repeats the same symptoms when accelerated or run at high speed, the preamplifier is probably defective. In such cases, the most practical solution is to substitute a known-good preamplifier to see if that solves the problem.

11. Turn the ignition OFF. Disconnect the battery cables, remove the voltmeter and reconnect the battery cables. Turn the ignition ON.

12. Grasp the end of the wire installed in Step 3 with a pair of insulated pliers and hold the end of it about 1/2 inch from the center of the distributor rotor, then spin the distributor pulley. Replace the rotor and shaft assembly if a strong spark is noted, as the rotor is grounding out to the shaft.

13. Reinstall the distributor cap and disconnect one spark plug cable from the cap. Repeat Step 12 holding the wire about 1/4 inch from the disconnected spark plug wire terminal in the distributor cap.

14. Reinstall the wire and test each remaining spark plug terminal base in the same manner. Remove a wire, perform the test, note the spark quality and reinstall the wire before testing the next terminal base. If a strong spark is noted at any of the terminal bases, replace the cap.

15. Reinstall the original coil wire to the distributor cap and ground the coil end of the wire to the engine with a jumper lead.

16. Grasp the end of the wire installed in Step 3 with insulated pliers, hold it about 1/4

inch from the distributor cap center terminal base and spin the distributor pulley. If a good spark does not jump the gap, replace the cap.

Delta and Motorola Magnapower Ignition Coil Test

Test the coils as described for magneto ignition systems in this chapter.

MAGNAPOWER II IGNITION SYSTEM

This ignition system was used on 1977-1980 55 hp, 1974-1976 60 hp, 1977-1978 65 hp and 1975-1977 105, 120 and 135 hp engines. It functioned erratically and Chrysler Marine offered an ignition retrofit kit part No. K1124 (55-65 hp) or part No. K1054 (105-135 hp) under warranty installation during 1978. Manufacture of the CD module (part No. S474301-1) was discontinued in 1981. Other system components are available but should the module fail, the only repair option is to install the retrofit kit, which is still available from some Chrysler/U.S. Marine dealers.

Circuit Test

Refer to **Figure 37** (2-cylinder) or **Figure 38** (4-cylinder) for this procedure.

1. Connect a power timing light to the No. 1 cylinder according to manufacturer's instructions.

2. Start the engine, let it run at idle, then activate the timing light. If the timing light comes on, the cylinder is firing.

3. Connect the timing light to each remaining cylinder and repeat Step 2 to determine which cylinder or cylinders are not firing.

NOTE
If the ignition is turned OFF to shut the engine off in Step 4, the capacitors will fire to ground and the system will not respond to troubleshooting.

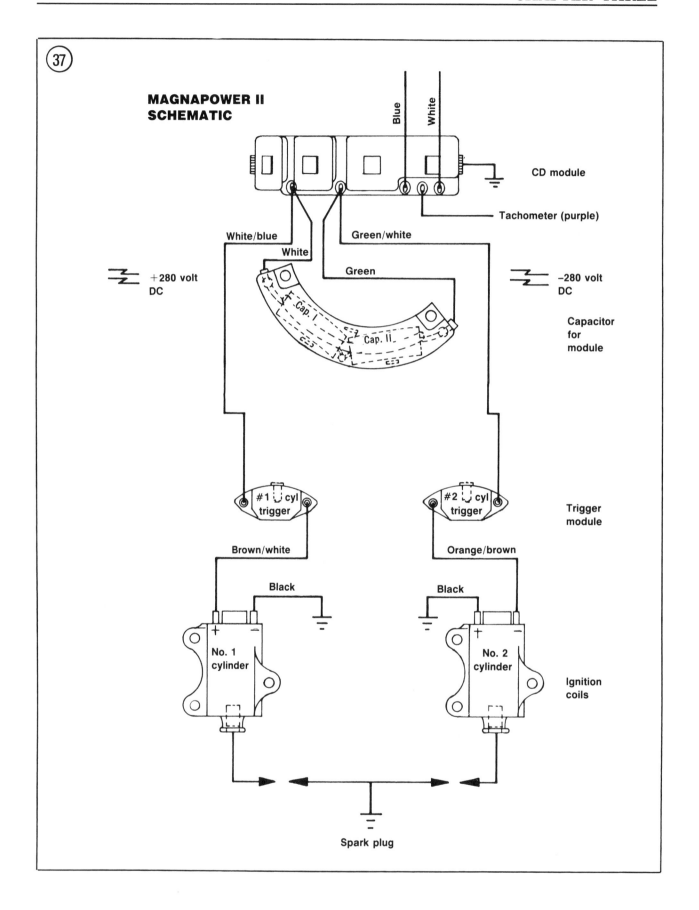

MAGNAPOWER II SCHEMATIC

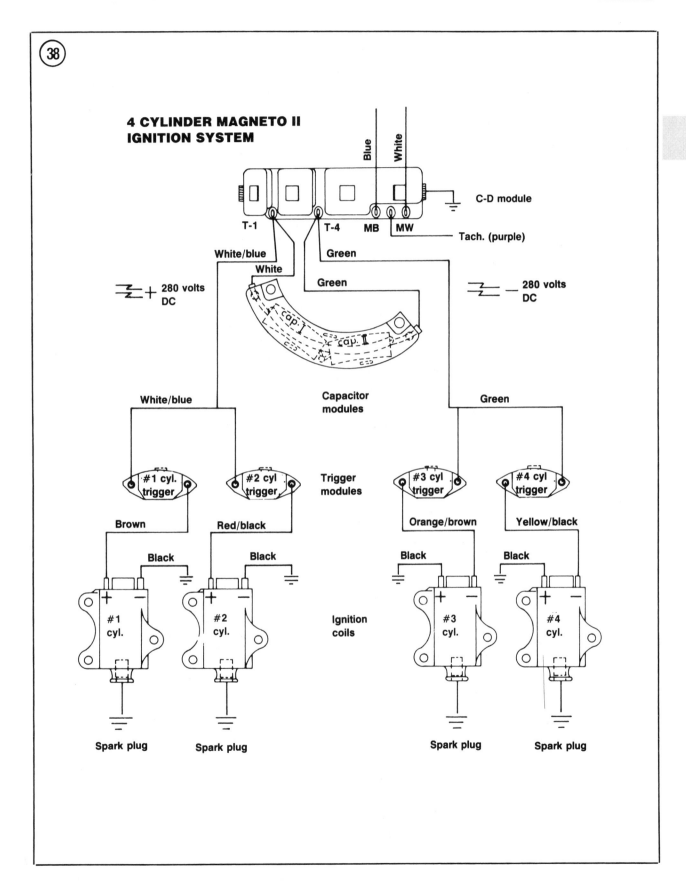

4 CYLINDER MAGNETO II IGNITION SYSTEM

3

4. Choke the engine to shut it off. This will retain a full charge in the capacitors, which is required for troubleshooting.

5. Disconnect all spark plug leads to prevent the engine from starting accidentally.

6. Ground the shaft of an insulated screwdriver to the flywheel and touch the screwdriver tip to the CD module T-1 terminal (**Figure 39**). Repeat this step to test the T-4 terminal.

7. If there is a good arc between the screwdriver tip and each terminal, proceed with Step 18.

8. If there is not a good arc at both terminals, disconnect the trigger wire at the terminal with the weakest arc and crank the engine to recheck the arc.

9. If there is a good arc with the trigger disconnected in Step 8, check the CD module-to-trigger and trigger-to-coil wiring for cracks, frayed insulation or other defects and correct as required. If wiring is satisfactory, replace the trigger.

10. If the arc is still poor with the trigger disconnected, the CD module or capacitor is probably defective.

NOTE
Steps 11-14 apply only to 55-60 hp engines.

11. Remove the flywheel. See Chapter Eight.
12. Disconnect all trigger module leads. Remove the trigger assembly bolts and rotate the assembly 180°. Reconnect the leads.

13. Install the flywheel and start the engine briefly to charge the capacitors. Choke the engine to shut it off.

14. Repeat Step 6 and Step 7. If the poor arc transfers to the other terminal, replace the trigger. If the poor arc remains on the same terminal as in Step 6 or Step 7, the CD module is probably defective.

NOTE
The remaining steps apply to all engines as specified.

15. Disconnect the capacitor and trigger leads from the malfunctioning terminal and install a spade terminal to provide a good test connection.

16. Connect CD tester T8953 as follows according to the terminal being tested:
   a. T-1 terminal—COIL P lead to T-1 terminal; COIL N lead to a good engine ground.
   b. T-4 terminal—COIL N lead to T-4 terminal; COIL P lead to a good engine ground.

17. Set CD tester indicator knob to "50" on the tester scale and crank the engine. If the tester light comes on, replace the capacitor module. If the tester light does not come on, replace the CD module.

18. Connect the CD tester COIL N lead to a good engine ground and the COIL P lead to the positive (+) terminal of the No. 1 cylinder coil.

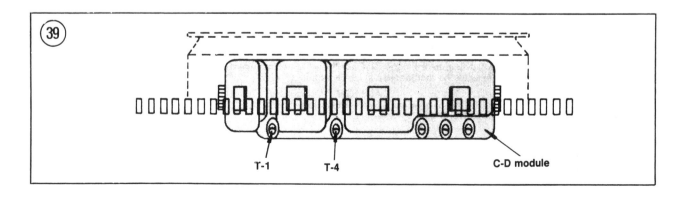

19. Set CD tester indicator knob to "65" on the tester scale and crank the engine. If the tester light does not come on, replace the No. 1 cylinder trigger module.

20. 4-cylinder engine—Repeat Step 18 and Step 19 with the tester connected to the No. 2 cylinder coil. If the tester light does not come on, replace the No. 2 cylinder trigger module.

21. Connect the CD tester COIL P lead to a good engine ground and the COIL N lead to the negative (–) terminal of the No. 2 cylinder coil (2-cylinder) or No. 3 cylinder coil (4-cylinder).

22. Crank the engine. If the tester light does not come on, replace the No. 2 (2-cylinder) or No. 3 (4-cylinder) cylinder trigger module.

23. 4-cylinder engine—Repeat Step 21 and Step 22 with the tester connected to the No. 4 cylinder coil. If the tester light does not come on, replace the No. 4 cylinder trigger module.

24. Start the engine and run at idle. If the engine misfires, the problem is a defective spark plug, secondary wiring or ignition coil. If the spark plugs and secondary wiring are satisfactory, test the coils as described for magneto ignition systems in this chapter.

MAGNAPOWER III
IGNITION SYSTEM

This ignition system is used on 1974-1978 10 hp, 1979 12 hp and 1974 and later 15 hp engines. On 1980 and later models, it is called a magneto CD ignition. The circuit test consists of a preliminary check and 5 individual component tests. The test sequences require the use of Chrysler CD tester T8953 and ignition tester T8996, which includes a No. 22 trigger pulse detector, load coil and MM-1 clip.

Preliminary Check

This procedure determines whether there is sufficient voltage present in the secondary system to fire the spark plug and if the plug is firing properly.

1. Install the MM-1 clip over the No. 1 spark plug wire as close as possible to the plug boot.

2. Connect the yellow test lead of CD tester T8953 to the MM-1 clip lead. Connect the red test lead to the terminal block ground wire screw.

3. Set the tester indicator dial on "80" (1974-1978) or "75" (1979-on) and crank the engine. The tester indicator lamp will light if the cylinder is firing normally.

4. Depress tester reset button to turn indicator lamp out. Unclip MM-1 clip from No. 1 plug wire and install on No. 2 plug wire. Repeat Steps 1-3 to test the other cylinder.

5. If the engine does not fire on either cylinder, disconnect one wire at the stop switch and repeat Steps 1-4. If both cylinders now fire normally, replace the stop switch. If the engine still does not fire on either cylinder, check the ground wire at the terminal block. If the ground is good, reconnect the stop switch and perform tests 1-5 (in that order) as described in this chapter. If the system passes all 5 tests and the cylinders still do not fire, replace the ignition coils.

6. If only the No. 1 cylinder fires, install a new spark plug in the No. 2 cylinder and repeat Steps 1-3. If that does not correct the problem, perform test 5 and test 3 (in that order) as described in this chapter. If the system passes both tests, replace the No. 2 ignition coil.

7. If only the No. 2 cylinder fires, install a new spark plug in the No. 1 cylinder and repeat Steps 1-3. If that does not correct the problem, perform test 4 and test 2 (in that order) as described in this chapter. If the system passes both tests, replace the No. 1 ignition coil.

3

Test 1
(Stator Winding Output)

1. Disconnect the blue and yellow stator wires at the terminal block. Connect the yellow stator wire to the terminal block ground terminal.

2. Connect the trigger pulse detector red test lead to the disconnected blue wire and the yellow test lead to the ground terminal. See **Figure 40**.

3. Place the pulse detector switch in position 2 and crank the engine. The tester lamp should light.

4. Reconnect the blue wire to the terminal block and disconnect the yellow wire from the ground terminal.

5. Connect the red test lead of CD tester T8953 to the disconnected yellow wire and the yellow test lead to the ground terminal. See **Figure 41**.

6. Set the indicator knob to "10" on the tester scale and crank the engine. The tester lamp should light.

7. If the tester lamp does not light in Step 3 or Step 6, replace the stator.

Test 2
(No. 1 Cylinder
Trigger Coil Output)

1. Disconnect the trigger housing white/green wire at the terminal block.

2. Connect the trigger pulse detector red test lead to the disconnected white/green wire. Connect the yellow test lead to the terminal block red wire terminal. See **Figure 42**.

3. Place the pulse detector switch in position 1 and crank the engine. If the test lamp lights, replace the CD unit. If it does not light, replace the trigger stator.

Test 3
(No. 2 Cylinder
Trigger Coil Output)

1. Disconnect the trigger housing orange wire at the terminal block.

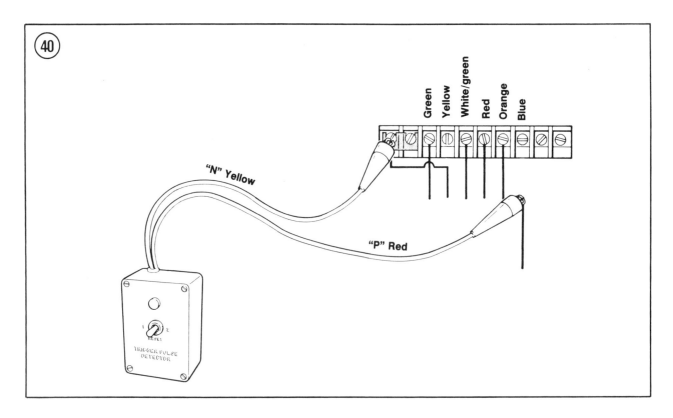

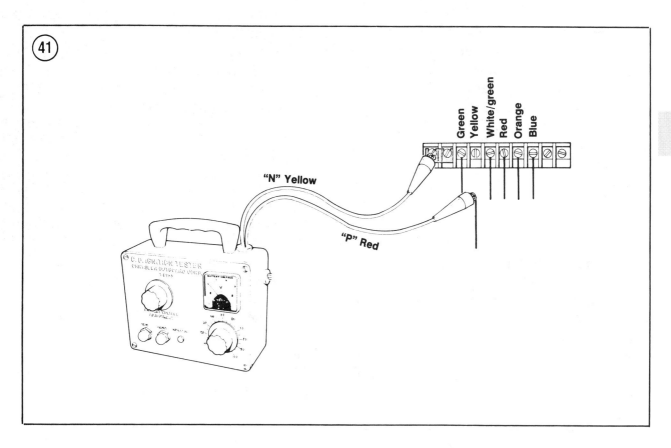

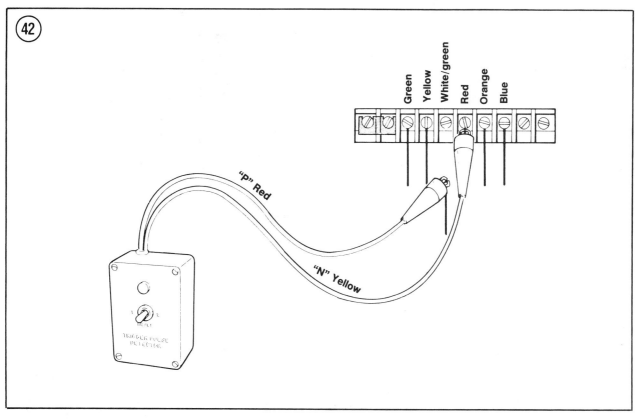

3

2. Connect the trigger pulse detector red test lead to the disconnected orange wire. Connect the yellow test lead to the terminal block green wire terminal. See **Figure 43**.

3. Place the pulse detector switch in position 1 and crank the engine. If the test lamp lights, replace the CD unit. If it does not light, replace the trigger stator.

Test 4
(CD Unit Output—No. 1 Coil)

Refer to **Figure 44** for this procedure.

1. Disconnect the CD unit brown wire at the No. 1 ignition coil.

2. Connect the load coil between the disconnected brown wire and the terminal block ground terminal.

3. Connect the red test lead of CD tester T8953 to the brown wire (same wire as in Step 2).

4. Connect the yellow test lead to the terminal block ground terminal.

5. Set the indicator knob to "50" on the tester scale and crank the engine. If the tester lamp does not light, replace the CD unit.

Test 5
(CD Unit Output—No. 2 Coil)

Refer to **Figure 45** for this procedure.

1. Disconnect the CD unit white wire at the No. 2 ignition coil.

2. Connect the load coil between the disconnected white wire and the terminal block ground terminal.

3. Connect the red test lead of CD tester T8953 to the white wire (same wire as in Step 2).

4. Connect the yellow test lead to the terminal block ground terminal.

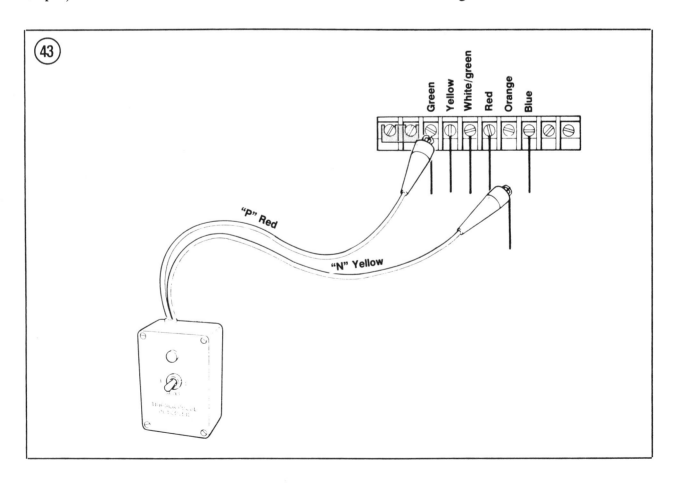

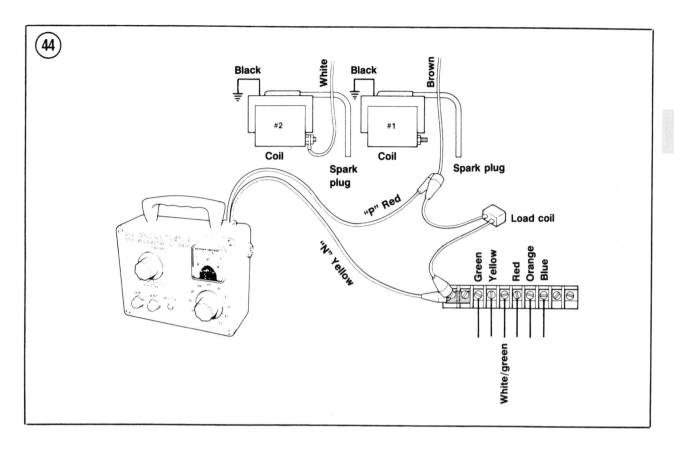

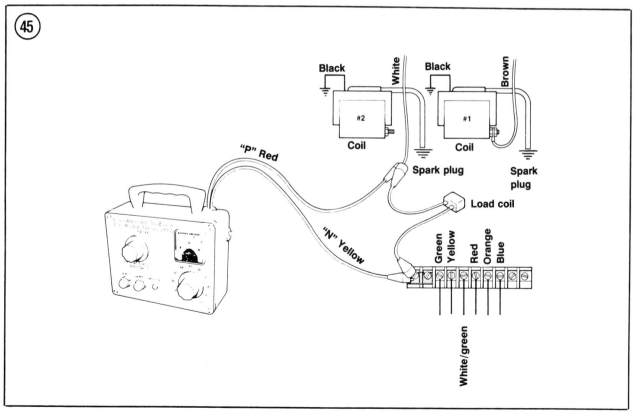

5. Set the indicator knob to "50" on the tester scale and crank the engine. If the tester lamp does not light, replace the CD unit.

Ignition Coil Test

1. Remove the ignition coil from the engine.
2. Connect an ohmmeter between either primary wire terminal and ground. The meter should show 0.03-0.05 ohm resistance.
3. Connect the ohmmeter between either primary wire terminal and the end of the secondary wire. The meter should show 230-260 ohms resistance.
4. If the meter readings are not as specified in Step 2 and Step 3, replace the coil.

MAGNETO CD SYSTEM

This ignition system is used on 1978-on 6, 7.5 and 8 hp engines. The circuit test consists of a preliminary check and 3 individual component tests. The test sequences require the use of Chrysler CD tester T8953, ignition tester T8996 (which includes a No. 22 trigger pulse detector and MM-1 clip) and plug adapter T11201.

Preliminary Check

This procedure determines whether there is sufficient voltage present in the secondary system to fire the spark plug and if the plug is firing properly.
1. Install the MM-1 clip over the No. 1 spark plug wire as close as possible to the plug boot.
2. Connect the yellow test lead of CD tester T8953 to the MM-1 clip lead. Connect the red test lead to the exhaust port cover ground wire screw.
3. Set the tester indicator dial on "75" and crank the engine. The tester indicator lamp will light if the cylinder is firing normally.
4. Depress tester reset button to turn indicator lamp out. Unclip MM-1 clip from No. 1 plug wire and install on No. 2 plug

wire. Repeat Steps 1-3 to test the other cylinder.
5. If the engine does not fire on either cylinder, disconnect one wire at the stop switch and repeat Steps 1-4. If both cylinders now fire normally, replace the stop switch. If the engine still does not fire on either cylinder, check the ground wire at the exhaust port cover. If the ground is good, reconnect the stop switch and perform tests 1-3 (in that order) as described in this chapter. If the system passes all 3 tests, replace the CD coil modules.
6. If only the No. 1 cylinder fires, install a new spark plug in the No. 2 cylinder and repeat Steps 1-3. If that does not correct the problem, perform test 3 and test 1 (in that order) as described in this chapter. If the system passes both tests, replace the No. 2 CD coil module.
7. If only the No. 2 cylinder fires, install a new spark plug in the No. 1 cylinder and repeat Steps 1-3. If that does not correct the problem, perform test 2 and test 1 (in that order) as described in this chapter. If the system passes both tests, replace the No. 1 CD coil module.

Test 1
(Stator Output)

Refer to **Figure 46** for this procedure.
1. Disconnect both single wire (blue) connector plugs between the stator and CD coil module. Connect the stator end of each plug to the single-wire plug connectors on plug adapter T11201.
2. Connect the alligator clip end of the black-sleeved wire on the plug adapter to a good engine ground.
3. Connect red test lead of CD tester T8953 to the red-sleeved wire marked "Charge +" on the plug adapter.
4. Connect the yellow test lead to a good ground.

5. Set the tester indicator dial on "45" and crank the engine. The tester lamp should light.

6. Remove the plug adapter and reconnect both stator connector plugs.

Test 2
(No. 1 Cylinder Trigger Output)

Refer to **Figure 47** for this procedure.

1. Disconnect the 2-wire (red and white/green) connector plug between the trigger housing and CD coil module. Connect the trigger housing end to the 2-wire plug on plug adapter T11201.

2. Connect the trigger pulse detector red test lead to the end of the red-sleeved plug adapter wire marked "plug 2."

3. Connect the yellow test lead to the end of the yellow-sleeved plug adapter wire marked "plug 1."

4. Place the pulse detector switch in position 1 and crank the engine. If the test lamp does not light, replace the trigger stator.

5. Remove the plug adapter and reconnect the 2-wire plug.

Test 3
(No. 2 Cylinder Trigger Output)

Refer to **Figure 48** for this procedure.

1. Disconnect the 2-wire (orange and green) connector plug between the trigger housing and CD coil module. Connect the trigger housing end to the 2-wire plug on plug adapter T11201.

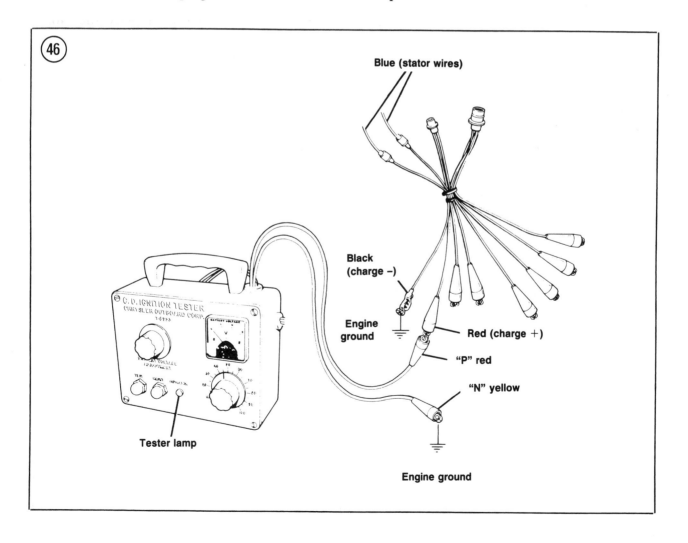

(46)

Blue (stator wires)

Black (charge –)

Engine ground

Red (charge +)

"P" red

"N" yellow

Tester lamp

Engine ground

3

2. Connect the trigger pulse detector red test lead to the end of the red-sleeved plug adapter wire marked "plug 2."

3. Connect the yellow test lead to the end of the yellow-sleeved plug adapter wire marked "plug 1."

4. Place the pulse detector switch in position 1 and crank the engine. If the test lamp does not light, replace the trigger stator.

5. Remove the plug adapter and reconnect the 2-wire plug.

MAGNAPOWER/MAGNETO CD IGNITION SYSTEM

This ignition system is used on 20, 25, 30 and 35 hp electric start engines. On models prior to 1980, it is called a Magnapower ignition. On 1980 and later models, it is called a magneto CD ignition. The circuit test consists of a preliminary check and 5 individual component tests. The test sequences require the use of Chrsyler CD tester T8953, ignition tester T8996 (which includes a No. 22 trigger pulse detector, load coil and MM-1 clip) and plug adapter T11201.

Preliminary Check

This procedure determines whether there is sufficient voltage present in the secondary system to fire the spark plug and if the plug is firing properly.

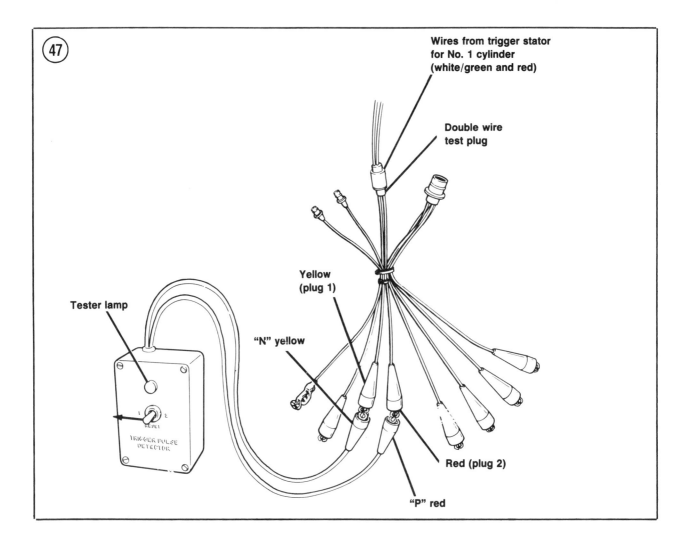

Wires from trigger stator for No. 1 cylinder (white/green and red)

Double wire test plug

Yellow (plug 1)

"N" yellow

Tester lamp

TRIGGER PULSE DETECTOR

Red (plug 2)

"P" red

1. Install the MM-1 clip over the No. 1 spark plug wire as close as possible to the plug boot.

2. Connect the yellow test lead of CD tester T8953 to the MM-1 clip lead. Connect the red test lead to the terminal block ground wire screw.

3. Set the tester indicator dial on "75" and crank the engine. The tester indicator lamp will light if the cylinder is firing normally.

4. Depress tester reset button to turn indicator lamp out. Unclip MM-1 clip from No. 1 plug wire and install on No. 2 plug wire. Repeat Steps 1-3 to test the other cylinder.

5. If the engine does not fire on either cylinder, perform tests 1-5 (in that order) as described in this chapter. If the system passes all 5 tests, remove either wire from the ignition switch M terminal. If both cylinders now fire, replace the ignition switch. If both cylinders still do not fire, replace both ignition coils.

6. If only the No. 1 cylinder fires, install a new spark plug in the No. 2 cylinder and repeat Steps 1-3. If that does not correct the problem, perform test 5 and test 3 (in that order) as described in this chapter. If the system passes both tests, replace the No. 2 ignition coil.

7. If only the No. 2 cylinder fires, install a new spark plug in the No. 1 cylinder and repeat Steps 1-3. If that does not correct the problem, perform test 4 and test 2 (in that order) as described in this chapter. If the system passes both tests, replace the No. 1 ignition coil.

3

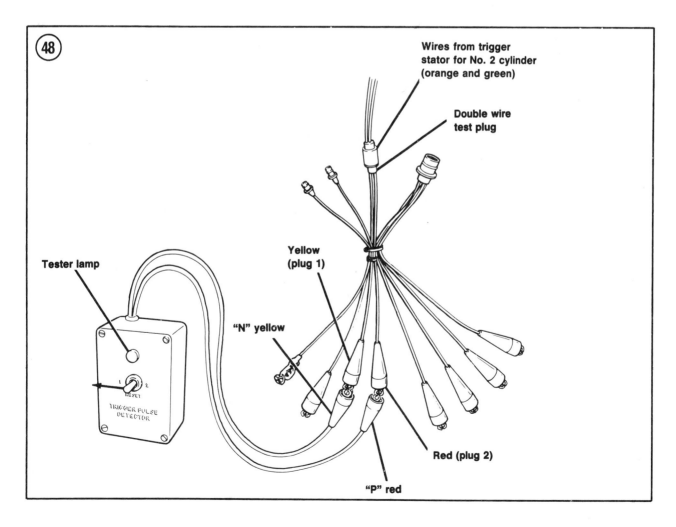

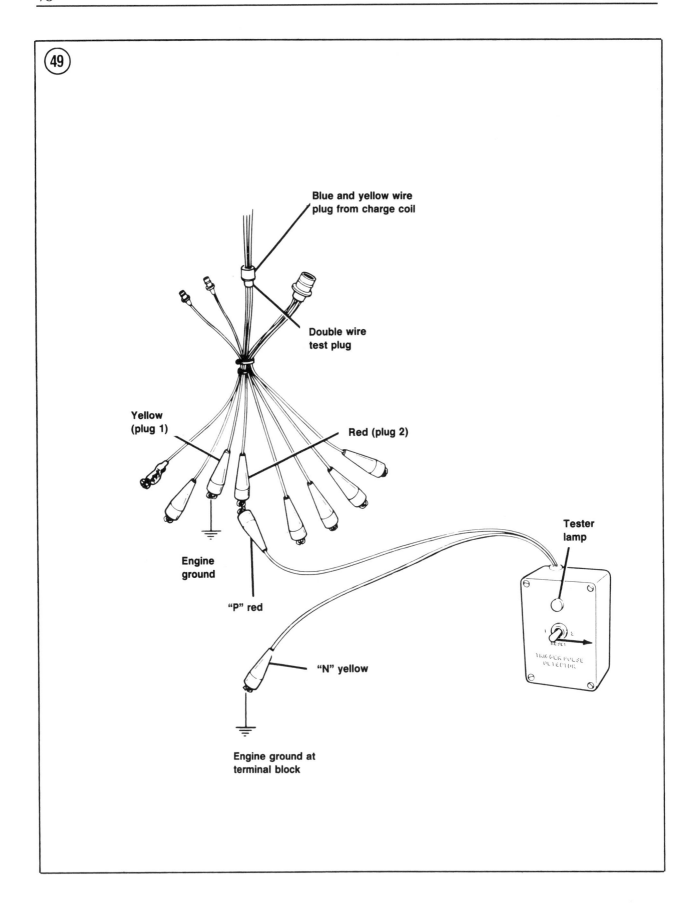

Blue and yellow wire
plug from charge coil

Double wire
test plug

Yellow
(plug 1)

Red (plug 2)

Engine
ground

Tester
lamp

"P" red

TRIGGER PULSE
DETECTOR

"N" yellow

Engine ground at
terminal block

Test 1
(Stator Winding Output)

1. Disconnect the 2-wire (blue and yellow) connector plug between the stator and CD unit. Connect the stator end of the plug to the 2-wire plug connector on plug adapter T11201. See **Figure 49**.

2. Connect the yellow-sleeved wire marked "plug 1" on the plug adapter to a good engine ground.

3. Connect red test lead of the trigger pulse detector to the end of the red-sleeved wire marked "plug 2" on the plug adapter.

4. Connect the yellow test lead to the terminal block ground terminal.

5. Set the tester switch in position 2 and crank the engine. The tester lamp should light.

6. Remove the trigger pulse detector from the circuit.

7. Disconnect the yellow-sleeved plug adapter wire from the engine ground and connect it to the red test lead of CD tester T8953. See **Figure 50**.

8. Connect the yellow test lead to the terminal block ground terminal.

9. Set the tester indicator knob on "10" and crank the engine. The test lamp should light.

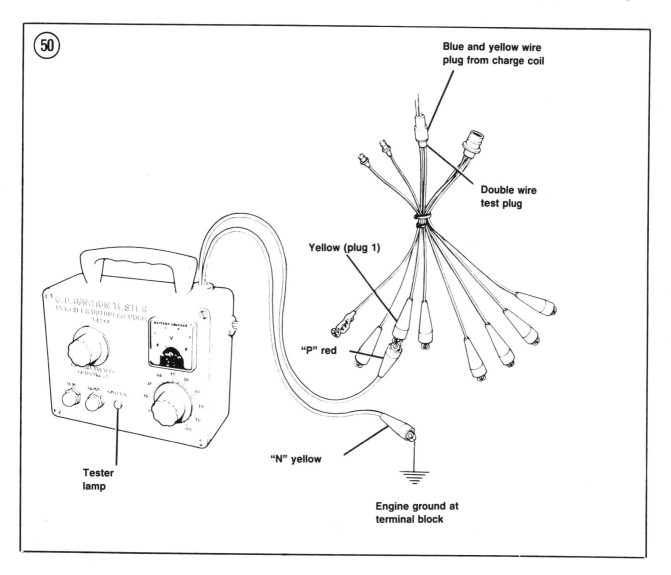

50

Blue and yellow wire plug from charge coil

Double wire test plug

Yellow (plug 1)

"P" red

Tester lamp

"N" yellow

Engine ground at terminal block

3

10. If the test lamp does not light in Step 5 or Step 9, replace the stator.

11. Remove the plug adapter and reconnect the 2-wire stator connector plug to the CD unit plug.

Test 2
(No. 1 Cylinder Trigger Coil Output)

Refer to **Figure 51** for this procedure.

1. Disconnect the 4-wire connector plug between the trigger housing and CD unit. Connect the trigger housing end to the 4-wire plug on plug adapter T11201.

2. Connect the trigger pulse detector red test lead to the end of the red-sleeved plug adapter wire marked "Trigger 1 Pos."

3. Connect the yellow test lead to the end of the yellow-sleeved plug adapter wire marked "Trigger 1 Neg."

4. Place the pulse detector switch in position 1 and crank the engine. If the test lamp lights, replace the CD unit. If it does not light, replace the trigger housing assembly.

5. Remove the plug adapter and reconnect the 4-wire plug.

Test 3
(No. 2 Cylinder Trigger Coil Output)

Refer to **Figure 52** for this procedure.

1. Disconnect the 4-wire connector plug between the trigger housing and CD unit.

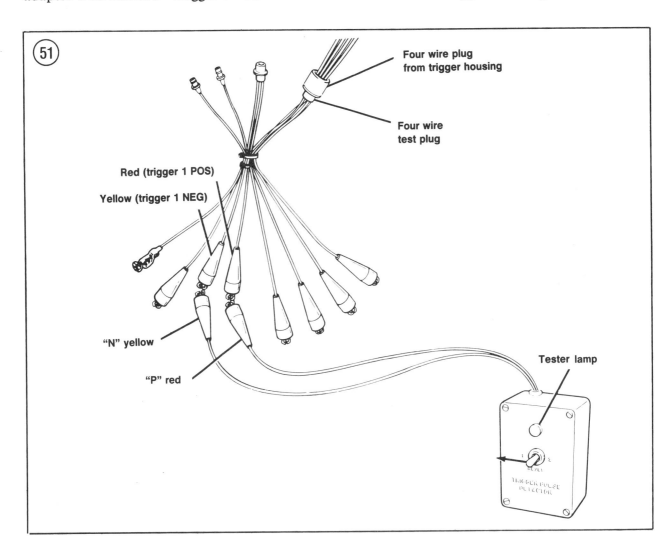

Connect the trigger housing end to the 4-wire plug on plug adapter T11201.

2. Connect the trigger pulse detector red test lead to the end of the red-sleeved plug adapter wire marked "Trigger 2 Pos."

3. Connect the yellow test lead to the end of the yellow-sleeved plug adapter wire marked "Trigger 2 Neg."

4. Place the pulse detector switch in position 1 and crank the engine. If the test lamp lights, replace the CD unit. If it does not light, replace the trigger housing assembly.

5. Remove the plug adapter and reconnect the 4-wire plug.

Test 4
(CD Unit Output—No. 1 Coil)

Refer to **Figure 53** for this procedure.

1. Disconnect the CD unit brown wire at the No. 1 ignition coil.

2. Connect the load coil between the disconnected brown wire and the terminal block ground terminal.

3. Connect the red test lead of CD tester T8953 to the brown wire (same wire as in Step 2).

4. Connect the yellow test lead to the terminal block ground terminal.

3

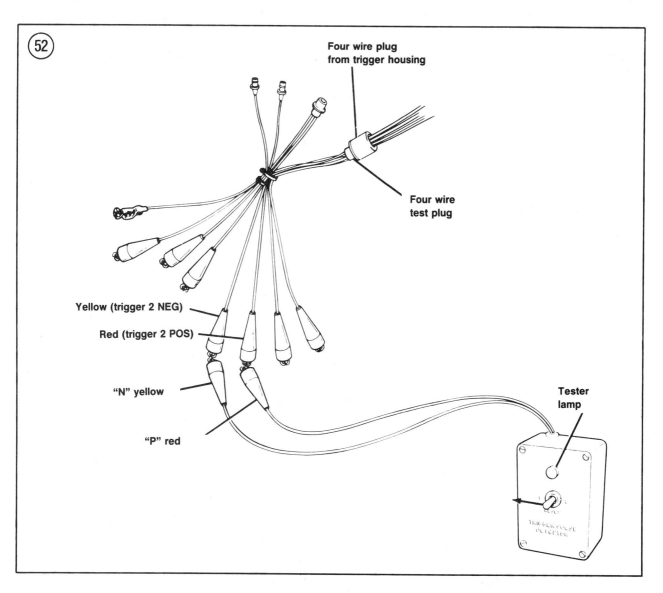

(52)

Four wire plug
from trigger housing

Four wire
test plug

Yellow (trigger 2 NEG)

Red (trigger 2 POS)

"N" yellow

"P" red

Tester
lamp

TRIGGER PULSE
DETECTOR

5. Set the indicator knob to "50" on the tester scale and crank the engine. If the tester lamp does not light, replace the CD unit.

Test 5
(CD Unit Output—No. 2 Coil)

Refer to **Figure 54** for this procedure.
1. Disconnect the CD unit white wire at the No. 2 ignition coil.
2. Connect the load coil between the disconnected white wire and the terminal block ground terminal.
3. Connect the red test lead of CD tester T8953 to the white wire (same wire as in Step 2).
4. Connect the yellow test lead to the terminal block ground terminal.

5. Set the indicator knob to "50" on the tester scale and crank the engine. If the tester lamp does not light, replace the CD unit.

Ignition Coil Test

1. Remove the ignition coil from the engine.
2. Connect an ohmmeter between either primary wire terminal and ground. The meter should show 0.03-0.05 ohm resistance.
3. Connect the ohmmeter between either primary wire terminal and the end of the secondary wire. The meter should show 230-260 ohms resistance.
4. If the meter readings are not as specified in Step 2 and Step 3, replace the coil.

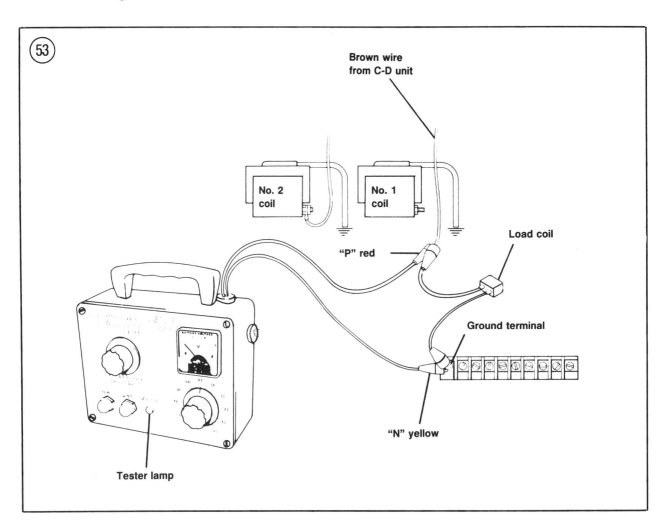

PRESTOLITE CD IGNITION SYSTEM

This alternator CD system was introduced on some 1980 55 hp engines. It is also used on 1982-1984 125 hp and some 1983-1984 75 hp and 85 hp engines. The circuit test consists of a preliminary check and 5 individual component tests (3 on the 75-85 hp). The test sequences require the use of Chrysler CD tester T8953, ignition tester T8996 (which includes a No. 22 trigger pulse detector, load coil and MM-1 clip) and plug adapter T11237.

CAUTION
If a tachometer is installed, Chrysler recommends the use of its 5H167 model or one wired in the same manner. Failure to use a correctly wired tachometer will damage the ignition system.

Preliminary Check

This procedure determines whether there is sufficient voltage present in the secondary system to fire the spark plug and if the plug is firing properly.
1. Install the MM-1 clip over the No. 1 spark plug wire as close as possible to the plug boot.
2. Connect the yellow test lead of CD tester T8953 to the MM-1 clip lead. Connect the red test lead to the terminal block ground wire screw.
3. Set the tester indicator dial on "80" and crank the engine. The tester indicator lamp will light if the cylinder is firing normally.
4. Depress tester reset button to turn indicator lamp out. Repeat Steps 1-3 on each remaining cylinder.

CAUTION
If the ignition switch is replaced in Step 5, it must be wired as magneto CD rather than battery CD.

3

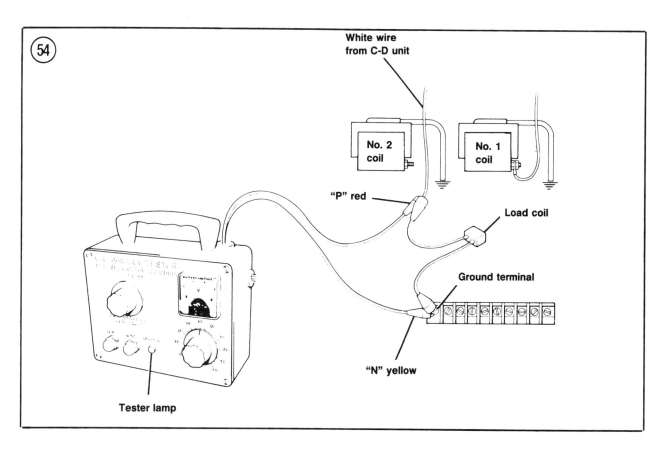

(54)

White wire
from C-D unit

No. 2 coil

No. 1 coil

"P" red

Load coil

Ground terminal

Tester lamp

"N" yellow

5. If the engine does not fire on any cylinder (55-85 hp) or does not fire on the No. 1 and 2 or the No. 3 and 4 cylinders (125 hp):

  a. Check the ground wire at the terminal block.

  b. If the ground is good, perform tests 1-5 (in that order) as described in this chapter.

  c. If the system passes all 5 tests, remove either wire from the ignition switch M terminal. If all cylinders now fire, replace the ignition switch. If the cylinders still do not fire, replace the ignition coils.

6. If only the No. 1 (55 hp) or the No. 1 and 2 (75, 85 and 125 hp) cylinders fire:

  a. Install a new spark plug in the No. 2 (55 hp), No. 3 (75-85 hp) or No. 3 and 4 (125 hp) cylinders and repeat Steps 1-3.

  b. If that does not correct the problem, perform test 5 and test 3 (for 2- and 4-cylinder engines) or test 6 and test 7 (for 3-cylinder engines).

  c. If the system passes both tests, replace the No. 2 (55 hp), No. 3 (75-85 hp) or No. 3 and 4 (125 hp) ignition coils.

7. If only the No. 2 (55 hp), No. 3 (75-85 hp) or No. 3 and 4 (125 hp) cylinders fire:

  a. Install a new spark plug in the No. 1 (55 hp) or the No. 1 and 2 (75, 85 and 125 hp) cylinders and repeat Steps 1-3.

  b. If that does not correct the problem, perform test 5 and test 3 (for 2- and 4-cylinder engines) or test 6 and test 7 (for 3-cylinder engines).

  c. If the system passes both tests, replace the No. 1 (55 hp) or No. 1 and 2 (75, 85 and 125 hp) ignition coils.

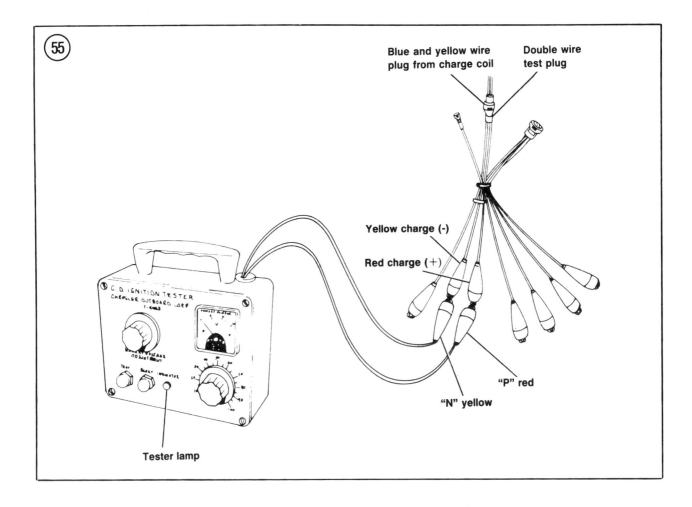

(55)

Blue and yellow wire plug from charge coil

Double wire test plug

Yellow charge (-)

Red charge (+)

"P" red

"N" yellow

C D IGNITION TESTER
CHRYSLER OUTBOARD

Tester lamp

Test 1
(Stator Winding Output)

Refer to **Figure 55** for this procedure.

1. Disconnect the 2-wire (blue and yellow) connector plug between the stator and CD unit.

2. Connect the stator end of the plug to the 2-wire plug connector on plug adapter T11237.

3. Connect the red test lead of CD tester T8953 to the red-sleeved plug adapter wire marked "charge +."

4. Connect the yellow test lead to the yellow-sleeved plug adapter wire marked "charge –."

5. Set the tester indicator knob on "60" (55 hp) or "50" (75-125 hp) and crank the engine. The test lamp should light.

6. If the test lamp does not light in Step 5, replace the stator.

7. Remove the plug adapter and reconnect the 2-wire stator connector plug to the CD unit plug.

8. On 3- and 4- cylinder engines, disconnect the other 2-wire (blue and yellow) connector plug and repeat the test to check the second stator coil.

Test 2
(Trigger Output)

This test checks the No. 1 (55 hp) or No. 1 and 2 (125 hp) cylinder output. Refer to **Figure 56** for this procedure.

1. Disconnect the 4-wire connector plug between the trigger housing and CD unit. Connect the trigger housing end to the 4-wire plug on plug adapter T11237.

2. Connect the trigger pulse detector red test lead to the end of the red-sleeved plug adapter wire marked "trigger 1 pos."

3. Connect the yellow test lead to the end of the yellow-sleeved plug adapter wire marked "trigger 1 neg."

4. Place the pulse detector switch in position 1 and crank the engine. If the test lamp lights, replace the CD unit. If the test lamp does not light, replace the trigger housing assembly.

5. Remove the plug adapter and reconnect the 4-wire plug.

3

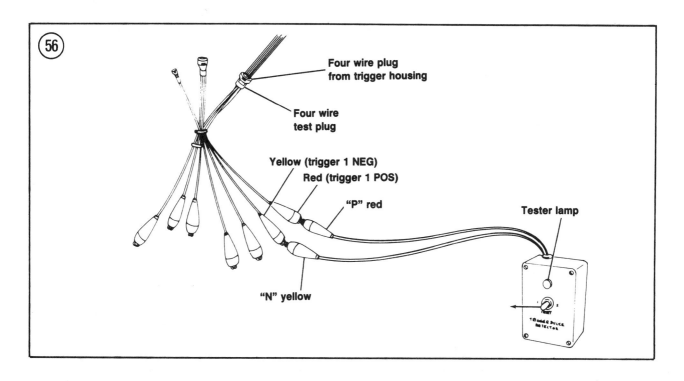

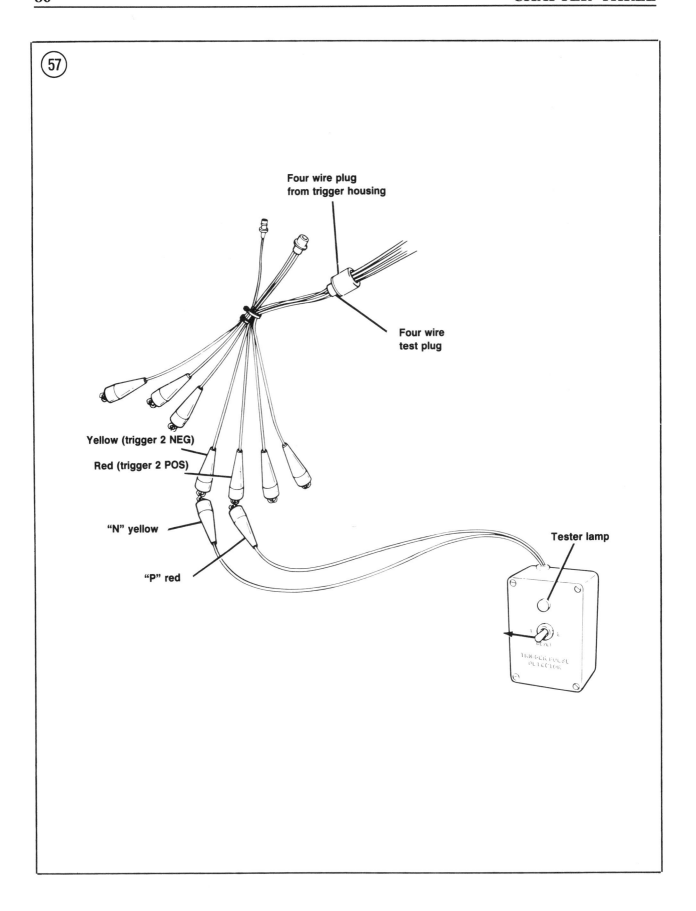

Four wire plug
from trigger housing

Four wire
test plug

Yellow (trigger 2 NEG)

Red (trigger 2 POS)

"N" yellow

"P" red

Tester lamp

TRIGGER PULSE
DETECTOR

Test 3
(Trigger Output)

This test checks the No. 2 (55 hp) or No. 3 and 4 (125 hp) cylinder output. Refer to **Figure 57** for this procedure.

1. Disconnect the 4-wire connector plug between the trigger housing and CD unit. Connect the trigger housing end to the 4-wire plug on plug adapter T11237.

2. Connect the trigger pulse detector red test lead to the end of the red-sleeved plug adapter wire marked "trigger 2 pos."

3. Connect the yellow test lead to the end of the yellow-sleeved plug adapter wire marked "trigger 2 neg."

4. Place the pulse detector switch in position 1 and crank the engine. If the test lamp lights, replace the CD unit. If the test lamp does not light, replace the trigger housing assembly.

5. Remove the plug adapter and reconnect the 4-wire plug.

Test 4
(CD Unit Output)

This test checks the CD unit output to the No. 1 (55 hp) or No. 1 and 3 (125 hp) ignition coils. Refer to **Figure 58** for this procedure.

NOTE
On 125 hp engines, the white coil lead may be disconnected at either the No. 1 or No. 3 coil in Step 1.

1. Disconnect the CD unit orange wire. Disconnect the white coil lead at the No. 1 ignition coil.

2. Connect the orange wire from the CD unit to the single wire plug on plug adapter T11237.

3. Connect the yellow test lead of CD tester T8953 to the yellow-sleeved wire on the plug adapter marked "CD."

4. Connect the red test lead to the terminal block ground terminal.

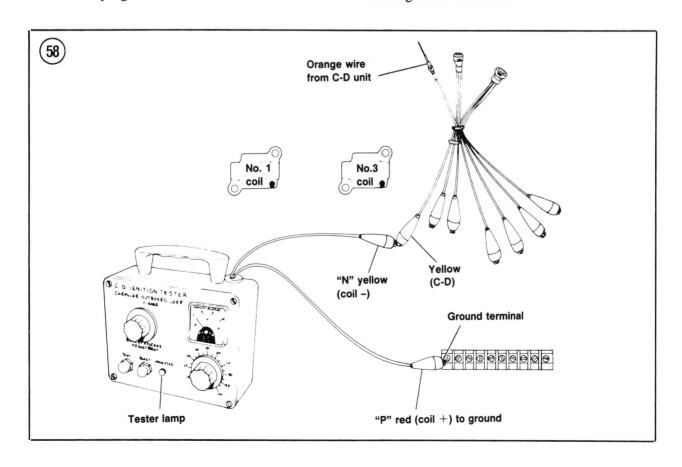

5. Set the indicator knob to "60" on the tester scale and crank the engine. If the test lamp lights, replace the No. 1 (55 hp) or No. 1 and 3 (125 hp) ignition coils. If the lamp does not light, replace the CD unit.

6. Reconnect the orange CD unit and white coil leads.

Test 5
(CD Unit Output)

This test checks the CD unit output to the No. 2 (55 hp) or No. 2 and 4 (125 hp) ignition coils. Refer to **Figure 59** for this procedure.

NOTE
On 125 hp engines, the white coil lead
may be disconnected at either the No. 2
or No. 4 coil in Step 1.

1. Disconnect the CD unit red wire. Disconnect the white coil lead at the No. 2 ignition coil.

2. Connect the red wire from the CD unit to the single wire plug on plug adapter T11237.

3. Connect the yellow test lead of CD tester T8953 to the yellow-sleeved wire on the plug adapter marked "CD."

4. Connect the red test lead to the terminal block ground terminal.

5. Set the indicator knob to "60" on the tester scale and crank the engine. If the test lamp lights, replace the No. 2 (55 hp) or No. 2 and 4 (125 hp) ignition coils. If the lamp does not light, replace the CD unit.

6. Reconnect the red CD unit and white coil leads.

Test 6
(75-85 hp Trigger Output)

This test checks the trigger coil output on each cylinder.

1. Disconnect the 4-wire (orange and green) connector plug between the trigger housing

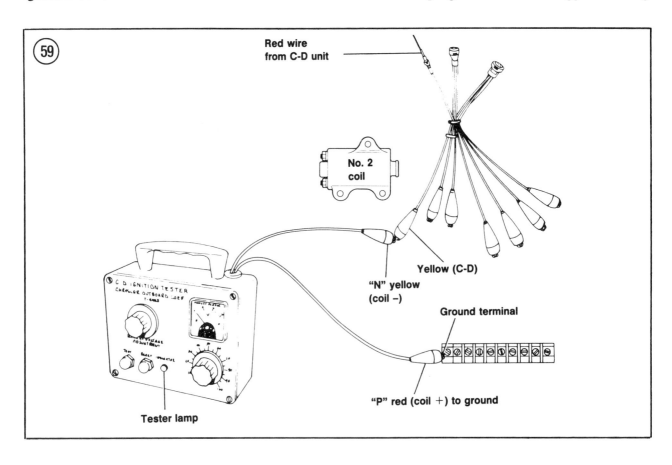

and CD unit. Connect the trigger housing end to the 4-wire plug on plug adapter T11237.

2. Connect the trigger pulse detector red test lead to the end of the red-sleeved plug adapter wire marked "trigger 1 pos."

3. Connect the yellow test lead to the end of the yellow-sleeved plug adapter wire marked "trigger 1 neg."

4. Place the pulse detector switch in position 1 and crank the engine. If the test lamp lights, replace the CD unit. If the lamp does not light, replace the trigger housing assembly.

5. Connect the trigger pulse detector red test lead to the end of the red-sleeved plug adapter wire marked "trigger 2 pos."

6. Connect the yellow test lead to the end of the yellow-sleeved plug adapter wire marked "trigger 2 neg."

7. Place the pulse detector switch in position 1 and crank the engine. If the test lamp lights, replace the CD unit. If the lamp does not light, replace the trigger housing assembly.

8. Remove the plug adapter and reconnect the 4-wire plug.

Test 7
(75-85 hp CD Unit Output)

1. Disconnect the white No. 1 ignition coil lead from the forward CD unit orange lead. Connect the orange lead to the single wire plug on plug adapter T11237 marked "CD."

2. Connect the yellow test lead of CD tester T8953 to the yellow-sleeved plug adapter lead.

3. Connect the red test lead to the terminal block ground terminal.

4. Set the indicator knob to "60" on the tester scale and crank the engine. The tester lamp should light.

5. Disconnect the white No. 2 ignition coil lead from the forward CD unit red lead.

Connect the red lead to the single wire plug on plug adapter T11237 marked "CD."

6. Repeat Step 4.

NOTE
The orange lead at the rear CD unit is not used on 3-cylinder engines.

7. Disconnect the white No. 3 ignition coil lead from the rear CD unit red lead. Connect the red lead to the single wire plug on plug adapter T11237 marked "CD."

8. Repeat Step 4.

9. If the test lamp fails to light in Step 4 or Step 6, replace the forward CD unit. If it does not light in Step 8, replace the rear CD unit.

IGNITION SWITCH

The ignition switch should be removed and all leads disconnected for testing.

Continuity and Resistance Check

Refer to **Figure 60** (Motorola) or **Figure 61** (Prestolite) for this procedure.

1. With the switch in the OFF position, connect a self-powered test lamp between terminals M and M. The lamp should light.

2. Turn the switch to the RUN position and connect the test lamp leads between terminals B and I. The lamp should light. Turn the key to the START position. The lamp should remain on.

3. Repeat Step 2 with the key depressed to engage the CHOKE position. The lamp should light and remain on as the switch is moved from the RUN to START position.

4. Turn the switch to the RUN position and connect the test lamp leads between terminals B and C. Depress key to engage CHOKE position. The lamp should light. Turn the key to the START position. The lamp should remain on.

5. With the switch in START, connect the test lamp between terminals B and S. The lamp should light. Hold switch in START and depress key to engage CHOKE position. The lamp should remain on.

6. Disconnect the spark plug leads to prevent the engine from starting.

7. Connect a negative voltmeter lead to ground and the positive lead to switch terminal B. Turn switch to START to crank the engine. Note voltmeter reading.

8. Move the positive voltmeter lead to terminal I and repeat Step 7. If voltage drops more than one volt, internal resistance is too great. Replace the switch.

9. If the switch does not perform as indicated in any step, replace the switch.

FUEL SYSTEM

Many outboard owners automatically assume that the carburetor is at fault when the engine does not run properly. While fuel system problems are not uncommon, carburetor adjustment is seldom the answer. In many cases, adjusting the carburetor only compounds the problem by making the engine run worse.

Fuel system troubleshooting should start at the gas tank and work through the system, reserving the carburetor(s) as the final point. The majority of fuel system problems result from an empty fuel tank, sour fuel, a plugged fuel filter or a malfunctioning fuel pump. **Table 4** provides a series of symptoms and causes that can be useful in locating fuel system problems.

Troubleshooting

As a first step, check the fuel flow. Remove the fuel tank cap and look into the tank. If there is fuel present, disconnect and ground the spark plug lead(s) as a safety precaution. Disconnect the fuel line at the carburetor (**Figure 62**, typical) and place it in a suitable container to catch any discharged fuel. See if gas

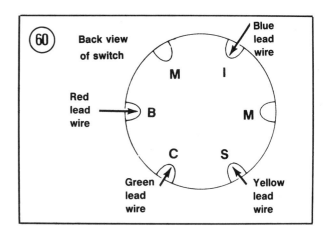

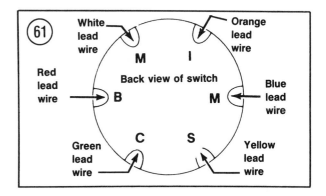

flows freely from the line when the primer bulb is squeezed.

If there is no fuel flow from the line, the fuel petcock may be shut off or blocked by rust or foreign matter, the fuel line may be stopped up or kinked or a primer bulb check valve may be defective. If a good fuel flow is present, crank the engine 10-12 times to check fuel pump operation. A pump that is operating satisfactorily will deliver a good, constant flow of fuel from the line. If the amount of flow varies from pulse to pulse, the fuel pump is probably failing.

Carburetor chokes can also present problems. A choke that sticks open will show up as a hard starting problem; one that sticks closed will result in a flooding condition.

During a hot engine shut-down, the fuel bowl temperature can rise above 200°, causing the fuel inside to boil. While Chrysler carburetors are vented to atmosphere to prevent this problem, there is a possibility that some fuel will percolate over the high-speed nozzle.

A leaking inlet needle and seat or a defective float will allow an excessive amount of fuel into the intake manifold. Pressure in the fuel line after the engine is shut down forces fuel past the leaking needle and seat. This raises the fuel bowl level, allowing fuel to overflow into the manifold.

Excessive fuel consumption may not necessarily mean an engine or fuel system problem. Marine growth on the boat's hull, a bent or otherwise damaged propeller or a fuel line leak can cause an increase in fuel consumption. These areas should all be checked *before* blaming the carburetor.

Spark plug wet fouling can occur on 3- and 4-cylinder engines if the fuel recirculating system is not functioning properly. If wet fouling proves a chronic problem, remove the recirculation hose from the carburetor adapter and try to blow through it. If you can,

one or more of the reed valves on the cylinder drain reed plate are not sealing properly.

ENGINE

Engine problems are generally symptoms of something wrong in another system, such as ignition, fuel or starting. If properly maintained and serviced, the engine should experience no problems other than those caused by age and wear.

Overheating and Lack of Lubrication

Overheating and lack of lubrication cause the majority of engine mechanical problems. Outboard motors create a great deal of heat and are not designed to operate at a standstill for any length of time. Using a spark plug of the wrong heat range can burn a piston. Incorrect ignition timing, a propeller that is too large (over-propping) or an excessively lean fuel mixture can also cause the engine to overheat.

Preignition

Preignition is the premature burning of fuel and is caused by hot spots in the combustion chamber (**Figure 63**). The fuel actually ignites before it is supposed to. Glowing deposits in the combustion chamber, inadequate cooling or overheated spark plugs can all cause preignition. This is first noticed in the form of a power loss but will eventually result in extensive damage to the internal parts of the engine because of higher combustion chamber temperatures.

63 Ignited by hot deposit   Regular ignition spark   Ignites remaining fuel   Flame fronts collide

Detonation

Commonly called "spark knock" or "fuel knock," detonation is the violent explosion of fuel in the combustion chamber prior to the proper time of combustion (**Figure 64**). Severe damage can result. Use of low octane gasoline is a common cause of detonation.

Even when high octane gasoline is used, detonation can still occur if the engine is improperly timed. Other causes are over-advanced ignition timing, lean fuel mixture at or near full throttle, inadequate engine cooling, cross-firing of spark plugs, excessive accumulation of deposits on piston and combustion chamber or the use of a prop that is too large (over-propping).

Since outboard motors are noisy, engine knock or detonation is likely to go unnoticed by owners, especially at high engine rpm when wind noise is also present. Such inaudible detonation, as it is called, is usually the cause when engine damage occurs for no apparent reason.

Poor Idling

A poor idle can be caused by improper carburetor adjustment, incorrect timing or ignition system malfunctions. Check the gas cap vent for an obstruction.

Misfiring

Misfiring can result from a weak spark or a dirty spark plug. Check for fuel contamination. If misfiring occurs only under heavy load, as when accelerating, it is usually caused by a defective spark plug. Run the motor at night to check for spark leaks along the plug wire and under spark plug cap or use a spark leak tester.

> *WARNING*
> *Do not run engine in dark garage to check for spark leak. There is considerable danger of carbon monoxide poisoning.*

Water Leakage in Cylinder

The fastest and easiest way to check for water leakage in a cylinder is to check the spark plugs. Water will clean a spark plug. If one of the 2 plugs on a multi-cylinder engine is clean and the other is dirty, there is most likely a water leak in the cylinder with the clean plug.

To remove all doubt, install a dirty plug in each cylinder. Run the engine in a test tank or on the boat in water for 5-10 minutes. Shut the engine off and remove the plugs. If one plug is clean and the other dirty (or if both plugs are clean), a water leak in the cylinder(s) is the problem.

Flat Spots

If the engine seems to die momentarily when the throttle is opened and then recovers, check for a dirty main jet in the carburetor, water in the fuel or an excessively lean mixture.

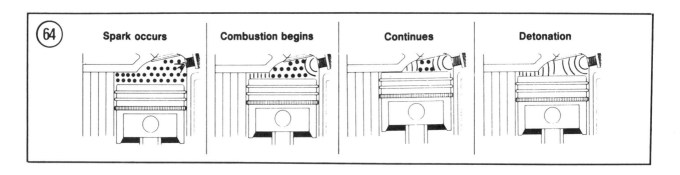

| Spark occurs | Combustion begins | Continues | Detonation |

Power Loss

Several factors can cause a lack of power and speed. Look for air leaks in the fuel line or fuel pump, a clogged fuel filter or a choke/throttle valve that does not operate properly. Check ignition timing.

A piston or cylinder that is galling, incorrect piston clearance or a worn/sticky piston ring may be responsible. Look for loose bolts, defective gaskets or leaking machined mating surfaces on the cylinder head, cylinder or crankcase. Also check the crankcase oil seal; if worn, it can allow gas to leak between cylinders.

Piston Seizure

This is caused by one or more pistons with incorrect bore clearances, piston rings with an improper end gap, the use of an oil-fuel mixture containing less than 1 part oil to 50 parts of gasoline or an oil of poor quality, a spark plug of the wrong heat range or incorrect ignition timing. Overheating from any cause may result in piston seizure.

Excessive Vibrations

Excessive vibrations may be caused by loose motor mounts, worn bearings or a generally poor running motor.

3

Engine Noises

Experience is needed to diagnose accurately in this area. Noises are difficult to differentiate and even harder to describe. Deep knocking noises usually mean main bearing failure. A slapping noise generally comes from a loose piston. A light knocking noise during acceleration may be a bad connecting rod bearing. Pinging should be corrected immediately or damage to the piston will result. A compression leak at the head-cylinder joint will sound like a rapid on-off squeal.

Table 1 STARTER TROUBLESHOOTING

Trouble	Cause	Remedy
Starter motor has low no-speed and high-current draw	Armature may be dragging on pole shoes from bent shaft, worn bearings or loose pole shoes. Tight or dirty bearings.	Replace shaft or bearings and/or tighten pole shoes. Loosen or clean bearings.
Low no-load speed and high current draw	Armature may be dragging on pole shoes. Tight or dirty bearings.	Remove armature and test on growler for short. Loosen or clean bearings.
High current draw with no armature rotation	A direct ground switch, at terminal, brushes or field connections.	Replace defective parts.
Starter motor has grounded armature or field winding	Field and/or armature is burned or lead is thrown out of commutator due to excess leakage.	Raise grounded brushes from commutator and insulate them with cardboard. Use Magneto Analyzer (part No. C-91-25213) (Selector No. 3) and test points to check between insulated terminal or starter motor and starter motor frame. (Remove ground connection of shunt coils on motors with this feature.) If analyzer shows resistance (meter needle moves to right), there is a ground. Raise other brushes from armature and check armature and fields separately to locate ground.
Starter motor has grounded armature or field winding ground field	Current passes through armature first, then to grounds in starter motor.	Disconnect grounded leads, then locate any abnormal windings.
Starter motor fails to operate and draws no current and/or high resistance	Open circuit in fields or armature, at connections or brushes or between brushes and commutator.	Repair or adjust broken or weak brush springs, worn brushes, high insulation between commutator bars or a dirty, gummy or oily commutator.
High resistance in starter motor	Low no-load speed and a low-current draw and low developed torque.	Close "open" field winding on unit which has 2 or 3 circuits in starter motor (unit in which current divides as it enters, taking 2 or 3 parallel paths).
High free speed and high current draw	Shorted fields in starter motor.	Install new fields and check for improved performance. (Fields normally have very low resistance, thus it is difficult to detect shorted fields, since difference in current draw between normal starter motor field windings would not be very great.)
Excessive voltage drop	Cables too small.	Install larger cables to accomodate high current draw.
High circuit resistance	Dirty connections.	Clean connections.

(continued)

Table 1 STARTER TROUBLESHOOTING (continued)

Trouble	Cause	Remedy
Starter does not operate	Run-down battery.	(1) Check battery with hydrometer. If reading is below 1.230, recharge or replace battery.
	Poor contact at terminals.	(2) Remove terminal clamps. Scrape terminals and clamps clean and tighten bolts securely.
	Wiring or key switch	Coat with sealer to protect against further corrosion.
	Starter solenoid.	Check for resistance between: (a) positive (+) terminal of battery and large input terminal of starter solenoid, (b) large wire at top of starter motor and negative (–) terminal of battery, and (c) small terminal of starter solenoid and positive battery terminal. Key switch must be in START position. Repair all defective parts.
	Starter motor.	(3) With a fully charged battery, connect a negative (–) jumper wire to upper terminal on side of starter motor and a positive jumper to large lower terminal of starter motor. If motor still does not operate, remove for overhaul or replacement.
Starter turns over too slowly	Low battery.	See "Starter does not operate."
	Poor contact at battery terminal.	See "Starter does not operate."
	Poor contact at starter . solenoid or starter motor.	Check all terminals for looseness and tighten all nuts securely.
	Starter mechanism.	Disconnect positive (+) battery terminal. Rotate pinion gear in disengaged position. Pinion gear and motor should run freely by hand. If motor does not turn over easily, clean starter and replace all defective parts.
	Starter motor.	See "Starter does not operate."
Starter spins freely but does not engage engine	Low battery.	See "Starter does not operate."
	Poor contact at battery terminal.	See "Starter does not operate."
	Poor contact @ starter solenoid or starter motor.	See "Starter does not operate."
	Dirty or corroded pinion drive.	Clean thoroughly and lubricate the spline underneath the pinion with multipurpose lubricant.

(continued)

3

Table 1 STARTER TROUBLESHOOTING (continued)

Trouble	Cause	Remedy
Starter does not engage freely	Pinion or flywheel gear.	Inspect mating gears for excessive wear. Replace all defective parts.
	Small anti-drift spring.	If drive pinion interferes with flywheel gear after engine has started, inspect anti-drift spring located under pinion gear. Replace all defective parts. If drive pinion tends to stay engaged in flywheel gear when starter motor is in idle position, start motor at 1/4 throttle to allow starter pinion gear to release flywheel ring gear instantly.
Starter keeps on spinning after key is turned to ON	Key not fully returned.	Check that key has returned to normal ON from START. Replace switch if key constantly stays in START.
	Starter solenoid.	Inspect starter solenoid to see if contacts have become stuck in closed position. If starter does not stop running with small yellow lead disconnected from starter solenoid, replace starter solenoid.
	Wiring or key switch.	Inspect all wires for defects. Open remote control box and inspect wiring at switches. Repair or replace all defective parts.
Wires overheat	Battery terminals improperly connected.	Check that negative marking on harness matches that of battery. If battery is connected improperly, red wire to rectifier will overheat.
	Short circuit in wiring system.	Inspect all connections and wires for looseness or defects. Open remote control box and inspect wiring at switches.
	Short circuit in choke solenoid.	Repair or replace all defective parts. Check for high resistance. If blue choke wire heats rapidly when choke is used, choke solenoid may have internal short. Replace if defective.
	Short circuit in starter solenoid.	If yellow starter solenoid lead overheats, there may be internal short (resistance) in starter solenoid. Replace if defective.
	Battery voltage.	Battery voltage is checked with an ampere-volt tester only when battery is under a starting load. Battery must be recharged if it registers under 9.5 volts. If battery is below specified hydrometer reading of 1.230, it will not turn engine fast enough to start it.

Table 2 STARTER GENERATOR TROUBLESHOOTING CHART

Problem	Cause
Test results show no-load speed and higher than normal current draw	Loose pole shoes Grounded armature Shorted armature
Test results indicate very high current draw and no armature rotation	Direct ground at the key switch Direct ground at the starter brushes Direct ground on the field connections
Test results indicate no current draw and the starter fails to operate	Open circuit at key switch Open circuit breaker Defective starter relay Broken or loose wire connections Defective interlock switch Incorrectly adjusted remote control Dirty brushes or dirty commutator
Test results in a low no-load speed and lower than normal current draw—high resistance in starter	Dirty or loose connections Dirty brushes Dirty commutator and brush spring
Test results show a higher than normal no-load speed and higher than normal current draw	Shorted fields

3

Table 3 IGNITION TROUBLESHOOTING

Symptom	Probable cause
Engine won't start, but fuel and spark are okay	Defective spark plugs. Spark plug gap set too wide. Improper spark timing.
Engine misfires at idle	Incorrect spark plug gap. Defective or loose spark plugs. Spark plugs of incorrect heat range. Cracked distributor cap. Leaking or broken high tension wires. Weak armature magnets. Defective coil or condenser. Defective ignition switch. Spark timing out of adjustment.
Engine misfires at high speed	Check all of above. Coil breaks down. Coil shorts through insulation. Spark plug gap too wide. Wrong type spark plugs. Too much spark advance.

(continued)

Table 3 IGNITION TROUBLESHOOTING (continued)

Symptom	Probable cause
Engine backfires	
Through exhaust	Cracked spark plug insulator.
	Carbon path in distributor cap.
	Improper timing.
	Crossed spark plug wires.
Through carburetor	Improper ignition timing.
Engine preignition	Spark advanced too far.
	Incorrect type spark plug.
	Burned spark plug electrodes.
Engine noises (knocking at power head)	Spark advanced too far.
Ignition coil fails	Extremely high voltage.
	Moisture formation.
	Excessive heat from engine.
Spark plugs burn and foul	Incorrect type plug.
	Fuel mixture too rich.
	Inferior grade of gasoline.
	Overheated engine.
	Excessive carbon in combustion chambers.
Ignition causing high fuel consumption	Incorrect spark timing.
	Leaking high tension wires.
	Incorrect spark plug gap.
	Fouled spark plugs.
	Incorrect spark advance.
	Weak ignition coil.
	Preignition.

Table 4 FUEL SYSTEM TROUBLESHOOTING

Symptom	Probable cause
No fuel at carburetor	No gas in tank.
	Air vent in gas cap not open.
	Air vent in gas cap clogged.
	Fuel tank sitting on fuel line.
	Fuel line fittings not properly connected to engine or fuel tank.
	Air leak at fuel connection.
	Fuel pickup clogged.
	Defective fuel pump.
Flooding at carburetor	Choke out of adjustment.
	High float level.
	Float stuck.
	Excessive fuel pump pressure.
	Float saturated beyond buoyancy.
	(continued)

Table 4 FUEL SYSTEM TROUBLESHOOTING (continued)

Symptom	Probable cause
Rough operation	Dirt or water in fuel. Reed valve open or broken. Incorrect fuel level in carburetor bowl. Carburetor loose at mounting flange. Throttle shutter not closing completely. Throttle shutter valve installed incorrectly.
Engine misfires at high speed	Dirty carburetor. Lean carburetor adjustment. Restriction in fuel system. Low fuel pump pressure.
Engine backfires	Poor quality fuel. Air/fuel mixture too rich or too lean.
Engine preignition	Excessive oil in fuel. Inferior grade of gasoline. Lean carburetor mixture.
Spark plugs burn and foul	Fuel mixture too rich. Inferior grade of gasoline.
High gas consumption 　Flooding or leaking	Cracked carburetor casting. Leaks at line connections. Defective carburetor bowl gasket. High float level. Plugged vent hole in cover. Loose needle and seat. Defective needle valve seat gasket. Worn needle valve and seat. Foreign matter clogging needle valve. Worn float pin or bracket. Float binding in bowl. High fuel pump pressure.
Overrich mixture	Choke lever stuck. High float level. High fuel pump pressure.
Abnormal speeds	Carburetor out of adjustment. Too much oil in fuel.

Table 5 CHARGING SYSTEM SPECIFICATIONS (1980-ON)

Year/ model	Minimum amperage @ rpm	Cut-in rpm	Stator ohms	Maximum current draw
1980-on 9.9-15 hp	1.0 @ 2,000 3.5 @ 4,000	1,000	1.0-2.0	4

(continued)

Table 5 CHARGING SYSTEM SPECIFICATIONS (1980-ON) (continued)

Year/model	Minimum amperage @ rpm	Cut-in rpm	Stator ohms	Maximum current draw
1980-on 20 and 30 hp; 1982-on 25 hp; 1983-on 35 hp	1.0 @ 2,000 2.0 @ 3,500	1,800	1.0-2.0	3
1980-1981 35 hp; 1980-on 45 and 50 hp	6.0 @ 3,500	1,100	0.1-0.4	6
1980-on 55 hp	0.5 @ 1,000 7.0 @ 4,000	800	1.0	6
1980-on 70-140 hp	7.0 @ 4,000	800	0.5-1.0	7

Table 6 CONDENSER SPECIFICATIONS

Chrysler No.	Capacity reading in microfarads	
	Min.	Max.
12042, 12042-1	0.28	0.33
12181, 12181-2, 12117, 12143, 12178, 12119, 12139, 1239-1,	0.16	0.20
12196	0.26	0.30
14022-1	0.36	0.44

Table 7 COIL SPECIFICATIONS

Chrysler Part No.	Operating amperage	Primary resistance		Secondary continuity	
		Min.	Max.	Min.	Max.
12057	1.5	0.62	0.68	45	55
12211	1.5	0.50	0.65	50	60
12226	1.4	0.48	0.54	50	60
12348	1.5	0.50	0.65	50	60
A12345, A12231	1.4	0.48	0.54	50	60
A316475	1.0	1.7	2.3	65	75
321475-1	2.35	*	*	10	14
345475-1	1.7	0.20	0.24	7	11
474475	1.7	*	*	7	11
510475	1.7	0.03	0.05	—	—
A85475-2	1.0	1.6	2.0	63	72
A91475-1	1.0	3.0	3.6	50	60
2A91475-2	1.2	3.0	4.0	53	63

* Too low to read.

Chapter Four

Lubrication, Maintenance and Tune-up

The modern outboard motor delivers more power and performance than ever before, with higher compression ratios, new and improved electrical systems and other design advances. Proper lubrication, maintenance and tune-ups have thus become increasingly important as ways in which you can maintain a high level of performance, extend engine life and extract the maximum economy of operation.

You can do your own lubrication, maintenance and tune-ups if you follow the correct procedures and use common sense. The following information is based on recommendations from Chrysler that will help you keep your outboard motor operating at its peak performance level.

Table 1 provides a complete model history. **Tables 1-7** are at the end of the chapter.

LUBRICATION

Proper Fuel Selection

Two-stroke engines are lubricated by mixing oil with the fuel. The various components of the engine are thus lubricated as the fuel-oil mixture passes through the crankcase and cylinders. Since outboard fuel serves the dual function of producing ignition and distributing the lubrication, the use of low octane marine white gasolines should be avoided. Such gasolines also have a tendency to cause ring sticking and port plugging.

All Chrysler outboards will use any gasoline with a minimum posted pump octane rating of 85 that works satisfactorily in an automotive engine. Lead-free gasoline is preferabler to leaded gasoline, as it offers longer spark plug life.

Sour Fuel

Fuel should not be stored for more than 60 days (under ideal conditions). Gasoline forms gum and varnish deposits as it ages. Such fuel will cause starting problems. A good grade of gasoline stabilizer and conditioner additive may be used to prevent gum and varnish formation during storage or prolonged periods of non-use but it is always better to drain the tank in such cases. Always use fresh gasoline when mixing fuel for your outboard.

Gasohol

Some gasolines sold for marine use now contain alcohol, although this fact may not be advertised. A mixture of 10 percent ethyl alcohol and 90 percent unleaded gasoline is called gasohol. While Chrysler *does not* recommend gasohol for use in its outboards, testing to date has found that it causes no major deterioration of the fuel system or its component parts when consumed immediately after purchase.

Fuels with an alcohol content tend to slowly absorb moisture from the air. When the moisture content of the fuel reaches approximately one percent, it combines with the alcohol and separates from the fuel. This separation does not normally occur when gasohol is used in an automobile, as the tank is generally emptied within a few days after filling it.

The problem does occur in marine use, however, because boats often remain idle between start-ups for days or even weeks. This length of time permits separation to take place. The alcohol-water mixture settles at the bottom of the fuel tank. Since outboard motors will not run on this mixture, it is necessary to drain the fuel tank, flush out the fuel system with clean gasoline and then remove, clean and reinstall the spark plugs before the engine can be started.

Continued use of fuels containing alcohol can "melt" the fuel level indicator lens in portable fuel tanks. Many late-model replacement tanks now contain an alcohol-resistant lens.

The major danger of using gasohol in an outboard motor is that a shot of the water-alcohol mix may be picked up and sent to one of the carburetors of a multicylinder engine. Since this mixture contains no oil, it will wash oil off the bore of any cylinder it enters. The other carburetor receiving good fuel-oil mixture will keep the engine running while the cylinder receiving the water-alcohol mixture can suffer internal damage.

The problem of unlabeled gasohol has become so prevalent around the United States that Miller Tools (32615 Park Lane, Garden City, MI 48135) now offers an Alcohol Detection Kit (part No. C-4846) so that owners and mechanics can determine the quality of fuel being used.

The kit cannot differentiate between types of alcohol (ethanol, methanol, etc.) nor is it considered to be absolutely accurate from a scientific standpoint, but it is accurate enough to determine whether or not there is sufficient alcohol in the fuel to cause the user to take precautions.

Recommended Fuel Mixture

Use the specified gasoline for your Chrysler outboard and mix with Chrysler Outboard Oil in the following ratios:

CAUTION
Do not, under any circumstances, use multigrade or other high detergent automotive oils, or oils containing metallic additives. Such oils are harmful to 2-stroke engines. Since they do not mix properly with gasoline, do not burn as 2-cycle oils do and leave an ash residue, their use may result in piston scoring, bearing failure or other engine damage.

a. During the break-in period, thoroughly mix two 16-ounce cans of Chrysler Outboard Oil with each 6 gallons of gasoline (or 16 ounces with each 3 gallons) in your remote fuel tank. This provides a 25:1 mixture.

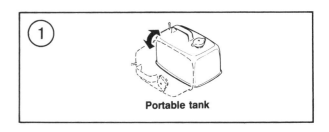

Portable tank

each 3 gallons) in your remote fuel tank. This provides a 25:1 mixture.

b. After engine break-in, mix one 16-ounce can with each 6 gallons of gasoline (8 ounces with each 3 gallons) in your remote fuel tank. This provides a 50:1 mixture.

c. Operation in Canada requires mixing 16 U.S. ounces of Chrysler Outboard Oil to each 5 Imperial gallons of gasoline in the remote fuel tank.

CAUTION
*There are a number of oil products on the market which specify use at 100:1. They are **not** BIA TC-W approved and should **not** be used.*

If Chrysler Outboard Oil is not available, any high-quality 2-stroke oil intended for outboard use may be substituted provided the oil meets BIA rating TC-W and specifies so on the container. Follow the manufacturer's mixing instructions on the container but do not exceed a 50:1 ratio (25:1 during break in).

Correct Fuel Mixing

Mix the fuel and oil outdoors or in a well-ventilated indoor location. Mix the fuel directly in the remote tank.

WARNING
Gasoline is an extreme fire hazard. Never use gasoline near heat, sparks or flame. Do not smoke while mixing fuel.

② Fuel nozzle
must contact
funnel

Measure the required amounts of gasoline and oil accurately. Pour a small amount of oil into the remote tank and add a small amount of gasoline. Mix it thoroughly by shaking or stirring vigorously; then add the balance of the gasoline and oil and mix again.

Using less than the specified amount of oil can result in insufficient lubrication and serious engine damage. Using more oil than specified causes spark plug fouling, erratic carburetion, excessive smoking and rapid carbon accumulation.

Cleanliness is of prime importance. Even a very small particle of dirt can cause carburetion problems. Always use fresh gasoline. Gum and varnish deposits tend to form in gasoline stored in a tank for any length of time. Use of sour fuel can result in carburetor problems and spark plug fouling.

Above 32° F (0° C)

Measure the required amounts of gasoline and Chrysler Outboard Oil accurately. Pour the oil into the portable tank and add the fuel. Install the tank filler cap and mix the fuel by tipping the tank on its side and back to an upright position several times. See **Figure 1**.

If a built-in tank is used, insert a large metal filter funnel in the tank filler neck. Slowly pour the Chrysler Outboard Oil into the funnel at the same time the tank is being filled with gasoline. See **Figure 2**.

Below 32° F (0° C)

Measure the required amounts of gasoline and Chrysler Outboard Oil accurately. Pour about one gallon of gasoline in the tank and then add the required amount of oil. Install the tank filler cap and shake the tank to thoroughly mix the fuel and oil. Remove the cap and add the balance of the gasoline.

4

If a built-in tank is used, insert a large metal filter funnel in the tank filler neck. Mix 1/6th pint of Chrysler Outboard Oil with one gallon of gasoline in a separate container. Slowly pour the mixture into the funnel at the same time the tank is being filled with gasoline.

Consistent Fuel Mixtures

The carburetor idle adjustment is sensitive to fuel mixture variations which result from the use of different oils and gasolines or from inaccurate measuring and mixing. This may require readjustment of the idle needle. To prevent the necessity for constant readjustment of the carburetor from one batch of fuel to the next, always be consistent. Prepare each batch of fuel exactly the same as previous ones.

Pre-mixed fuels sold at some marinas are not recommended for use in Chrysler outboards, since the quality and consistency of pre-mixed fuels can vary greatly. The possibility of engine damage resulting from use of an incorrect fuel mixture outweighs the convenience offered by pre-mixed fuel.

Gearcase Lubrication

Check the gearcase lubricant after the first 30 hours of operation and replace at 100 hour or 6 month intervals (at least once per season). Check the gearcase at 50 hour intervals between lubricant changes. Use Chrysler Gear Lube. If this is not available, use a high quality non-corrosive E.P. 90 outboard gear lubricant.

> *CAUTION*
> *Do not use regular automotive grease in the gearcase. Its expansion and foam characteristics are not suitable for marine use.*

Gearcase Lubricant Check

To assure a correct level check, the engine must be in the upright position and not run for at least 2 hours before performing this procedure. Refer to **Figure 3** or **Figure 4** as required.

1. Remove the engine cover and disconnect the spark plug lead(s) as a safety precaution to prevent any accidental starting of the engine.
2. Locate and loosen (but do not remove) the gearcase drain or fill/drain plug. Allow a small amount of lubricant to drain. If there is water in the gearcase, it will drain before the lubricant. Retighten the drain plug securely.
3. If water is noted in Step 2, retighten the drain plug securely and pressure test the gearcase to determine if a seal has failed or if the water is simply condensation in the gearcase. See Chapter Nine.
4. Remove the vent plug. Do not lose the accompanying washer. The lubricant should be level with the bottom of the vent plug hole.

> *CAUTION*
> *Never lubricate the gearcase without first removing the vent plug, as the injected lubricant displaces air which must be allowed to escape. The gearcase cannot be completely filled otherwise.*

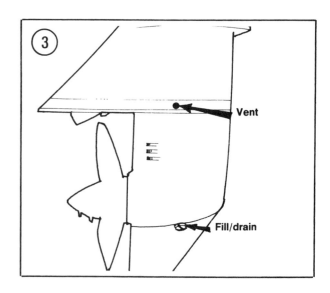

5. If the lubricant level is low, remove the fill plug on models so equipped. See **Figure 4**. Remove the fill/drain plug on models without a separate fill plug.

6. Inject lubricant into the fill/drain hole (**Figure 3**) or fill hole (**Figure 4**) until excess fluid flows out the vent plug hole.

7. Install the vent plug, then the fill/drain or fill plug. Be sure the washers are in place under the head of each, so that water will not leak past the threads into the housing.

8. Wipe any excess lubricant off the gearcase exterior.

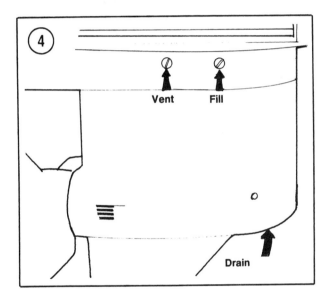

Vent   Fill

Drain

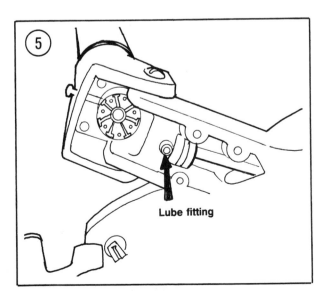

Lube fitting

9. On models with separate fill and drain plugs (**Figure 4**), remove the vent plug and washer. Let gearcase stand upright for a minimum of 1/2 hour, then recheck the lubricant level. Top up if necessary, then reinstall vent plug and washer.

Gearcase Lubricant Change

Refer to **Figure 3** or **Figure 4** for this procedure.

1. Remove the engine cover and disconnect the spark plug lead(s) as a safety precaution to prevent any accidental starting of the engine.

2. Place a container under the fill/drain (**Figure 3**) or drain plug (**Figure 4**) and remove it. Remove the vent plug. Drain the lubricant from the gearcase.

NOTE
If the lubricant is creamy in color or metallic particles are found in Step 3, remove and disassemble the gearcase to determine and correct the cause of the problem.

3. Wipe a small amount of lubricant on a finger and rub the finger and thumb together. Check for the presence of metallic particles in the lubricant. Note the color of the lubricant. A white or creamy color indicates water in the lubricant. Check the drain container for signs of water separation from the lubricant.

4. Perform Steps 6-9 of *Gearcase Lubricant Check* in this chapter.

Other Lubrication Points

Refer to **Figures 5-18** (typical) and **Table 2** for other lubricant points, frequency of lubrication and lubricant to be used. In addition to these lubrication points, some motors may also have grease fittings provided at critical points where bearing surfaces are not externally exposed. These fittings should be lubricated at least once each season with

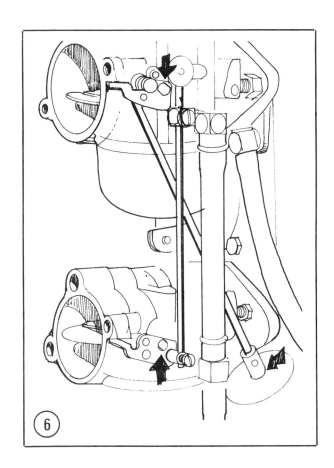

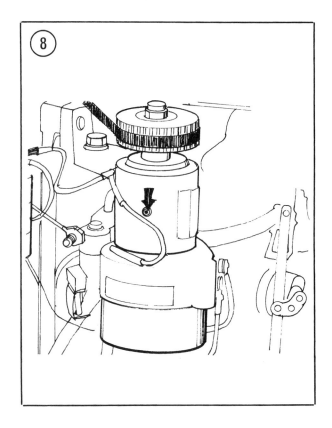

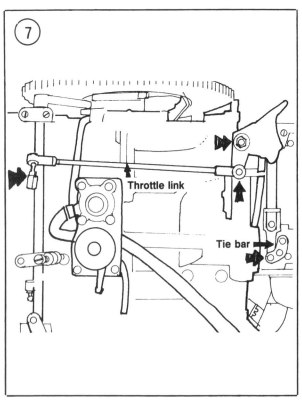

Throttle link

Tie bar

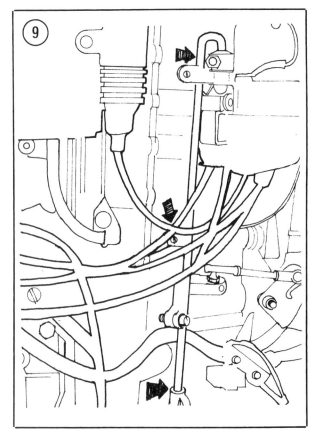

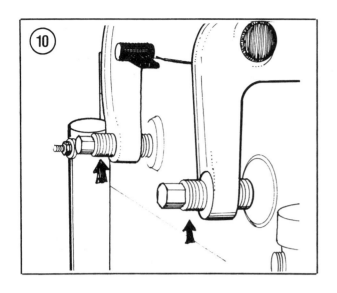

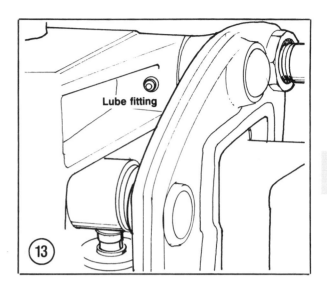

4

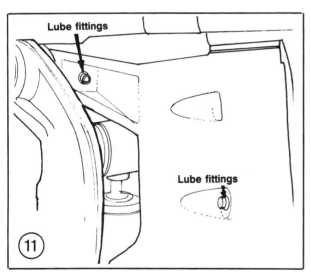

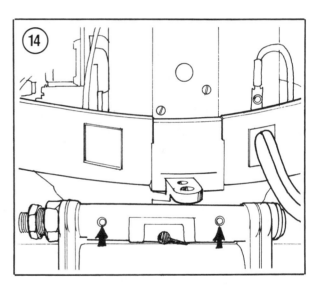

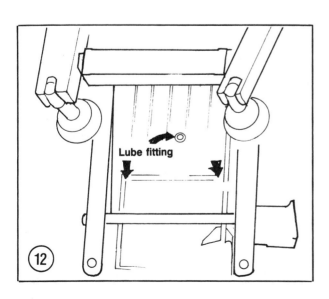

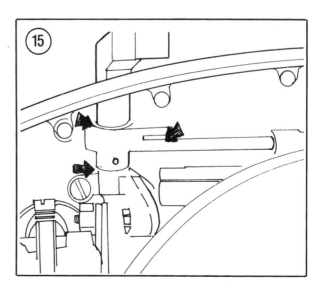

an automotive type grease gun and Rykon No. 2EP grease.

> *CAUTION*
> *When lubricating the steering cable on models so equipped, make sure its core is fully retracted into the cable housing. Lubricating the cable while extended can cause a hydraulic lock to occur.*

Salt Water Corrosion of Gearcase Bearing Cage or Spool

Salt water corrosion that is allowed to build up unchecked can eventually split the gearcase and destroy the lower unit. If the motor is used in salt water, remove the propeller assembly, zinc anode or bearing cage cap and the bearing cage or spool (**Figure 19**, typical) at least once a year after the initial 30-hour inspection. Clean all corrosive deposits and dried-up lubricant from each end of the cage or spool.

Install new O-rings on cage or spool, wipe O-rings with Chrysler Lubricant part No. T 2961 and install spool with new O-ring seals on spool bolts. If a zinc anode cover is used, apply a liberal amount of Loctite No. 75 (part No. T 2963-1) to the anode screw threads. On models without the anode, coat the cap screw threads with anti-seize compound part No. T 2987-1.

Coat propeller shaft splines with anti-seize compound and install propeller.

STORAGE

The major consideration in preparing an outboard motor for storage is to protect it from rust, corrosion and dirt. Chrysler recommends the following procedure.

1. Remove the engine cover.
2. Operate the motor in a test tank or on the boat in the water. Start the engine and run at fast idle until warmed up.
3. Disconnect the fuel line and let engine run while pouring 2 oz. of a good rust preventive

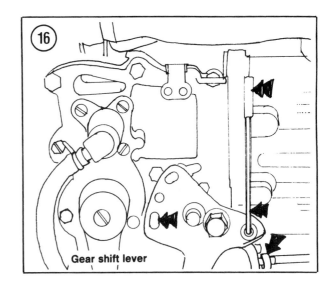
(16) Gear shift lever

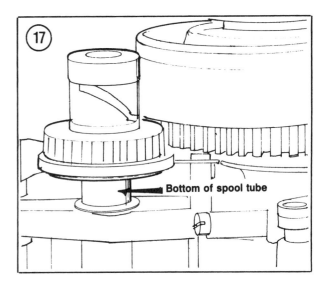
(17) Bottom of spool tube

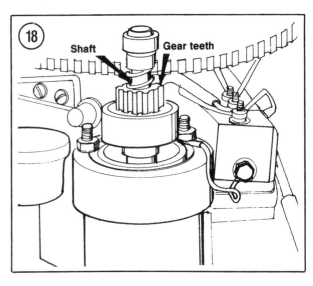
(18) Shaft   Gear teeth

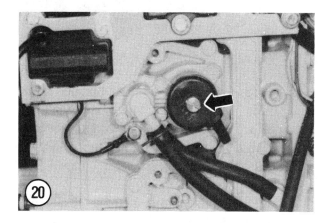

compressed air through the screen in the direction of the plastic collar.

  e. Reinstall filter screen to the tank adapter. Thread adapter in place and connect the fuel line.

7. Service 4.4-15 hp fuel pump filters as follows:

  a. Disconnect the fuel line at the lower pump fitting.

  b. Remove the lower pump fitting with an open-end wrench.

  c. Remove and discard the filter installed on the end of the pump fitting.

  d. Install a new filter on the pump fitting.

  e. Reinstall fitting to pump and reconnect fuel line to fitting.

8. Service Type A fuel pump filters (**Figure 20**) as follows:

  a. Remove the fuel pump sediment bowl screw.

  b. Remove the sediment bowl and filter from the pump.

  c. Remove and clean or replace the filter screen.

  d. Install the filter screen in the sediment bowl with its turned edge facing the engine.

  e. Install sediment bowl and filter assembly to fuel pump with a new gasket. Tighten bowl screw securely.

9. Service Type B fuel pump filters (**Figure 21**) as follows:

  a. Remove the fuel pump bracket from the power head.

  b. Loosen the knurled nut on the wire bail and remove the fuel bowl from the pump.

  c. Remove the bowl gasket and filter screen from the pump.

  d. Clean or replace the filter screen.

  e. Install filter screen in fuel pump counterbore.

  f. Install fuel bowl with a new gasket. Position wire bail under bowl and tighten knurled nut securely.

oil into each carburetor throat. Let engine stall out, indicating that the carburetor(s) have run completely dry.

4. Remove motor from water. Drain all fuel lines and carburetor(s).

5. Remove spark plug(s) as described in this chapter. Pour about one ounce of Chrysler Outboard Oil into each spark plug hole. Turn engine over by hand several times to distribute the oil throughout the cylinder(s). Reinstall spark plugs.

6. Service the fuel tank filter as follows:

  a. Disconnect the fuel line at the tank adapter.

  b. Unthread and remove the tank adapter.

  c. Remove the hose clamp holding the filter screen to the pick-up end of the adapter.

  d. Remove the screen, rinse it in clean benzine and blow low-pressure

g. Install pump bracket to power head.

10. Check fuel filter installation for leakage by priming fuel system with fuel line primer bulb.

11. Drain and refill gearcase as described in this chapter. Check condition of vent, fill or fill/drain plug gaskets. Replace as required.

12. Refer to **Figures 5-18** and **Table 2** as appropriate and lubricate motor at all specified points. See **Table 3** for recommended lubricants.

13. Clean the motor, including all accessible power head parts. Coat with a good marine-type wax. Install the engine cover.

14. Remove the propeller and lubricate propeller shaft splines with anti-seize compound (part No. T 2987-1). Reinstall the propeller.

15. Store the motor upright in a dry and well-ventilated area.

16. Service the battery as follows:

   a. Disconnect the negative battery cable, then the positive battery cable.

   b. Remove all grease, corrosion and dirt from the battery surface.

   c. Check the electrolyte level in each battery cell and top up with distilled water, if necessary. Fluid level in each cell should not be higher than 3/16 in. above the perforated baffles.

   d. Lubricate the terminal bolts with grease or petroleum jelly.

CAUTION
A discharged battery can be damaged by freezing.

   e. With the battery in a fully charged condition (specific gravity 1.260-1.275), store in a dry place where the temperature will not drop below freezing.

   f. Recharge the battery every 45 days or whenever the specific gravity drops below 1.230. Before charging, cover the

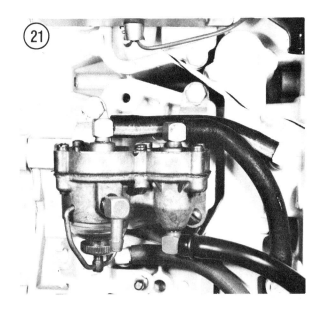

plates with distilled water, but not more than 3/16 in. above the perforated baffles. The charge rate should not exceed 6 amps. Discontinue charging when the specific gravity reaches 1.260 at 80° F (27° C).

   g. Before placing the battery back into service after winter storage, remove the excess grease from the terminals, leaving a small amount on. Install battery in a fully charged state.

COMPLETE SUBMERSION

An outboard motor which has been lost overboard should be recovered as quickly as possible. If the motor was running when submerged, disassemble and clean it immediately—any delay will result in rust and corrosion of internal components once it has been removed from the water. If the motor was not running and appears to be undamaged mechanically with no abrasive dirt or silt inside, take the following emergency steps immediately.

1. Wash the outside of the motor with clean water to remove weeds, mud and other debris.

2. Remove the engine cover.

3. If recovered from salt water, flush motor completely with fresh water and spray entire power head with LPSI Lubricant (part No. T 8969).

4. Remove the spark plug(s) as described in this chapter.

CAUTION
Do not force the motor if it does not turn over freely by hand in Step 5. This may be an indication of internal damage such as a bent connecting rod or broken piston.

5. Drain as much water as possible from the power head by placing the motor in a horizontal position. Manually rotate the flywheel by hand with the spark plug hole(s) facing downward.

6. Dry and reinstall the spark plugs.

7. Dry all ignition components and spray with LPSI Lubricant (part No. T 8969).

8. Drain the fuel lines and carburetor(s).

9. On models with an integral fuel tank, drain the tank and flush with fresh gasoline until all water has been removed.

CAUTION
If there is a possibility that sand may have entered the power head, do not try to start the motor or severe internal damage may occur.

10. Try starting the motor with a fresh fuel source. If motor will start, let it run at least one hour to eliminate any water remaining inside.

CAUTION
If it is not possible to disassemble and clean the motor immediately in Step 11, resubmerge it in water to prevent rust and corrosion formation until it can be properly serviced.

11. If the motor will not start in Step 10, try to diagnose the cause as fuel, electrical or mechanical then correct. If the engine cannot be started within 2 hours, disassemble, clean and oil all parts thoroughly as soon as possible.

ANTI-CORROSION MAINTENANCE

1. Flush the cooling system with fresh water as described in this chapter after each time motor is used in salt water. Wash exterior with fresh water.

2. Dry exterior of motor and apply primer over any paint nicks and scratches. Use only Chrysler Boat White Lacquer (part No. 5H134). Do not use paints containing mercury or copper. Do not paint sacrificial anodes.

3. Spray power head and all electrical connections with a good quality corrosion and rust preventive.

4. Check sacrifical anodes. Replace any that are less than half their original size.

5. Lubricate more frequently than specified in **Table 2**. If used consistently in salt water, reduce lubrication intervals by one-half.

ENGINE FLUSHING

Periodic engine flushing will prevent salt or silt deposits from accumulating in the water passageways. This procedure should also be performed whenever an outboard motor is operated in salt water or polluted water.

Keep the motor in an upright position during and after flushing. This prevents water from passing into the power head through the drive shaft housing and exhaust ports during the flushing procedure. It also eliminates the possibility of residual water being trapped in the drive shaft housing or other passageways.

1. Attach a flushing device according to manufacturer's instructions.

2. Connect a garden hose between a water tap and the flushing device.

3. Open the water tap partially—do not use full pressure.

4. Shift into NEUTRAL, then start motor. Keep engine speed between 800-1,100 rpm.

5. Adjust water flow so that there is a slight loss of water around the rubber cups of the flushing device.

6. Check the motor to make sure that water is being discharged from the "tell-tale" nozzle. If it is not, stop the motor immediately and determine the cause of the problem.

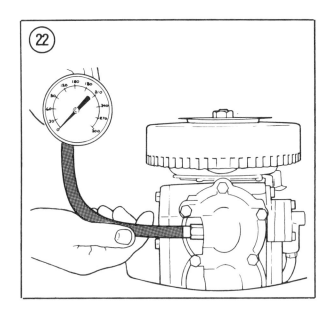

CAUTION
Flush the motor for at least 5 minutes if
used in salt water.

7. Flush motor until discharged water is clear. Stop motor.

8. Close water tap and remove flushing device from lower unit.

TUNE-UP

A tune-up consists of a series of inspections, adjustments and parts replacements to compensate for normal wear and deterioration of outboard motor components. Regular tune-ups are important for power, performance and economy. Chrysler recommends that outboards be serviced every 6 months or 100 hours of operation, whichever comes first. If subjected to limited use, the engine should be tuned at least once a year.

Since proper outboard motor operation depends upon a number of interrelated system functions, a tune-up consisting of only one or two corrections will seldom give lasting results. For best results, a thorough and systematic procedure of analysis and correction is necessary.

Prior to performing a tune-up, it is a good idea to flush the motor as described in this chapter and check for satisfactory water pump operation.

The tune-up sequence recommended by Chrysler includes the following:

   a. Compression check.

   b. Cylinder head bolt torque.

   c. Spark plug service.

   d. Gearcase and water pump check.

   e. Fuel system service.

   f. Ignition system service.

   g. Battery, starter motor and solenoid check (if so equipped).

   h. Wiring harness check.

   i. Timing, synchronization and adjustment.

   j. Performance test (on boat).

Any time the fuel or ignition systems are adjusted or defective parts replaced, the engine timing, synchronization and adjustment *must* be checked. These procedures are described in Chapter Five. Perform the timing, synchronization and adjustment procedure for your engine *before* running the performance test.

COMPRESSION CHECK

An accurate cylinder compression check gives a good idea of the condition of the basic working parts of the engine. It is also an important first step in any tune-up, as a motor with low or unequal compression between cylinders *cannot* be satisfactorily tuned. Any compression problem discovered during this check must be corrected before continuing with the tune-up procedure.

1. With the engine warm, disconnect the spark plug wire(s) and remove the plug(s) as described in this chapter.

2. Ground the spark plug wire(s) to the engine to disable the ignition system.

3. Connect the compression tester to the top spark plug hole according to manufacturer's instructions (**Figure 22**).

4. Make sure the throttle is set to the wide open position and crank the engine through at least 4 compression strokes. Record the gauge reading.

5. Repeat Step 3 and Step 4 on each cylinder of multi-cylinder engines.

While minimum cylinder compression should not be less than shown in **Table 4**, the actual readings are not as important as the differences between cylinders when interpreting the results. A variation of more than 10-15 psi between 2 cylinders indicates a problem with the lower reading cylinder, such as worn or sticking piston rings and/or scored pistons or cylinders. In such cases, pour a tablespoon of engine oil into the suspect cylinder and repeat Step 3 and Step 4. If the compression is raised significantly (by 10 psi in an older engine), the rings are worn and should be replaced.

Many outboard engines are plagued by hard starting and generally poor running for which there seems to be no good cause. Carburetion and ignition check out satisfactorily and a compression test may show that everything is well in the engine's upper end.

What a compression test does *not* show is lack of primary compression. In a 2-stroke engine, the crankcase must be alternately under high pressure and low pressure. After the piston closes the intake port, further downward movement of the piston causes the entrapped mixture to be pressurized so that it can rush quickly into the cylinder when the scavenging ports are opened. Upward piston movement creates a lower pressure in the crankcase, enabling fuel-air mixture to pass in from the carburetor.

When the crankshaft seals or case gaskets leak, the crankcase cannot hold pressure and proper engine operation becomes impossible. Any other source of leakage, such as defective cylinder base gaskets or a porous or cracked crankcase casting, will result in the same conditions.

If the power head shows signs of overheating (discolored or scorched paint) but the compression test turns up nothing abnormal, check the cylinder(s) visually through the transfer ports for possible scoring. A cylinder can be slightly scored and still deliver a relatively good compression reading. In such a case, it is also a good idea to double-check the water pump operation as a possible cause for overheating.

CYLINDER HEAD BOLT TORQUE

Loosen each bolt and retorque to specifications in **Table 5**. Refer to Chapter Nine for the proper tightening sequence according to engine.

CAUTION
Excessive torque will distort the cylinder bores or head. Insufficient torque will allow the cylinder head gasket to leak.

SPARK PLUGS

Chrysler outboards are equipped with Champion spark plugs selected for average use conditions. Under adverse use conditions, the recommended spark plug may foul or overheat.

SPARK PLUG ANALYSIS
(CONVENTIONAL GAP SPARK PLUGS)

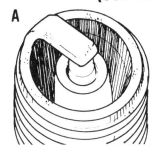

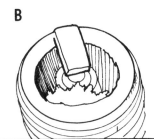

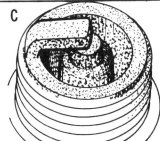

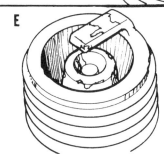

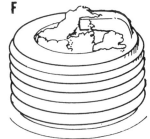

A. Normal—Light tan to gray color of insulator indicates correct heat range. Few deposits are present and the electrodes are not burned.

B. Core bridging—These defects are caused by excessive combustion chamber deposits striking and adhering to the firing end of the plug. In this case, they wedge or fuse between the electrode and core nose. They originate from the piston and cylinder head surfaces. Deposits are formed by one or more of the following:
a. Excessive carbon in cylinder.
b. Use of non-recommended oils.
c. Immediate high-speed operation after prolonged trolling.
d. Improper fuel-oil ratio.

C. Wet fouling—Damp or wet, black carbon coating over entire firing end of plug. Forms sludge in some engines. Caused by one or more of the following:
a. Spark plug heat range too cold.
b. Prolonged trolling.
c. Low-speed carburetor adjustment too rich.

d. Improper fuel-oil ratio.
e. Induction manifold bleed-off passage obstructed.
f. Worn or defective breaker points.

D. Gap bridging—Similar to core bridging, except the combustion particles are wedged or fused between the electrodes. Causes are the same.

E. Overheating—Badly worn electrodes and premature gap wear are indicative of this problem, along with a gray or white "blistered" appearance on the insulator. Caused by one or more of the following:
a. Spark plug heat range too hot.
b. Incorrect propeller usage, causing engine to lug.
c. Worn or defective water pump.
d. Restricted water intake or restriction somewhere in the cooling system.

F. Ash deposits or lead fouling—Ash deposits are light brown to white in color and result from use of fuel or oil additives. Lead fouling produces a yellowish brown discoloration and can be avoided by using unleaded fuels.

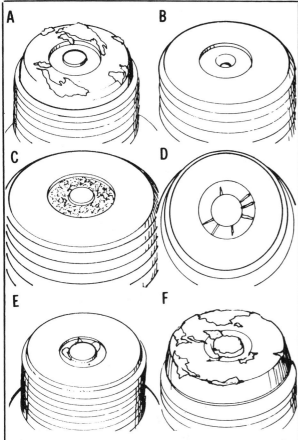

**SPARK PLUG ANALYSIS
(SURFACE GAP SPARK PLUGS)**

A. **Normal**—Light tan or gray colored
   deposits indicate that the engine/ignition
   system condition is good. Electrode wear
   indicates normal spark rotation.

B. **Worn out**—Excessive electrode wear
   can cause hard starting or a misfire
   during acceleration.

C. **Cold fouled**—Wet oil or fuel deposits
   are caused by "drowning" the plug with raw
   fuel mix during cranking, overrich carburetion
   or an improper fuel-oil ratio. Weak ignition
   will also contribute to this condition.

D. **Carbon tracking**—Electrically conductive
   deposits on the firing end provide a
   low-resistance path for the voltage.
   Carbon tracks form and can cause misfires.

E. **Concentrated arc**—Multi-colored appearance
   is normal. It is caused by electricity
   consistently following the same firing path.
   Arc path changes with deposit conductivity
   and gap erosion.

F. **Aluminum throw-off**—Caused by preignition.
   This is not a plug problem but the result of
   engine damage. Check engine to determine
   cause and extent of damage.

⑳₄

In such cases, check the ignition and carburetion systems to make sure they are operating correctly. If no defect is found, replace the spark plug with one of a hotter or colder heat range as required. **Table 6** contains the recommended spark plugs for all models covered in this book. **Table 7** provides a cross-reference for use when Champion spark plugs are not available.

Spark Plug Removal

> *CAUTION*
> *Whenever the spark plugs are removed, dirt around them can fall into the plug holes. This can cause engine damage that is expensive to repair.*

1. Blow out any foreign matter from around the spark plugs with compressed air. Use a compressor if you have one. If you do not, use a can of compressed inert gas, available from photo stores.

2. Disconnect the spark plug wires by twisting the wire boot back and forth on the plug insulator while pulling outward. Pulling on the wire instead of the boot may cause internal damage to the wire.

3. Remove the plugs with a 13/16 in. spark plug socket or box end wrench. Keep the plugs in order so you know which cylinder they came from.

4. Examine each spark plug. Compare its condition with **Figure 23** (conventional gap) or **Figure 24** (surface gap). Spark plug condition indicates engine condition and can warn of developing trouble.

5. Check each plug for make and heat range. All should be of the same make and number or heat range.

6. Discard the plugs. Although they could be cleaned and reused if in good condition, they seldom last very long. New plugs are inexpensive and far more reliable.

Spark Plug Gapping
(Conventional Gap Only)

New plugs should be carefully gapped to ensure a reliable, consistent spark. Use a special spark plug tool with a wire gauge. See **Figure 25** for one common type.

1. Remove the plugs and gaskets from the boxes. Install the gaskets.

> *NOTE*
> *Some plug brands may have small end pieces that must be screwed on (**Figure 26**) before the plugs can be used.*

2. Insert a 0.030 in. wire gauge between the electrodes. If the gap is correct, there will be a slight drag as the wire is pulled through. If there is no drag or if the wire will not pull through, bend the side electrode with the gapping tool (**Figure 27**) to change the gap. Remeasure with the wire gauge.

> *CAUTION*
> *Never try to close the electrode gap by tapping the spark plug on a solid surface. This can damage the plug internally. Always use the gapping and adjusting tool to open or close the gap.*

Spark Plug Installation

Improper installation is one of the most common causes of poor spark plug performance in outboard motors. The gasket on the plug must be fully compressed against a clean plug seat in order for heat transfer to take place effectively. This requires close attention to proper tightening during installation.

1. Inspect the spark plug hole threads and clean them with a thread chaser (**Figure 28**). Wipe cylinder head seats clean before installing the new plugs.

2. Screw each plug in by hand until it seats. Very little effort is required. If force is necessary, the plug is cross-threaded. Unscrew it and try again.

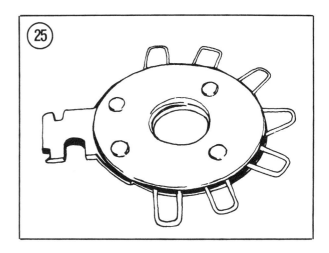

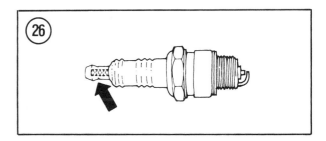

3. Tighten the spark plugs. If you have a torque wrench, tighten to 10-15 ft.-lb. If not, seat the plug finger-tight on the gasket, then tighten an additional 1/4 turn with a wrench.

4. Inspect each spark plug wire before reconnecting it to its cylinder. If insulation is damaged or deteriorated, install a new plug wire. Push wire boot onto plug terminal and make sure it seats fully.

GEARCASE AND
WATER PUMP CHECK

A faulty water pump or one that performs below specifications can result in extensive engine damage. Thus, it is a good idea to replace the water pump impeller, seals and gaskets once a year or whenever the lower unit is removed for service. See Chapter Nine.

FUEL SYSTEM SERVICE

The clearance between the carburetor and choke shutter should not be greater than

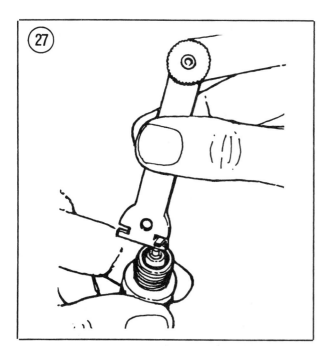

0.015 in. when the choke is closed or a hard starting condition will result. When changing from one brand of gasoline to another, it may be necessary to readjust the carburetor idle mixture needle slightly (1/4 turn) to accomodate the variations in volatility.

Fuel Lines

1. Visually check all fuel lines for kinks, leaks, deterioration or other damage.
2. Disconnect fuel lines and blow out with compressed air to dislodge any contamination or foreign material.

3. Coat fuel line fittings sparingly with Permatex and reconnect the lines.

Engine Fuel Filter

Models with an integral fuel tank that utilize gravity fuel feed instead of a fuel pump have a filter screen attached to the fuel line fitting on the bottom of the tank. All 4.4-15 hp models have the filter screen installed inside the fuel pump on the lower fuel line fitting. All other models use a filter screen installed in the sediment bowl attached to the fuel pump.

4.4-15 hp models

1. Disconnect the fuel line at the lower pump fitting.
2. Remove the lower pump fitting with an open-end wrench.
3. Remove and discard the filter installed on the end of the pump fitting.
4. Install a new filter on the pump fitting.
5. Reinstall fitting to pump and reconnect fuel line to fitting.
6. Check filter installation for leakage by priming fuel system with the fuel line primer bulb.

Type A fuel pump filter

1. Remove the fuel pump sediment bowl screw.
2. Remove the sediment bowl and filter from the pump.
3. Remove and clean or replace the filter screen.
4. Install the filter screen in the sediment bowl with its turned edge facing the engine.
5. Install sediment bowl and filter to fuel pump with a new gasket. Tighten bowl screw securely.
6. Check filter installation for leakage by priming fuel system with the fuel line primer bulb.

Type B fuel pump filter

1. Remove the fuel pump bracket from the power head.
2. Loosen the knurled nut on the wire bail and remove the fuel bowl from the pump.
3. Remove the bowl gasket and filter screen from the pump.
4. Clean or replace the filter screen.
5. Install filter screen in fuel pump counterbore.
6. Install fuel bowl with a new gasket. Position wire bail under bowl and tighten knurled nut securely.
7. Install pump bracket to power head.
8. Check filter installation for leakage by priming fuel system with the fuel line primer bulb.

Fuel Pump

The fuel pump does not generally require service during a tune-up. However, if the engine has more than 100 hours on it since the fuel pump was last serviced, it is a good idea to remove and disassemble the pump, inspect each part carefully for wear or damage and reassemble it with a new diaphragm. See Chapter Six.

Fuel pump diaphragms are fragile and one that is defective often produces symptoms that are diagnosed as an ignition system problem. A common malfunction results from a tiny pinhole or crack in the diaphragm caused by an engine backfire. This defect allows gasoline to enter the crankcase and wet-foul the spark plug at idle speed, causing hard starting and engine stall at low rpm. The problem disappears at higher speeds, as fuel quantity is limited. Since the plug is not fouled by excess fuel at higher speeds, it fires normally.

Fuel Pump Pressure Test

On single fuel pump installations, use a tee-fitting to connect a fuel pressure gauge into the line between the carburetor(s) and fuel pump. If equipped with twin fuel pumps, connect the pressure gauge into the line between the 2 pumps at the lower pump to test the lower pump. To test the upper pump, disconnect the fuel inlet line at the lower pump and the line between the 2 pumps at the upper pump. Move the disconnected fuel inlet line from the lower pump to the upper pump and connect it to the pump with the tee and pressure gauge in place of the line between the 2 pumps.

Before starting the engine, loosen the fuel tank vent cap to relieve any pressure in the system. With the engine running in a test tank or on the boat in the water, fuel pump pressure must be at least:

    a. 1 psi at 600 rpm.
    b. 1.5 psi at 2,500-3,000 rpm.
    c. 2.5 psi at 4,500 rpm.

If not, rebuild the fuel pump with a new diaphragm and gaskets. See Chapter Six.

BREAKER POINT IGNITION SYSTEM SERVICE

The condition and gap of the breaker points will greatly affect engine operation. Burned or badly oxidized points will allow little or no current to pass. A gap that is too narrow will not allow the coil to build up sufficient voltage and will result in a weak spark. An excessive point gap will allow the points to open before the primary current flow reaches its maximum.

While slightly pitted points can be dressed with a file, this should be done only as a temporary measure, as the points may arc after filing. Oxidized, dirty or oily points can be cleaned with alcohol but new points are inexpensive and always preferable for efficient engine operation.

The condenser absorbs the surge of high voltage from the coil and prevents current from arcing across the points when they open.

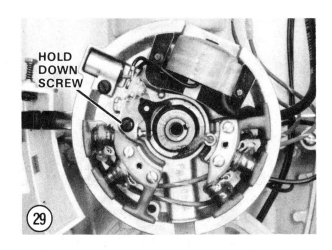

Condensers can be tested as described in Chapter Three, but are also inexpensive and should be replaced as a matter of course whenever new breaker points are installed.

NOTE
Breaker points must be adjusted correctly. An error in gap of 0.0015 in. will change engine timing by as much as one degree. With multi-cylinder models, variations in the gap between point sets can throw timing off by several degrees.

All breaker point sets installed on a breaker plate under the flywheel are set to 0.020 in. regardless of model. Breaker points used in a distributor ignition are set to 0.014 in. (3-cylinder) or 0.010 in. (4-cylinder) regardless of model. The following procedures provided by Chrysler involve setting breaker point gap with a flat feeler gauge. Some marine mechanics prefer to set the gap with a test light to indicate the precise position at which they open and close.

CAUTION
Always rotate the crankshaft in a clockwise direction in the following procedures. If rotated in a counter-clockwise direction, the water pump impeller may be damaged.

Magneto Plate Breaker Point Replacement (1-cylinder Models)

Refer to **Figure 29** for this procedure.
1. Disconnect the negative battery cable, if so equipped.
2. Remove the engine cover.
3. Remove the flywheel. See Chapter Eight.
4. Remove the 2 screws holding the breaker point set to the stator plate.
5. Disconnect the coil and condenser leads at the breaker point set. Remove the breaker point set.
6. Remove the screw holding the condenser to the stator plate. Remove the condenser.
7. Install a new breaker point set on the stator plate. Install but do not tighten the hold-down screw.
8. Install a new condenser on the stator plate and tighten attaching screw securely, then connect the coil and condenser leads to the breaker point set.

CAUTION
Do not over-lubricate the wick in Step 9. Excessive lubrication will cause premature point set failure.

9. Squeeze the cam lube wick to see if it has sufficient lubrication. If dry, work a small amount of Rykon No. 2EP (part No. 2961) into the wick with your fingers.
10. Adjust the breaker point gap as described in this chapter.
11. Reverse Steps 1-3 to complete installation.

Magneto Plate Breaker Point Adjustment (1-cylinder Models)

1. Install the flywheel screw on the crankshaft and rotate the magneto stator to the wide-open throttle position.
2. Place a wrench on the flywheel screw and rotate the crankshaft clockwise until the cam index mark aligns with the breaker point rubbing block. See **Figure 30**.

3. Turn the point set adjusting screw to obtain a gap of 0.020 in. when measured with a flat feeler gauge. The gap is correct when the feeler gauge offers a slight drag as it is slipped between the points. When the gap is correct, tighten the hold-down screw securely and recheck the point gap.

4. Attach a spring tension gauge (part No. T 8974) and measure the breaker point spring tension (**Figure 31**) at right angles to the surface at the center of contact. If the tension is not 14-28 oz. (3.6 hp) or 16-32 oz. (all others) install another set of points.

Magneto Plate Breaker Point Replacement (2-cylinder Models)

1. Disconnect the negative battery cable.
2. Remove the engine cover.
3. Remove the flywheel. See Chapter Eight.
4. Remove the 2 screws holding the breaker point set to the stator plate. See **Figure 32**.
5. Disconnect the coil and condenser leads at the breaker point set. Remove the breaker point set.
6. Remove the screw holding the condenser to the stator plate. Remove the condenser.
7. Repeat Steps 4-6 to remove the other point set and condenser.
8. Install a new breaker point set on the stator plate. Install but do not tighten the hold-down screw.
9. Install a new condenser on the stator plate and tighten attaching screw securely, then connect the coil and condenser leads to the breaker point set.
10. Repeat Step 8 and Step 9 to install the other point set and condenser.

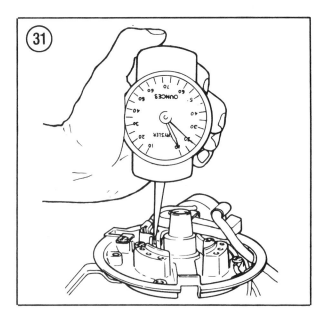

> *CAUTION*
> *Do not over-lubricate the wick in Step 11. Excessive lubrication will cause premature point set failure.*

11. Squeeze the cam lube wick to see if it has sufficient lubrication. If dry, work a small

amount of Rykon No. 2EP (part No. 2961) into the wick with your fingers.

12. Adjust the breaker point gaps as described in this chapter.

13. Reverse Steps 1-3 to complete installation.

Magneto Plate Breaker Point Adjustment (2-cylinder Models)

1. Move the magneto stator ring to the wide-open throttle position.

2. Install the flywheel knock-off nut (used during flywheel removal) on the crankshaft. Rotate the crankshaft clockwise on alternator

models until the breaker point rubbing block is about 10° beyond the top of the breaker cam ramp. On all other models, position the rubbing block at the highest point of the breaker cam. At this point, the breaker points will not open wider if the crankshaft is turned.

3. Mark the cam opposite the rubbing block to indicate the alignment point for adjusting the other set of points.

4. Turn the point set adjusting screw to obtain a gap of 0.020 in. when measured with a flat feeler gauge. The gap is correct when the feeler gauge offers a slight drag as it is slipped between the points. When the gap is correct, tighten the hold-down screw securely and recheck the point gap.

5. Rotate the crankshaft clockwise 2 full revolutions and recheck the point gap.

6. Rotate the crankshaft clockwise until the cam alignment mark made in Step 3 aligns with the rubbing block on the second set of breaker points.

7. Repeat Step 4 and Step 5 to set and check the gap on the second set of points.

8. Attach a spring tension gauge (part No. T 8974) and measure the breaker point spring tension (**Figure 31**) at right angles to the surface at the center of contact. The correct tension depends upon the breaker points used. If the tension is not 30-40 oz. (part No. 12044), 24-32 oz. (part No. 2329-2) or 16-32 oz. (all others), install another set of points.

Distributor Breaker Point Replacement (3- and 4-cylinder Models)

1. Disconnect the negative battery cable.

2. Remove the engine cover.

3. Remove the distributor cap by loosening the screws on the cap retaining clips. See **Figure 33**.

4. Remove the screws holding the distributor to the power head. See **Figure 34**. Slip

distributor drive belt from pulley and remove distributor from power head.

5. Turn the distributor over to expose points and remove the 2 screws holding the point set to the distributor housing.

6. Lift the point set from the housing and disconnect the lead wire.

7. Connect the lead wire to a new breaker point set.

8. Align the breaker point pivot with the distributor housing pivot hole. Put the point set in the housing.

9. Install the point set attaching screws but do not tighten.

10. Adjust the breaker point gap as described in this chapter.

11. Reverse Steps 1-4 to complete installation.

Distributor Breaker Point Adjustment (3- and 4-cylinder Models)

1. Turn the distributor rotor shaft until the breaker point rubbing block is on the high point of one of the cam lobes.

2. Turn the point set adjusting screw to obtain a gap of 0.014 in. (3-cylinder) or 0.010 in. (4-cylinder) when measured with a flat feeler gauge. The gap is correct when the feeler gauge offers a slight drag as it is slipped between the points. When the gap is correct, tighten the hold-down screw securely and recheck the point gap.

3. With the breaker points closed, connect a test lamp part No. T 2938-1 between the breaker point terminal screw and a good ground.

4. Switch test light on. Install spring tension gauge (part No. T 8974) and measure the breaker point spring tension. Note the gauge reading as the points open and the test light goes out. Tension should be 25-30 oz.

5. Repeat Step 4. The test light should go out at a tension within 2 oz. of that required in

Step 4. If it does not, repeat this step until the readings are consistent. If consistency cannot be achieved, install another set of points.

BATTERY AND STARTER MOTOR CHECK (ELECTRIC START MODELS ONLY)

1. Check the battery's state of charge. See Chapter Seven.

2. Connect a voltmeter between the starter motor positive terminal (**Figure 35**) and ground.

3. Turn ignition switch to START and check voltmeter scale:

    a. If voltage exceeds 9.5 volts and the starter motor does not operate, replace the motor.

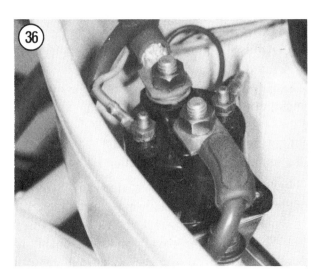

b. If voltage is less than 9.5 volts, recheck battery and connections. Charge battery, if necessary, and repeat procedure.

STARTER RELAY/ SOLENOID CHECK (ELECTRIC START MODELS ONLY)

1. Disconnect the negative battery cable.
2. Connect an ohmmeter between the small terminals on the starter relay or solenoid. See **Figure 36**.
3. If the meter does not read 2.2-2.8 ohms on the R×1 scale, replace the relay or solenoid.

WIRING HARNESS CHECK

Figure 37 (Motorola) and **Figure 38** (Prestolite) show typical wiring harness installations.

1. Check the wiring harness for signs of frayed or chafed insulation.
2. Check for loose connections between the wires and terminal ends.
3. Prestolite—Check the harness connectors for bent electrical pins. Check the harness connector and pin sockets for signs of corrosion. Clean as required.
4. If the harness is suspected of contributing to electrical malfunctions, check all wires for continuity and resistance between harness

4

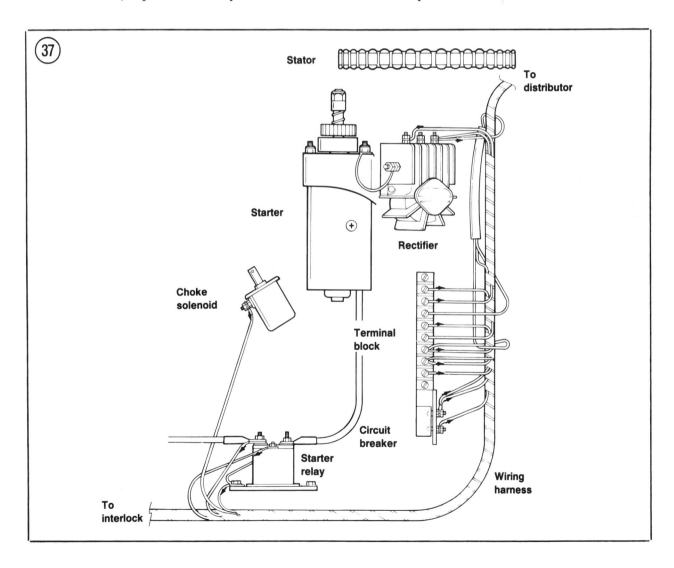

(37) Stator / To distributor / Starter / Rectifier / Choke solenoid / Terminal block / Circuit breaker / Starter relay / Wiring harness / To interlock

connection and terminal end. Repair or replace as required.

ENGINE SYNCHRONIZATION AND ADJUSTMENT

See Chapter Five.

PERFORMANCE TEST (ON BOAT)

Before performance testing the engine, make sure that the boat bottom is cleaned of all marine growth and that there is no evidence of a "hook" or "rocker" (**Figure 39**) in the bottom. Any of these conditions will reduce performance considerably. The boat should be performance tested with an average load and with the motor tilted at an angle that will allow the boat to ride on an even keel. If equipped with an adjustable trim tab, it should be properly adjusted to allow the boat to steer in either direction with equal ease.

Check engine rpm at full throttle. If not within the maximum rpm range for the motor as specified in Chapter Five, check the propeller pitch. A high pitch propeller will reduce rpm while a lower pitch prop will increase it.

Readjust the idle mixture and speed under actual operating conditions as required to obtain the best low-speed engine performance.

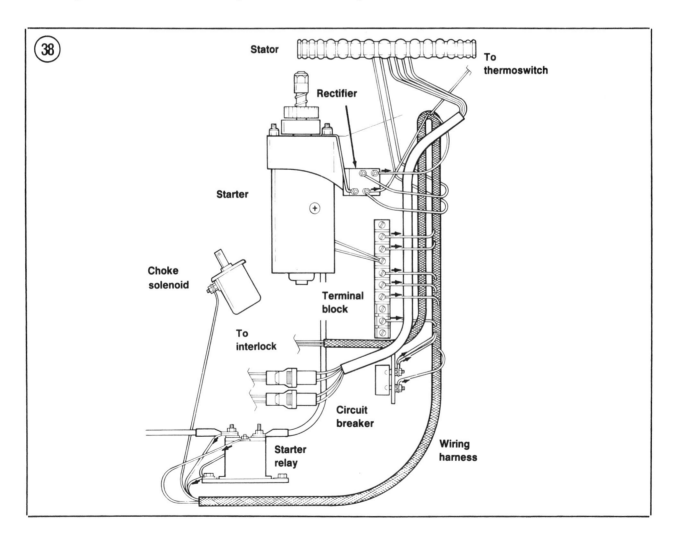

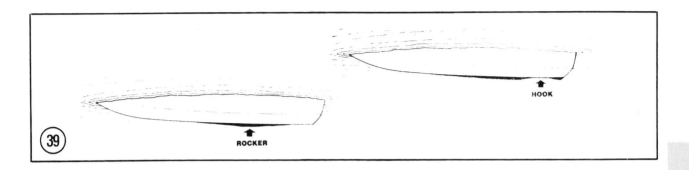

ROCKER HOOK

Table 1 CHRYSLER MODEL HISTORY

Model	1966	1967	1968	1969	1970	1971	1972	1973	1974	1975
3.5 hp	X	X	X	X						
3.6 hp					X	X	X	X	X	X
4 hp										
4.4 hp			X							
4.5 hp							X			
4.9 hp								X	X	X
5 hp				X	X				X	X
6 hp	X	X				X	X	X	X	X
6.6 hp			X							
7 hp				X	X					
7.5 hp										
8 hp						X	X	X	X	X
9.2 hp	X	X								
9.6 hp									X	X
9.9 hp			X	X	X	X	X	X		
10 hp									X	X
12 hp										
12.9 hp						X	X	X		
15 hp									X	X
20 hp	X	X	X	X	X	X	X	X	X	X
25 hp								X	X	X
30 hp								X	X	X
35 hp	X	X	X	X	X	X	X	X	X	X
45 hp	X	X	X	X	X	X	X	X	X	X
50 hp	X									
55 hp		X	X	X	X	X	X	X	X	X
60 hp									X	X
65 hp										
70 hp				X	X	X	X	X		
75 hp	X	X	X						X	X
85 hp				X	X	X	X	X		
90 hp									X	X
100 hp										
105 hp	X	X	X	X	X	X	X	X	X	X
115 hp										
120 hp					X	X	X	X	X	X
125 hp										
130 hp							X	X		
135 hp				X	X	X			X	X
140 hp										

(continued)

Table 1 CHRYSLER MODEL HISTORY (continued)

Model	1976	1977	1978	1979	1980	1981	1982	1983	1984
3.5 hp		X	X	X	X	X	X	X	X
3.6 hp	X	X							
4 hp	X	X	X	X	X	X	X	X	X
4.4 hp									
4.5 hp									
4.9 hp									
5 hp	X								
6 hp	X	X	X	X	X	X	X		
6.6 hp									
7 hp									
7.5 hp				X	X	X	X	X	X
8 hp	X	X	X	X					
9.2 hp									
9.6 hp	X	X	X						
9.9 hp				X	X	X	X	X	X
10 hp	X	X	X						
12 hp				X					
12.9 hp									
15 hp	X	X	X	X	X	X	X	X	X
20 hp	X			X	X	X	X		
25 hp	X	X	X				X	X	X
30 hp				X	X	X	X		
35 hp	X	X	X	X	X	X		X	X
45 hp	X	X	X	X	X	X	X	X	X
50 hp						X	X	X	X
55 hp	X	X	X	X	X	X	X	X	
60 hp	X	X	X						X
65 hp		X	X						
70 hp				X	X				
75 hp	X	X	X	X	X	X	X	X	X
85 hp		X	X	X	X	X	X	X	X
90 hp	X	X						X	X
100 hp				X	X	X	X	X	X
105 hp	X	X	X						
115 hp		X	X	X	X	X	X	X	X
120 hp	X	X							
125 hp								X	
130 hp									
135 hp	X	X							
140 hp			X	X	X	X	X	X	X

Table 2 LUBRICATION & MAINTENANCE[1]

Lubrication points	Figure No.
Steering arm	5
Carburetor and choke linkage	6
Throttle linkage	7
Distributor	8
Tower shaft	9
Clamp screws	10
Swivel bracket and tilt lock	11, 12
Stern bracket	13
Steering support tube	14
Shift linkage	15, 16
Manual starter pinion gear	17
Electric starter pinion gear	18
Propeller shaft splines[2]	
Drive shaft splines[3]	

1. Lubricate with Rykon No. 2EP (part No. T 2961) every 60 days (fresh water) or 30 days (salt water) as required. Complete list does not apply to all models. Perform only those items which apply to your model. Refer to owner's manual.
2. Use Chrysler Lubricant T 2987-1.
3. Lubricate once each season.

Table 3 RECOMMENDED LUBRICANTS AND SEALANTS

Chrysler lubricant	Use	Chrysler part No.
Marine Lubricant (Rykon No. 2EP)	All-purpose waterproof marine grease	T 2961
Marine Gear Lubricant	Extreme pressure E.P. 90 gearcase lubricant	597
Outboard Multi-Purpose Lubricant	Bearings and other moving parts	5H059
Anti-seize Compound	Propeller shaft splines	T 2987-1
LPSI Lubricant	Lubricating water displacer	T 8969
Loctite No. 75	—	T 8936-1
Loctite H (non-hardening)	Screw threads	T 2962-1

(continued)

Table 3 RECOMMENDED LUBRICANTS AND SEALANTS (continued)

Chrysler lubricant	Use	Chrysler part No.
Loctite D (hardening)	Screw threads	T 2963-2
Locquic	Cleaner	T 8935
RTV sealant	Power head assembly	T 8983
EC-750 Industrial sealant	Main bearing seal grooves	T 8955
Gasoila	Screw threads	T 2960
Sealant primer	—	T 8935

Table 4 POWER HEAD COMPRESSION SPECIFICATIONS*

Model	Compression (psi)
3.5 hp	
1965-1969	65-75
1977-on	110-125
3.6 hp	65-75
4 hp	110-125
4.4 hp	NA
4.5 hp	55-65
4.9 hp	NA
5 hp	NA
6 hp	115-130
6.6 hp	NA
7 hp	NA
7.5 hp	115-130
8 hp	110-130
9.2 hp	105-115
9.6 hp	105-115
9.9 hp	
1968-1973	105-115
1979-on	115-125
10 hp	105-125
12 hp	125-135
12.9 hp	105-125
15 hp	125-135
20 hp	
1965-1976	115-125
1979-on	95-105
25 hp	95-105
30 hp	125-135
35 hp	95-105
45 hp	115-130
50 hp	130-140
55 hp	150-165
60-65 hp	150-160
70 hp	130-140
75 hp	145-155
	(continued)

Table 4 POWER HEAD COMPRESSION SPECIFICATIONS* (continued)

Model	Compression (psi)
85 hp	
CD ignition and Charger	145-165
All others	145-155
90 hp	155-165
100 hp	145-165
105 hp	
CD ignition	145-165
Charger	150-165
All others	130-140
115 hp	125-145
120 hp	
CD ignition	150-165
All others	145-155
125 hp	145-165
130 hp	155-165
135 hp	
CD ignition	150-165
All others	155-165
140 hp	
Charger	155-165
All others	150-160

* NA = **Not available.**

Table 5 CYLINDER HEAD BOLT TORQUES*

Model	in.-lb.
3.5 hp	
1965-1969	NA
1977-on	130
3.6 hp	NA
4-4.5 hp	130
4.9-5 hp	80
6-8 hp	130
9.2 hp	65
9.9-15 hp	130
20 hp	
1965-1978	120
1979-on	190
25 hp	225
30 hp	
1965-1978	225
1979-on	190
35-55 hp	270
60-65 hp	225
70-140 hp	
3/8-16 bolts	265-275
5/16-18 bolts	220-230

* NA = **Not available.**

Table 6 SPARK PLUG RECOMMENDATIONS[1]

Model	Champion spark plug No.[2]
3.5 (1965-1969) and 3.6 hp	H8J
4.9 and Series 54/55 5 hp	L10
3.5 (1977-on), 4, 6 (1979-on) and 7.5 hp	L86
20 hp (Series 2000)	J4J
All Magnapower II models, 90, 1983 115, 125 and 140 hp Charger	UL18V
All other 2-cylinder models	L4J
All other 3-cylinder models	L20V

1. All conventional spark plugs are gapped to 0.030 in.
2. 50 hp factory recommendation is AC M42K plug.

Table 7 SPARK PLUG CROSS-REFERENCE CHART[1]

Champion	NGK	AC	Autolite
H8J	—	—	2675
J4J[2]	B8S	M42K	353
L4J	B8S	M42FF	2635
L10	B5HS	45F	536
L20V	BUH	V40FFM	2852
L86	B6HS	—	414
UL18V	BUHX	VB40FFM	2892

1. Spark plugs produced by different manufacturers do not have identical heat range characteristics. If the spark plug recommended by the engine manufacturer is not available, use this chart to determine a suitable equivalent plug.
2. Champion J4J and J6J plugs are superceded by J6C; no cross-reference is available for J6C.

Chapter Five

Engine Synchronization and Linkage Adjustments

If an engine is to deliver its maximum efficiency and peak performance, the ignition must be timed and the carburetor operation synchronized with the ignition. This procedure is the final step of a tune-up. It must also be performed whenever the fuel or ignition systems are serviced or adjusted.

Procedures for engine synchronization and linkage adjustment on Chrysler outboards differ according to model and ignition system. This chapter is divided into self-contained sections dealing with particular models/ignition systems for fast and easy reference.

Each section specifies the appropriate procedure and sequence to be followed and provides the necessary tune-up data. Read the general information at the beginning of the chapter and then select the section pertaining to your outboard.

ENGINE TIMING AND SYNCHRONIZATION

As engine rpm increases, the ignition system must fire the spark plug(s) more rapidly. Proper ignition timing synchronizes the spark plug firing with engine speed.

As engine speed increases, the carburetor must provide an increased amount of fuel for combustion. Synchronizing is the process of timing the carburetor operation to the ignition (and thereby the engine speed).

Required Equipment

Chrysler recommends that static timing of an engine with a breaker point ignition be done by setting the point gap with a feeler gauge. A test lamp or ohmmeter can also be used to set the breaker point gap.

A stroboscopic timing light can be used to check timing mark alignment on some models. As the engine is cranked or operated, the light flashes each time the spark plug fires. When the light is pointed at the moving flywheel, the mark on the flywheel appears to stand still. The flywheel mark should align with the stationary timing pointer on the engine.

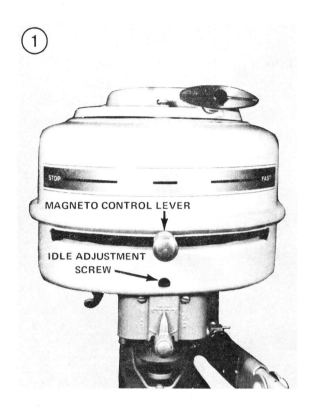

A tachometer connected to the engine is used to determine engine speed during idle and high-speed adjustments.

CAUTION
Never operate the engine without water circulating through the gearcase to the engine. This will damage the water pump and gearcase and can cause engine damage.

Some form of water supply is required whenever the engine is operated during the procedure. Using a test tank is the most convenient method, although the procedures may be carried out on land with a flushing device or with the boat in the water.

3.5 HP (1966-1969)

Throttle Pick-up
Point Adjustment

1. Align the magneto control lever with the carburetor idle adjustment screw. See **Figure 1**.

2. Make sure throttle link is installed in the throttle shaft lever hole nearest the throttle shaft.

3. Loosen the throttle bellcrank swivel screw and push the throttle bellcrank against the stator throttle cam.

4. Lightly depress the brass throttle link enough to take up all play in the swivel. Do not push down hard enough to move the throttle lever. Hold the pressure on the link and tighten the swivel screw.

5. Retard the magneto control lever and then slowly advance it until the throttle shaft lever starts to move. The magneto control lever should be aligned with the idle adjustment screw (**Figure 1**).

6. If the lever and screw are not properly aligned in Step 5:

   a. Move the swivel down on the brass link if the pick-up point is too early.

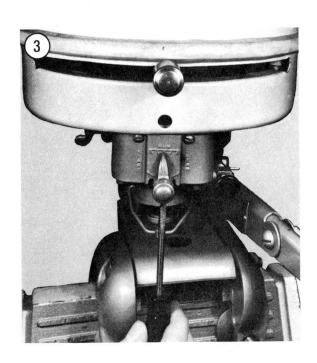

b. Move the swivel up on the brass link if the pick-up point is too late.

Carburetor Adjustment

1. Turn the idle adjustment screw clockwise until it lightly seats in the carburetor (**Figure 2**). Do not turn screw tightly into carburetor or the tip will be damaged.

2. Back the screw out one full turn.

3. Repeat Step 1 and Step 2 with the high-speed adjustment screw. Disregard the position of the high-speed adjusting screw knob at this time.

4. Start the motor and warm to normal operating temperature.

5. Set the throttle lever control to the lowest setting where the engine will run smoothly.

6. Turn the idle adjustment screw clockwise until the engine starts to misfire. Note the position of the adjustment screw slot.

7. Turn the idle adjustment screw counterclockwise until the engine speed becomes rough and starts to decrease. Note the screw slot setting.

8. Turn the idle adjustment screw to a midpoint between the settings noted in Step 6 and Step 7.

9. Move the magneto control lever to the wide-open throttle position.

10. Turn the high-speed adjustment screw clockwise until the engine starts to misfire. Note the position of the adjustment screw knob pointer.

11. Turn the high-speed adjustment screw counterclockwise until the engine speed becomes rough and starts to decrease. Note the adjustment screw knob pointer position.

12. Turn the-high speed adjustment screw to a midpoint between the settings noted in Step 10 and Step 11.

13. Loosen the adjusting knob setscrew and reposition the pointer so that it is vertical. See **Figure 3**. Tighten the setscrew.

3.5 HP (1980-ON) AND 4 HP (1979-ON)

Throttle Pick-up Point Adjustment

1. Remove the engine cover.

2. Advance the throttle control until the carburetor throttle plate starts to open. If adjustment is correct, the center of the throttle shaft follower should be aligned with the cam mark. See **Figure 4**.

3. If the cam mark did not intersect the follower at the point where the throttle started to open, loosen the screw holding the throttle cam to the stator plate.

4. Pivot the cam in or out as required to align the mark with the center of the follower. Tighten the screw securely.

5. Repeat Step 2 to recheck the adjustment. If still not correct, repeat Step 3 and Step 4.

5

Carburetor Adjustment

1. Turn the idle adjustment screw clockwise until it lightly seats in the carburetor (**Figure 5**). Do not turn screw tightly into carburetor or the tip will be damaged.

2. Back the screw out one full turn.

3. Start the motor and warm to normal operating temperature.

4. Move throttle to SHIFT position and shift engine into FORWARD.

5. Set the throttle lever control to the lowest setting where the engine will run smoothly.

6. Turn the idle adjustment screw clockwise until the engine starts to misfire. Note the position of the adjustment screw slot.

7. Turn the idle adjustment screw counterclockwise until the engine speed becomes rough and starts to decrease. Note the screw slot setting.

8. Turn the idle adjustment screw to a midpoint between the settings noted in Step 6 and Step 7.

9. Adjust the support plate idle stop to bring the idle speed to 800-1,000 rpm in gear.

10. Install the engine cover.

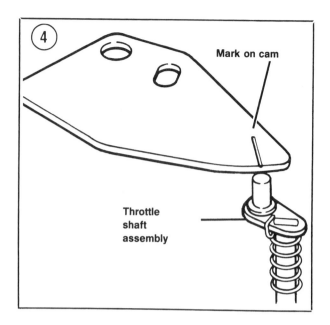

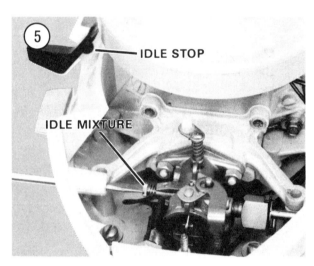

3.6 HP (1970-1977)
AND 4 HP (1976-1978)

1. Remove the engine cover.

2. Move the throttle cam forward until it just contacts the roller on the throttle shaft lever.

> *CAUTION*
> *Excessive bending of the throttle shaft lever in Step 3 may loosen the lever on the shaft. Do not bend throttle shaft under any circumstances.*

3. If the throttle cam mark does not align with the center of the roller at the point of initial contact, bend the throttle shaft lever with needlenose pliers until proper alignment is achieved. See **Figure 6**.

4. With the cam mark and roller aligned at the pick-up point, check the ball stud position

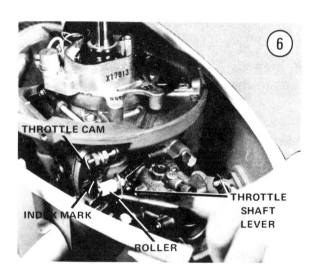

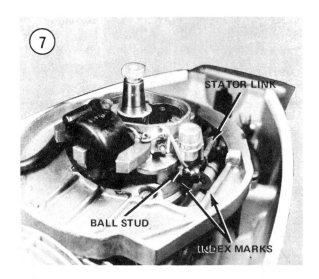

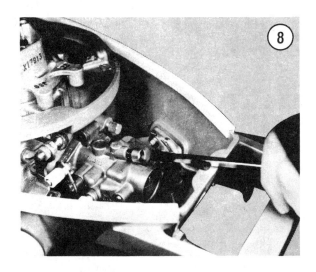

on the stator relative to the bearing cage index marks. If the ball stud is not centered between the index marks, disconnect the stator control rod link.

5. Loosen the ball connector locknuts and rotate the connector on the link. Initial center-to-center length of link, as measured from the center of the ball connector, should be about 3 1/2 in. There should also be an equal number of exposed threads on each end of the link rod. See **Figure 7**. Tighten the locknuts after adjustment.

6. Rotate the magneto control lever to its wide open throttle position. The carburetor shutter should be horizontal and the throttle

cam should be on the surface of the roller away from the end of the roller. If it is not, recheck any adjustment made in Step 3.

Carburetor Adjustment

1. Turn the idle adjustment screw clockwise until it lightly seats in the carburetor (**Figure 8**). Do not turn screw tightly into carburetor or the tip will be damaged.
2. Back the screw out one full turn.
3. Start the motor and warm to normal operating temperature.
4. Set the throttle lever control to the lowest setting where the engine will run smoothly.
5. Turn the idle adjustment screw clockwise until the engine starts to misfire. Note the position of the adjustment screw slot.
6. Turn the idle adjustment screw counterclockwise until the engine speed becomes rough and starts to decrease. Note the screw slot setting.
7. Turn the idle adjustment screw to a midpoint between the settings noted in Step 5 and Step 6.
8. Install the engine cover.

4.4-8 HP
(BREAKER POINT IGNITION)

Throttle Pick-up
Point Adjustment

1. Remove the engine cover.
2. Back off the idle stop screw until the throttle cam mark is on the starboard side of the carburetor follower roller. See **Figure 9**.
3. Advance the throttle until the cam contacts the follower roller. Then slowly advance the throttle until the throttle shaft arm just starts to move. The follower roller should be at or within 1/32 in. of the cam mark. See **Figure 10**.
4. If the roller and cam mark are not aligned as specified in Step 3, turn the carburetor

follower adjusting screw (**Figure 10**) until the cam mark and roller align properly.

Carburetor Adjustment

1. Turn the idle adjustment screw clockwise until it lightly seats in the carburetor (**Figure 11**). Do not turn screw tightly into carburetor or the tip will be damaged.
2. Back the screw out one full turn.
3. Shift the engine into NEUTRAL and turn twist grip throttle to START position.
4. Start the motor and warm to normal operating temperature.
5. Move throttle to SHIFT position and shift engine into FORWARD.
6. Turn the throttle twist grip clockwise to the lowest setting where the engine will run smoothly.
7. Turn the idle adjustment screw clockwise until the engine starts to misfire. Note the position of the adjustment screw slot.
8. Turn the idle adjustment screw counterclockwise until the engine speed

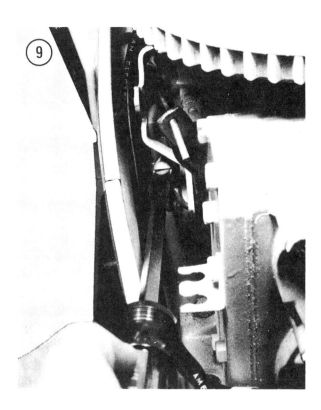

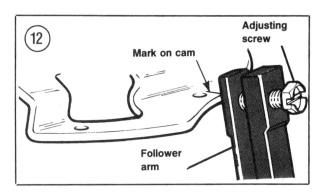

becomes rough and starts to decrease. Note the screw slot setting.

9. Turn the idle adjustment screw to a midpoint between the settings noted in Step 7 and Step 8.

10. Adjust the idle stop screw to bring the idle speed to 700 rpm in gear.

11. Install the engine cover.

6-8 HP (CD IGNITION)

Throttle Pick-up
Point Adjustment

1. Remove the engine cover.

2. Advance the throttle control until the edge of the follower aligns with the stator ring cam

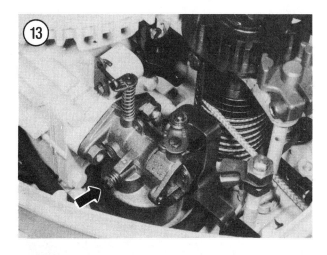

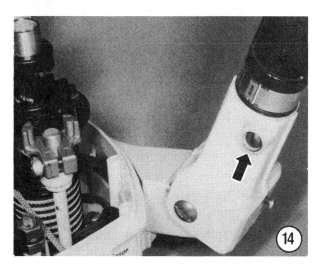

mark. See **Figure 12**. The throttle should just begin to move at this point.

3. If the throttle does not begin to move as the follower and cam mark align in Step 1, turn the follower adjusting screw in or out as required until the throttle starts to move.

Carburetor Adjustment

1. Turn the idle adjustment screw clockwise until it lightly seats in the carburetor (**Figure 13**). Do not turn screw tightly into carburetor or the tip will be damaged.

2. Back the screw out one full turn.

3. Start the motor and warm to normal operating temperature. Once it is warmed up, allow another 1-2 minutes of idling for the recirculation system to start functioning.

4. Move throttle to SHIFT position and shift engine into FORWARD.

5. Turn the throttle twist grip clockwise to the lowest setting where the engine will run smoothly.

6. Turn the idle adjustment screw counterclockwise until the engine speed becomes rough and starts to decrease. Note the screw slot setting.

7. Turn the idle adjustment screw clockwise until the engine starts to misfire. Note the position of the adjustment screw slot.

8. Turn the idle adjustment screw to a midpoint between the settings noted in Step 7 and Step 8.

9. Adjust the idle stop screw (**Figure 14**) to bring the idle speed to 600-750 rpm in gear.

10. Install the engine cover.

9.2-15 HP (1966-ON)

Throttle Pick-up Point Adjustment
(1966-1982)

1. Remove the engine cover.

2. Back off idle stop screw until throttle cam mark is on starboard side of carburetor follower roller. See **Figure 15**.

3. Advance throttle control until throttle cam just contacts follower roller. Then slowly advance the throttle cam until the throttle shaft arm just starts to move.

4. Check throttle cam mark and roller alignment. Roller should be at or within 1/32 in. of the mark. If not, turn adjusting screw on carburetor follower (**Figure 15**) until alignment is correct.

Throttle Pick-up
Point Adjustment
(1983-on)

1. Remove the engine cover.
2. Advance the throttle control until the center of the carburetor follower aligns with the stator ring cam mark. The throttle should just begin to move at this point.
3. If the throttle does not begin to move as the follower and cam mark align in Step 2, turn the follower adjusting screw in or out as required until the throttle starts to move.

Carburetor Adjustment

1. Turn the idle mixture screw clockwise until it lightly seats in the carburetor (**Figure 15**). Do not turn screw tightly into carburetor or the tip will be damaged.
2. Back the screw out 1 1/4 turns (1966-1982) or 1 turn (1983-on).
3. Start the motor and warm to normal operating temperature. Once it is warmed up, allow another 1-2 minutes of idling for the recirculation system to start functioning.
4. Move throttle to SHIFT position and shift engine into FORWARD.
5. Turn the throttle control clockwise to the lowest setting where the engine will run smoothly.
6. Turn the idle mixture screw counter-clockwise until the engine speed becomes

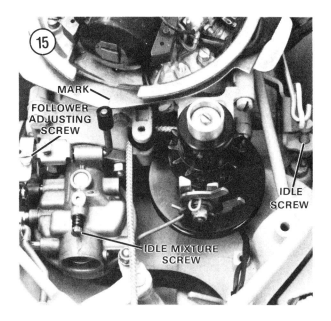

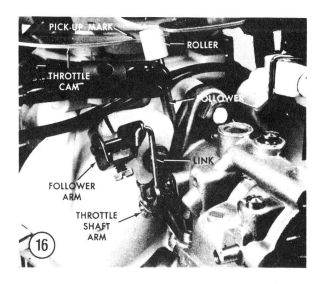

rough and starts to decrease. Note the screw slot setting.

7. Turn the idle adjustment screw clockwise until the engine starts to misfire. Note the position of the adjustment screw slot.

8. Turn the idle adjustment screw to a midpoint between the settings noted in Step 6 and Step 7.

9. Adjust the idle stop screw (**Figure 15**) to bring the idle speed to 700 rpm (1966-1982) or 600-750 rpm (1983-on) in gear.

10. Install the engine cover.

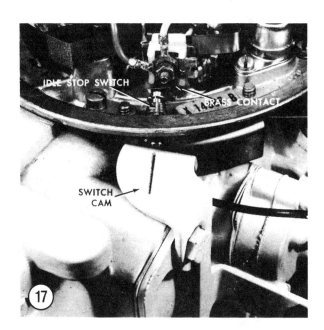

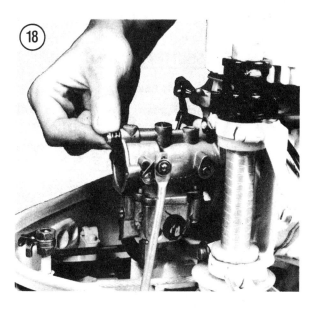

20 HP (1966-1976)

Throttle Pick-up Point Adjustment

1. Remove the engine cover.

2. Close the throttle until the throttle cam mark is on the starboard side of the carburetor follower roller. See **Figure 16**.

3. Advance throttle control until throttle cam just contacts follower roller. Then slowly advance the throttle cam until the throttle shaft arm just starts to move.

4. Check throttle cam mark and roller alignment. Roller should be at or within 1/32 in. of the mark. If not, loosen screw on carburetor follower, adjust follower arm position until alignment is correct and retighten screw.

5. Repeat Steps 2-4 to recheck pick-up point. Readjust if necessary.

Idle Stop Switch Check

Earlier 20 hp manual start engines use an idle stop switch to short out the ignition when the magneto is fully retarded. Idle stop switch adjustment should be checked on models so equipped.

1. Remove the flywheel. See Chapter Eight.

2. Rotate stator plate until centerline of idle stop switch aligns with the switch cam indent line. See **Figure 17**.

3. If the switch plunger does not touch the brass contact on the breaker point terminal screw when switch and cam are aligned, loosen the 2 screws holding the cam to the transfer port cover and align plunger and contact, then retighten the screws.

4. Install the flywheel. See Chapter Eight.

Carburetor Adjustment

1. Turn the idle mixture screw clockwise until it lightly seats in the carburetor (**Figure 18**). Do not turn screw tightly into carburetor or the tip will be damaged.

2. Back the screw out 1 1/4 turns.

3. Shift engine into NEUTRAL and rotate twist grip to START position.

4. Start the motor and warm to normal operating temperature.

5. Move throttle to SHIFT position and shift engine into FORWARD.

6. Turn the throttle control clockwise to the lowest setting where the engine will run smoothly.

5

7. Turn the idle mixture screw counterclockwise until the engine speed becomes rough and starts to decrease. Note the screw slot setting.

8. Turn the idle adjustment screw clockwise until the engine starts to misfire. Note the position of the adjustment screw slot.

9. Turn the idle adjustment screw to a midpoint between the settings noted in Step 7 and Step 8.

10. Adjust the idle stop screw (**Figure 19**) to bring the idle speed to 700 rpm in gear.

11. Install the engine cover.

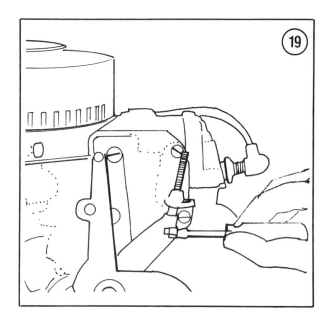

20 HP (1979-1982), 25 AND 30 HP (ALL) AND 35 HP (1976-1979)

Throttle Pick-up Point Adjustment

1. Remove the engine cover.

> *CAUTION*
> *Do not disconnect throttle cam rod at the throttle cam. The pressure required to do so can distort the cam.*

2. Disconnect the throttle cam rod at the magneto control lever. See **Figure 20**.

3. Loosen the locknut on the eccentric screw holding the nylon roller to the throttle shaft arm. See **Figure 21**.

4. Set the throttle cam mark at a point close to where it will touch the nylon roller.

5. Turn the eccentric screw until the roller is at the maximum distance from the cam mark, then turn the screw counterclockwise until the roller barely touches the cam mark.

6. Tighten eccentric screw locknut, taking care not to turn the screw.

7. Tear a small strip from a sheet of writing paper and insert it between the cam and roller from the top, then from the bottom. Note the gap between the cam mark and paper edge. If both gaps are equal, the cam mark is at the center of its contact with the roller. See

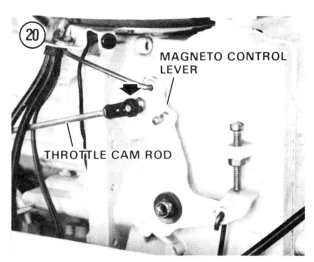

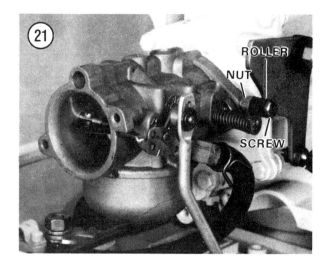

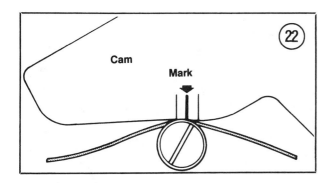

Cam

Mark

STATOR RING STOP

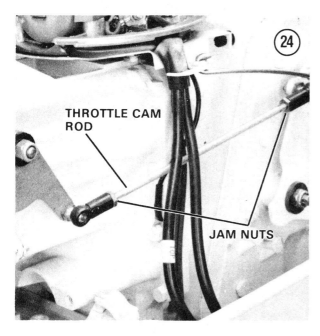

THROTTLE CAM ROD

JAM NUTS

Figure 22. If it is not, repeat the eccentric screw adjustment.

8. Reconnect throttle cam rod to magneto control lever. Rotate magneto until the nylon throttle stop on the stator ring rests against the cylinder. See **Figure 23.** Wide-open throttle mark on cam should intersect the roller.

9. If cam mark does not intersect roller in Step 8, disconnect the throttle cam rod from the magneto control lever. Loosen the jam nuts and rotate the rod to lengthen or shorten it as required. See **Figure 24.** Reconnect rod and recheck cam mark/roller intersection. Repeat this step until the mark intersects the roller when the rod is installed, then tighten jam nuts.

10. Check the throttle shutter position with the magneto at wide-open throttle position and the wide-open throttle mark on the cam intersecting the roller. If the shutter is not horizontal, repeat Steps 1-9.

**Throttle Stop Adjustment
(Manual Start Models)**

The setscrew should touch the steering handle wall when the wide-open throttle mark on the cam intersects the roller. If it does not, adjust the screw in the throttle stop. See **Figure 25.**

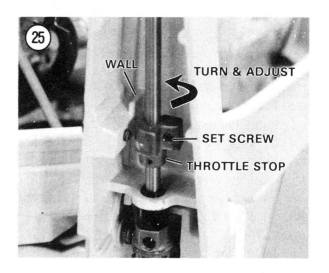

WALL   TURN & ADJUST

SET SCREW

THROTTLE STOP

5

Carburetor Adjustment

1. Connect a tachometer according to manufacturer's instructions.

2. Turn the idle adjustment screw clockwise until it lightly seats in the carburetor (**Figure 26**). Do not turn screw tightly into carburetor or the tip will be damaged.

3. Back the screw out 1 1/4 turns.

4. Shift engine into NEUTRAL and rotate twist grip to START position.

5. Start the motor and run at 2,000 rpm until it reaches operating temperature. Once it is warmed up, allow another 1-2 minutes of idling for the recirculation system to start functioning.

6. Move throttle to SHIFT position and shift engine into FORWARD.

7. Turn the throttle control clockwise to the lowest setting where the engine will run smoothly.

8. Turn the idle adjustment screw (**Figure 26**) clockwise 1/8 turn at a time, pausing at least 10 seconds between turns and adjusting the throttle to keep the engine from exceeding 900 rpm. Continue this step until the tachometer shows that engine speed has peaked and is decreasing and the engine starts to misfire. Note the screw slot setting.

9. Return idle adjustment screw to original position and repeat Step 8, turning the screw in a *counterclockwise* direction until the engine speed becomes rough. Note the position of the adjustment screw slot.

10. Turn the idle adjustment screw to a midpoint between the settings noted in Step 8 and Step 9.

11. Quickly open the throttle. If the engine dies, it is too lean. Turn the adjustment screw counterclockwise. If the engine stumbles, it is too rich. Turn the screw clockwise.

12. Adjust the idle speed screw to bring the idle speed to 750 rpm in gear. See **Figure 27** (manual start) or **Figure 28** (electric start).

13. Install the engine cover.

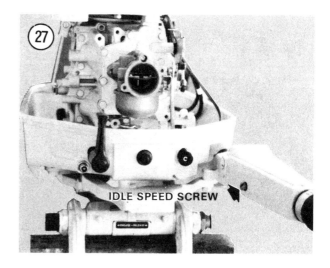

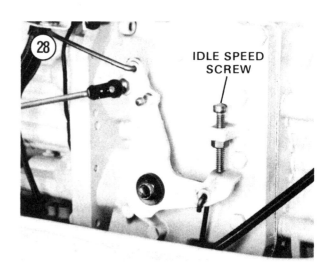

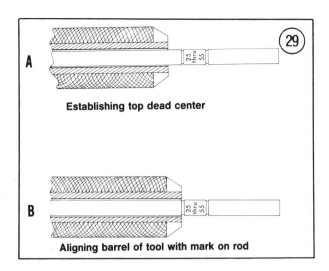

Establishing top dead center

Aligning barrel of tool with mark on rod

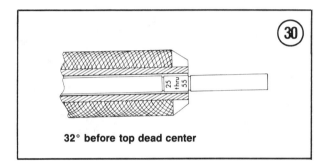

32° before top dead center

35 AND 45 HP (1966-ON),
50 HP (1966 AND 1981-ON), AND
55 HP (1967-1976)

Engine Timing

1. Disconnect the negative battery lead.

2. Remove the engine cover.

3. Remove all spark plugs from the power head. See Chapter Four.

4. Select the proper timing tool. There is one designed especially for 35 and 45 hp manual tiller engines beginning with the serial numbers shown in **Table 1** (part No. T 8938). The other (part No. T 2937-1) is used with all other 35-55 hp models.

5. Remove the flywheel. See Chapter Eight.

6. Check and adjust breaker points as required. An error of 0.0015 inch in the gap setting will change engine timing as much as 1 degree.

7. Thread the timing tool barrel in the top cylinder spark plug hole and screw it in completely.

8. Install the timing tool rod into the barrel with its 2 identification marks ("25-55 HP") toward the outside. The distance between the 2 marks equals 28° (part No. T 8938) or 32° (part No. T 2937-1).

9. While holding the rod tightly against the piston, rotate the crankshaft clockwise to locate top dead center (TDC). This is the point at which the marks on the timing tool rod are at a maximum distance from the barrel. See A, **Figure 29**.

10. Thread the timing tool barrel in or out as required to align the inner timing rod mark with the edge of the barrel. See B, **Figure 29**.

NOTE
If the crankshaft is rotated too far in Step 11, the outer timing rod mark will go beyond the edge of the timing tool barrel. If this happens, repeat the step—do not try to position the rod mark by rotating the crankshaft counterclockwise.

11. Apply slight finger pressure on the end of the rod to prevent it from shooting out during the piston compression stroke. Rotate crankshaft clockwise almost one full turn until the outer timing rod mark aligns with the edge of the barrel. See **Figure 30**. The piston is now positioned exactly 28° (part No. T 8938) or 32° (part No. T 2937-1) before TDC.

12. Disconnect the throttle link at the tower shaft. Place tower shaft in wide-open throttle position by rotating it until the nylon stop rests against the cylinder crankcase cover.

13A. Battery ignition—Disconnect the breaker point lead wires at the slide terminal.

13B. Magneto ignition—Remove the coil ground wire and condenser attaching screw. Make sure wire and condenser do not touch ground.

5

14. Connect a 12-volt test lamp between the No. 1 cylinder breaker point lead wire and a good engine ground.

15. Loosen the locknut on the spark control link (**Figure 31**):

   a. If test light is on, rotate spark control link counterclockwise until light dims or goes off.

   b. If test lamp is off, rotate spark control link clockwise until light just comes on.

16. Rotate crankshaft back and forth. The test light should go on and off as the timing rod mark passes the edge of the timing tool barrel.

17. Hold the spark control link in position and tighten the locknut.

18. Remove the timing tool. Reinstall the flywheel (Chapter Eight). Reinstall the spark plugs.

Throttle Pick-up
Point Adjustment

> *CAUTION*
> *Do not disconnect throttle link at the throttle cam. The pressure required to do so can distort the cam.*

1. Disconnect the throttle link ball-joint connector at the tower shaft. See **Figure 32**.

2. Loosen the locknut on the eccentric screw holding the nylon roller to the throttle shaft arm, then tighten locknut just enough to eliminate any clearance between the screw, roller, throttle cam and nut. See **Figure 33**.

3. Set the throttle cam mark at a point close to where it will touch the nylon roller.

4. Turn the eccentric screw until the roller is at the maximum distance from the cam mark, then turn the screw counterclockwise until the roller barely touches the cam mark.

5. Tighten eccentric screw locknut, taking care not to turn the screw.

6. Tear a small strip from a sheet of writing paper and insert it between the cam and roller from the top, then from the bottom. Note the

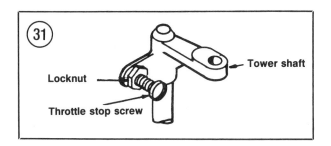

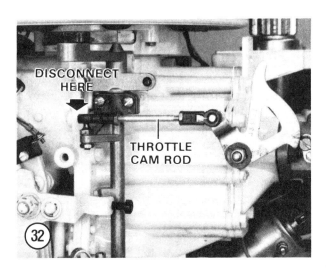

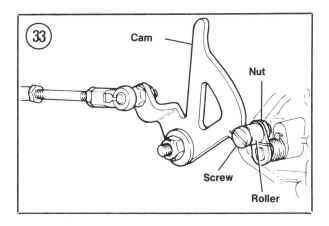

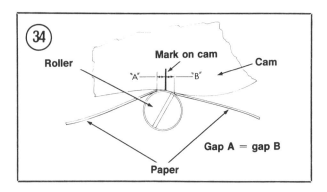

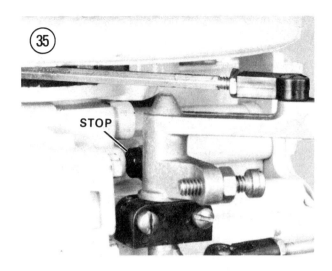

STOP

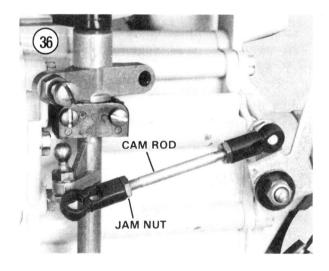

CAM ROD

JAM NUT

gap between the cam mark and paper edge. If both gaps are equal, the cam mark is at the center of its contact with the roller. See **Figure 34**. If it is not, repeat the eccentric screw adjustment.

7. Reconnect throttle link to tower shaft. Rotate magneto until the nylon throttle stop on the tower shaft rests against the cylinder. See **Figure 35**. The wide-open throttle mark on the cam should intersect the roller.

8. If cam mark does not intersect roller in Step 7, disconnect the throttle link from the tower shaft. See **Figure 36**. Loosen the jam nuts and rotate the rod to lengthen or shorten it as required. Reconnect rod and recheck cam mark/roller intersection. Repeat this step

until the mark intersects the roller when the rod is installed, then tighten jam nuts.

9. Check the throttle shutter position with the tower shaft at its wide-open throttle position and the wide-open throttle mark on the cam intersecting the roller. If the shutter is not horizontal, repeat Steps 1-8.

Carburetor Adjustment

1. Connect a tachometer according to manufacturer's instructions.

2. Remove the carburetor intake cover and gaskets, if so equipped.

3. Turn the idle adjustment screw clockwise until it lightly seats in the carburetor (**Figure 33**). Do not turn screw tightly into carburetor or the tip will be damaged.

4. Back the screw out 1 1/4 turns.

5. Shift engine into NEUTRAL and rotate twist grip to START position.

6. Start the motor and run at 2,000 rpm until it reaches operating temperature. Once it is warmed up, allow another 1-2 minutes of idling for the recirculation system to start functioning.

7. Move throttle to SHIFT position and shift engine into FORWARD.

8. Turn the throttle control clockwise to the lowest setting where the engine will run smoothly.

9. Turn the idle adjustment screw (**Figure 37**) clockwise 1/8 turn at a time, pausing at least 10 seconds between turns and adjusting the throttle to keep the engine from exceeding 900 rpm. Continue this step until the tachometer shows that engine speed has peaked and is decreasing and the engine starts to misfire. Note the screw slot setting.

10. Return idle adjustment screw to original position and repeat Step 9, turning the screw in a *counterclockwise* direction until the engine speed becomes rough. Note the position of the adjustment screw slot.

5

11. Turn the idle adjustment screw to a midpoint between the settings noted in Step 9 and Step 10.

12. Quickly open the throttle. If the engine dies, it is too lean. Turn the adjustment screw counterclockwise. If the engine stumbles, it is too rich. Turn the screw clockwise.

13. Adjust the idle speed screw to bring the idle speed to 750 ± 50 rpm in gear.

14. On models equipped with a neutral warm-up speed screw on the end of the gear shift arm, shift the engine into NEUTRAL and open the throttle fully. If the tachometer does not read 1,500-1,800 rpm, adjust the warm-up speed screw as required.

Choke Solenoid Adjustment

1. Loosen the screw holding the solenoid plunger to the choke lever.

2. Slip a small piece of writing paper about 1/2 in. wide into the carburetor air horn. The paper will provide sufficient clearance to prevent the choke shutter from sticking closed.

3. Hold the choke lever in the closed position and depress the solenoid plunger (not the rod) until it bottoms in the solenoid, then tighten the choke lever screw.

4. Pull the paper from the air horn. There should be a slight drag as the paper is removed. If drag is excessive, repeat Step 3.

5. Install the carburetor intake cover with new gaskets.

6. Install the engine cover.

**55 HP (1977-1980),
60 HP (1974-1976),
65 HP (1974-1978), AND
105, 120 AND 135
HP (1975-1977)
WITH MAGNAPOWER
II IGNITION**

Engine Timing

1. Disconnect the negative battery lead.

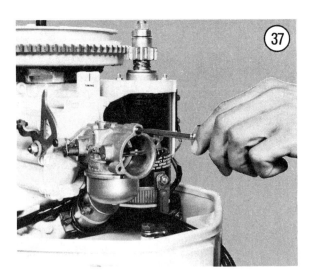

2. Remove the engine cover.

3. Remove all spark plugs from the power head. See Chapter Four.

4. 55-65 hp—Check and adjust breaker points as required. An error of 0.0015 inch in the gap setting will change engine timing as much as 1 degree.

5. Thread timing tool barrel (part No. T 2937-1) into the top cylinder spark plug hole and screw it in completely.

6. Install the timing tool rod in the barrel with its 2 identification marks ("25-55 HP") toward the outside.

7. While holding the rod tightly against the piston, rotate the crankshaft clockwise to locate top dead center (TDC). This is the point at which the marks on the timing tool rod are at a maximum distance from the barrel. See A, **Figure 29**.

8. Thread the timing tool barrel in or out as required to align the inner timing rod mark with the edge of the barrel. See B, **Figure 29**.

NOTE
If the crankshaft is rotated too far in Step 9, the outer timing rod mark will go beyond the edge of the timing tool barrel. If this happens, repeat the step—do not try to position the rod mark by rotating the crankshaft counterclockwise.

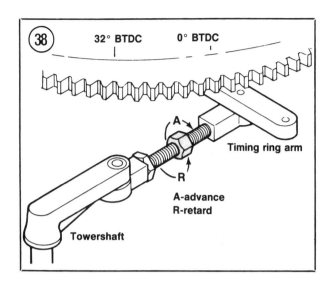

9. Apply slight finger pressure on the end of the rod to prevent it from shooting out during the piston compression stroke. Rotate crankshaft clockwise almost one full turn until the outer timing rod mark aligns with the edge of the barrel. See **Figure 30**. The piston is now positioned exactly 32° before TDC.

10. Make sure the 32 degree mark on the flywheel ring gear decal aligns with the index mark at the top of the carburetor adapter flange (55-65 hp) or timing pointer index mark (105, 120 and 135 hp). If it does not, install a new decal.

11. Remove the timing tool and reinstall the spark plugs. Connect a timing light according to manufacturer's instructions.

12. With the engine in a test tank or on the boat in the water, place the throttle at the full retard position and crank the engine over while pointing the timing light at the flywheel timing marks. If the TDC mark on the timing decal does not align ± 3° with the carburetor adapter flange index mark (55-65 hp) or timing pointer index mark (105, 120 and 135 hp), repeat Steps 3-11.

13. Start the engine and bring its speed up to 4,750-5,250 rpm (1980-on 55 hp) or 5,000-5,500 rpm (all others). Point the timing light at the flywheel timing marks. The 32°

mark on the decal should align with the carburetor adapter flange index mark (55-65 hp) or timing pointer index mark (105, 120 and 135 hp).

WARNING
Do not attempt to adjust the tower shaft link with the engine running. It is close to the flywheel ring gear and serious personal injury can result.

14. If the timing marks do not align in Step 13, retard the engine and shut it off. Loosen the locknut on the tower shaft link under the flywheel and rotate the link clockwise to advance or counterclockwise to retard the timing as required. See **Figure 38**. One full turn of the link equals about a 2 degree change in timing.

15. Restart the engine and recheck any adjustment made in Step 14. Repeat Step 14 as required until the timing marks align. When adjustment is correct, shut the engine off and tighten the tower shaft link locknut securely.

Throttle Shutter Synchronization

1. Remove the carburetor intake cover and gaskets, if so equipped.

2. Disconnect the tower shaft throttle link at the throttle cam. Move cam away from carburetor throttle roller.

3. Remove the retaining ring holding the throttle tie bar to the top carburetor throttle arm.

4. Loosen the throttle tie bar screw (**Figure 39**).

5. Make sure the throttle shutters are closed, then tighten the tie bar screw.

6. Operate the tie bar and make sure its pivot goes in and out of the carburetor throttle arm without binding.

7. Install the retaining ring and connect the link to the cam.

5

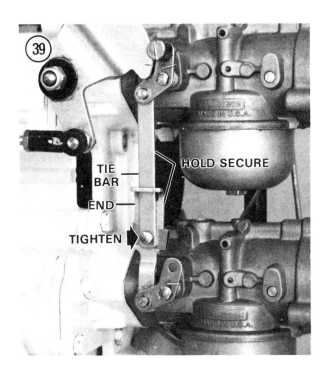

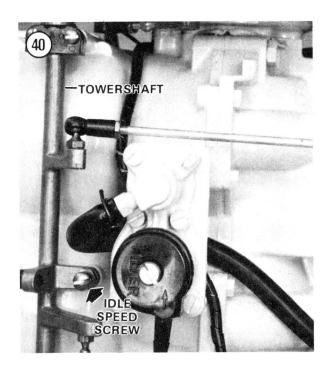

**Throttle Pick-up
Point Adjustment**

Adjust the throttle pick-up point as described under *35 and 45 hp (1966-on), 50 hp (1966 and 1981-on) and 55 hp (1967-1976)* in this chapter.

Carburetor Adjustment

1. Perform Steps 1-11 of *35 and 45 hp (1966-on), 50 hp (1966 and 1981-on) and 55 hp (1967-1976)* in this chapter to adjust the top carburetor.

2. Perform Steps 9-11 of *35 and 45 hp (1966-on), 50 hp (1966 and 1981-on) and 55 hp (1967-1976)* in this chapter to adjust the bottom carburetor.

3. Loosen the idle speed screw locknut on the tower shaft (**Figure 40**) and adjust stop screw as required until engine idles at 750-800 rpm in gear, then tighten the locknut.

Choke Shutter Synchronization

Refer to **Figure 41** for this procedure.

1. Loosen the screw on each choke swivel to free the choke link.

2. Adjust top end of link about 1/16-1/8 in. beyond the top of the choke swivel and tighten the screw to hold the link in that position.

3. Close top carburetor choke shutter completely and hold in that position. Close bottom carburetor choke completely and tighten bottom choke swivel screw.

4. Loosen the 2 solenoid attaching screws just enough to permit the solenoid to be moved.

5. Slip a small piece of writing paper about 1/2 in. wide into each carburetor air horn. The paper will provide sufficient clearance to prevent the choke shutters from sticking closed.

6. Depress the solenoid plunger (not the rod) to close the choke shutters. See **Figure 42**.

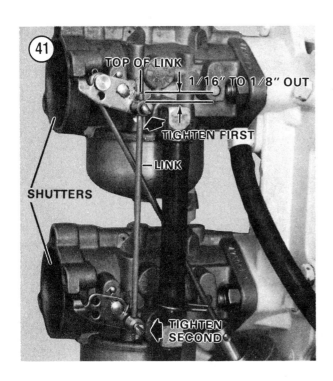

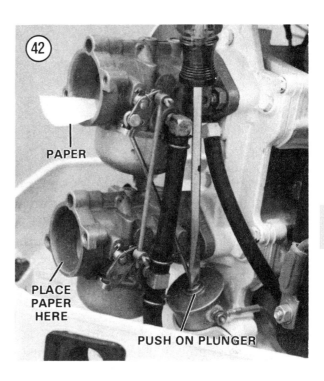

Slide the solenoid against the plunger and tighten the solenoid attaching screws.

7. Release the solenoid plunger and push plunger into solenoid. Pull the papers from the air horn. There should be a slight drag as the papers are removed. If drag is excessive, repeat Step 5 and Step 6.

8. Release the solenoid plunger. If the choke shutters are not horizontal, bend the shutter stop groove pin in the carburetor air horn(s) as required to bring the shutters into a horizontal position.

9. Install the carburetor intake cover with new gaskets.

10. Install the engine cover.

70-135 HP (DISTRIBUTOR BREAKER POINT IGNITION)

Engine Timing

1. Disconnect the negative battery lead.

2. Remove the engine cover.

3. Remove all spark plugs from the power head. See Chapter Four.

4. Check and adjust breaker points as required. An error of 0.0015 inch in the gap setting will change engine timing as much as 1 degree.

5. Thread timing tool barrel (part No. T 2937-1) into the top cylinder spark plug hole and screw it in completely.

6. Install the timing tool rod in the barrel with its 2 identification marks ("70 hp and up") toward the outside.

7. While holding the rod tightly against the piston, rotate the crankshaft clockwise to locate top dead center (TDC). This is the point at which the marks on the timing tool rod are at a maximum distance from the barrel. See **Figure 43**.

8. Thread the timing tool barrel in or out as required to align the inner timing rod mark with the edge of the barrel. See **Figure 43**.

NOTE
If the crankshaft is rotated too far in Step 9, the outer timing rod mark will go beyond the edge of the timing tool barrel. If this happens, repeat the step—do not try to position the rod mark by rotating the crankshaft counterclockwise.

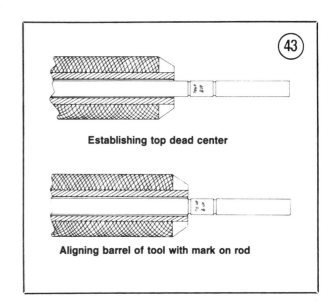

Establishing top dead center

Aligning barrel of tool with mark on rod

9. Apply slight finger pressure on the end of the rod to prevent it from shooting out during the piston compression stroke. Rotate crankshaft clockwise almost one full turn until the outer timing rod mark aligns with the edge of the barrel. See **Figure 44**. The piston is now positioned exactly 36° before TDC.

10. Make sure the 36 degree mark on the flywheel ring gear decal aligns with the timing pointer index line. If it does not, loosen the timing pointer screws, align the pointer index line with the decal mark and tighten the pointer screws.

11. Rotate the flywheel to align the 0 degree mark with the timing pointer index line.

12. Rotate the distributor pulley to align the pulley index mark with the outer diameter of the flywheel. See **Figure 45**.

13. Install distributor belt (removed when breaker points were checked) over pulley. Position the tip of a 0.008 in. feeler gauge against the mid-point of the belt length. Push against the belt until the feeler gauge bends without moving the belt further. This should provide the desired 3/16-1/4 in. belt deflection. See **Figure 46**. Tighten the distributor bracket screws.

14. Disconnect the throttle link at the tower shaft. Move tower shaft to wide-open throttle position with the nylon stop resting against the cylinder crankcase cover.

15. Rotate the flywheel clockwise to align the 36 degree mark with the appropriate mark on the timing pointer as follows:
    a. -4 degree mark (32° BTDC) for all 3-cylinder and 105 hp. See **Figure 47**.

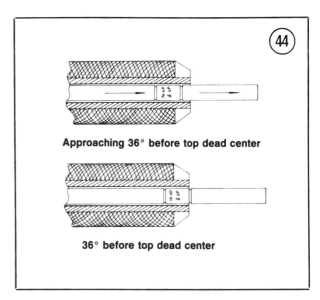

Approaching 36° before top dead center

36° before top dead center

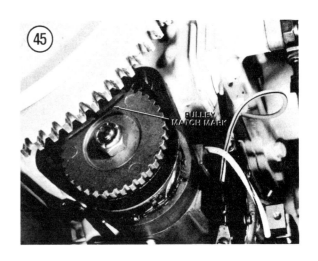

PULLEY MATCH MARK

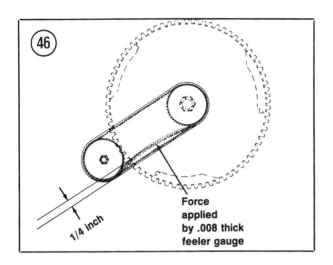

Force applied by .008 thick feeler gauge

1/4 inch

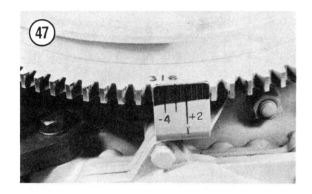

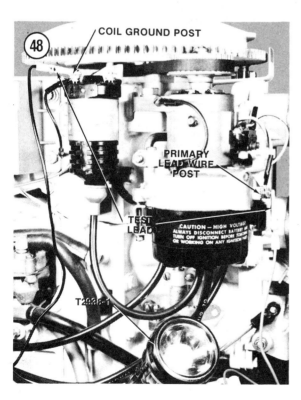

COIL GROUND POST

PRIMARY LEAD WIRE POST

TEST LEAD

CAUTION — HIGH VOLTAGE
ALWAYS DISCONNECT BATTERY or
TURN OFF IGNITION BEFORE TOUCHING
OR WORKING ON ANY IGNITION PART

T2038-1

b. -2 degree mark (second to left of pointer index mark or 34° BTDC) for 120 and 130 hp.

c. -6 degree mark (left edge of pointer or 30° BTDC) for 135 hp.

16. Connect a 12-volt test light between the distributor primary lead wire post and the coil ground post. See **Figure 48**.

17. Loosen distributor control rod locknut (**Figure 49**):

a. If test light is off, rotate control rod clockwise until light just comes on.

b. If test light is on, rotate control rod counterclockwise until light just goes off.

18. Depress distributor belt lightly several times. Test light should come on and go off each time.

19. Check the length of the control rod between the tower shaft swivel and distributor. It should be 3/8-1/2 in. See **Figure 49**. If it is not within specifications, one of the following mistakes was made. Correct as required:

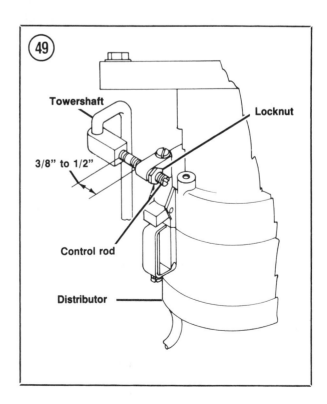

Towershaft

Locknut

3/8" to 1/2"

Control rod

Distributor

5

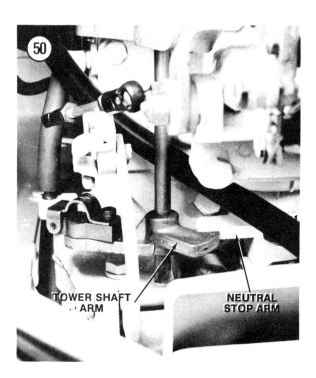

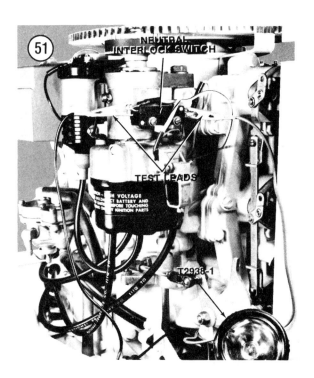

a. The flywheel was turned in the wrong direction to align the 36 degree mark with the timing pointer.

b. The breaker point gap is not correct.

c. The No. 1 cylinder was not at top dead center when the distributor drive belt was installed.

20. When control rod length is correct, hold rod in position and tighten locknut securely.

Neutral Interlock
Switch Adjustment

1. Shift engine to NEUTRAL.

2. Advance tower shaft to neutral stop arm as shown in **Figure 50**.

3. Connect a 12-volt test light to the neutral interlock switch as shown in **Figure 51**.

4. Loosen the screws holding the switch cam to the distributor housing.

5. Move cam upward until test light just comes on and tighten cam screws.

Neutral Interlock Switch Cam
(Shift Arm) Adjustment

1. Shift engine to NEUTRAL.

2. Advance tower shaft to neutral stop arm as shown in **Figure 50**.

3. Connect a 12-volt test light to the neutral interlock switch as shown in **Figure 52**.

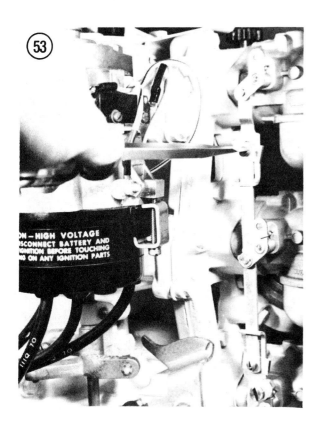

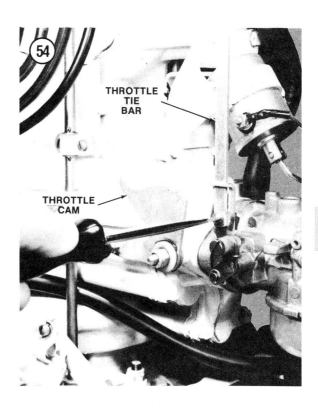

5

4. Loosen the screws holding the switch cam to the shift arm.

5. Move cam upward until test light just comes on and tighten cam screws.

Throttle Shutter Synchronization (3-cylinder Engines)

1. Disconnect throttle link from tower shaft. Pivot throttle cam away from carburetor throttle roller.

2. Loosen the throttle tie bar screws. See **Figure 53**.

3. Make sure throttle shutters are fully closed and tighten tie bar screws (**Figure 53**).

4. Connect throttle link to tower shaft. Set engine at wide-open throttle and make sure throttle shutters are horizontal. If they are not, adjust the throttle cam-to-tower shaft link as required.

Throttle Shutter Synchronization (4-cylinder Engines)

1. Disconnect throttle link from tower shaft. Pivot throttle cam away from carburetor throttle roller.

2. Remove retaining ring holding throttle tie bar to the top carburetor throttle arm.

3. Loosen the throttle tie bar screws. See **Figure 54**.

4. Make sure throttle shutters are fully closed and tighten tie bar screws (**Figure 54**).

5. Check to see that the throttle tie bar end pivot goe in and out of carburetor throttle arm without binding.

6. Install retaining ring to hold throttle tie bar to the top carburetor throttle arm.

7. Connect throttle link to tower shaft.

Throttle Pick-up Point Adjustment

1. Disconnect the throttle link from the tower shaft.

2. Loosen throttle roller shaft locknut. Rotate shaft until throttle roller just touches the throttle cam at the index line as shown in **Figure 55**, then tighten locknut.

3. Reconnect throttle link at the tower shaft.

4. Move tower shaft to wide-open throttle position.

5. Loosen throttle link jam nuts and adjust link until throttle shutters are horizontal, then tighten the jam nuts.

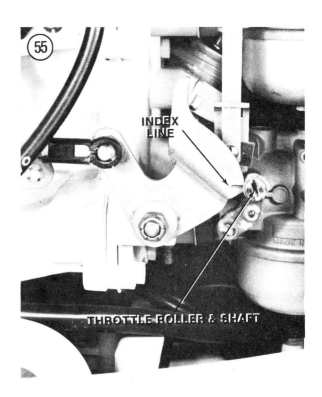

**Carburetor Adjustment
(3-cylinder Engines)**

1. Perform Steps 1-11 of *35 and 45 hp (1966-on), 50 hp (1966 and 1981-on) and 55 hp (1967-1976)* in this chapter to adjust the top carburetor.

2. Perform Steps 9-11 of *35 and 45 hp (1966-on), 50 hp (1966 and 1981-on) and 55 hp (1967-1976)* in this chapter to adjust the middle carburetor.

3. Shut the engine off and turn the top and middle carburetor idle adjustment needles clockwise until they lightly seat. Note the number of turns required to seat the needle on each carburetor.

4. Take the average of the settings noted in Step 3 and set all carburetors accordingly. For example, if the top carburetor required 7/8 turn and the middle carburetor required 5/8 turn, the average of the 2 is 6/8 or 3/4. Back all 3 carburetor needles out 3/4 turn from a lightly seated position.

5. Adjust the idle stop screw on the tower shaft as required until engine idles at 700-900 rpm in NEUTRAL.

**Carburetor Adjustment
(4-cylinder Engines)**

1. Perform Steps 1-11 of *35 and 45 hp (1966-on), 50 hp (1966 and 1981-on) and 55 hp (1967-1976)* in this chapter to adjust the top carburetor.

2. Perform Steps 9-11 of *35 and 45 hp (1966-on), 50 hp (1966 and 1981-on) and 55 hp (1967-1976)* in this chapter to adjust the bottom carburetor.

3. Adjust the idle stop screw on the tower shaft as required until engine idles at 700-900 rpm in NEUTRAL.

**Choke Shutter Synchronization
(3-cylinder Engines)**

Refer to **Figure 56** for this procedure.

1. Loosen the screw on each choke swivel to free the choke link.

2. Adjust top end of link about 1/16-1/8 in. beyond the top of the choke swivel and tighten the screw to hold the link in that position.

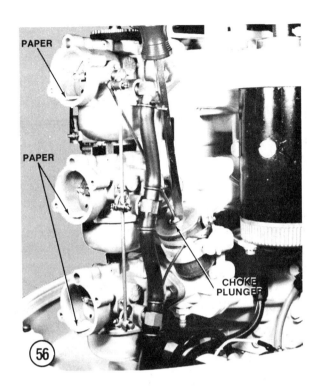

PAPER

PAPER

CHOKE PLUNGER

56

3. Close the top and middle carburetor choke shutters completely and hold in that position. Tighten middle choke swivel screw.

4. Close bottom carburetor choke completely and tighten bottom choke swivel screw.

5. Loosen the 2 solenoid attaching screws just enough to permit the solenoid to be moved.

6. Slip a small piece of writing paper about 1/2 in. wide into each carburetor air horn. The paper will provide sufficient clearance to prevent the choke shutters from sticking closed.

7. Depress the solenoid plunger (not the rod) to close the choke shutters. Slide the solenoid against the plunger and tighten the solenoid attaching screws.

8. Release the solenoid plunger and push plunger into solenoid. Pull the papers from the air horn. There should be a slight drag as the papers are removed. If drag is excessive, repeat Steps 5-7.

9. Release the solenoid plunger. If the choke shutters are not horizontal, bend the shutter stop groove pin in the carburetor air horn(s) as required to bring the shutters into a horizontal position.

10. Install the carburetor intake cover with new gaskets.

11. Install the engine cover.

Choke Shutter Synchronization (4-cylinder Engines)

The choke shutters are synchronized as described under *Chrysler outboards 35 and 45 hp (1966-on), 50 hp (1966 and 1981-on) and 55 hp (1967-1976)* in this chapter.

70-140 HP (MOTOROLA ELECTRONIC DISTRIBUTOR IGNITION)

Engine Timing Preparation

1. Disconnect the negative battery lead.
2. Remove the engine cover.
3. Remove all spark plugs from the power head. See Chapter Four.
4. Check position of timing decal on flywheel ring gear. The TDC and 36° marks on the decal must align with the top of the 0° and 36° flywheel lines. If they do not, align and install a new decal.
5. Thread timing tool barrel (part No. T 2937-1) into the top cylinder spark plug hole and screw it in completely.

NOTE
The 85 hp high performance engine has a short crankshaft throw. Top dead center must be found with a dial indicator instead of the timing tool described below. After locating TDC, zero the dial indicator and rotate the flywheel counterclockwise until the indicator gauge reads 0.266 in. Proceed with Step 10.

5

6. Install the timing tool rod in the barrel with its 2 identification marks ("25-55 hp") toward the outside.

7. While holding the rod tightly against the piston, rotate the crankshaft clockwise to locate top dead center (TDC). This is the point at which the marks on the timing tool rod are at a maximum distance from the barrel. See A, **Figure 29**.

8. Thread the timing tool barrel in or out as required to align the inner timing rod mark with the edge of the barrel. See B, **Figure 29**.

NOTE
If the crankshaft is rotated too far in Step 9, the outer timing rod mark will go beyond the edge of the timing tool barrel. If this happens, repeat the step—do not try to position the rod mark by rotating the crankshaft counterclockwise.

9. Apply slight finger pressure on the end of the rod to prevent it from shooting out during the piston compression stroke. Rotate crankshaft clockwise almost one full turn until the outer timing rod mark aligns with the edge of the barrel. See **Figure 30**. The piston is now positioned exactly 32° before TDC.

10. Make sure the 32 degree mark on the flywheel ring gear decal aligns with the timing pointer index line. See **Figure 57**. If it does not, loosen the timing pointer screws, align the pointer index line with the decal mark and tighten the pointer screws.

11. Rotate the flywheel to align the 0 degree mark with the timing pointer index line.

12. Carefully pry drive belt off distributor pulley.

13. Rotate the distributor pulley to align the pulley index mark with the outer diameter of the flywheel. See **Figure 45**.

14. Install drive belt over distributor pulley. Position the tip of a 0.008 in. feeler gauge against the mid-point of the belt length. Push

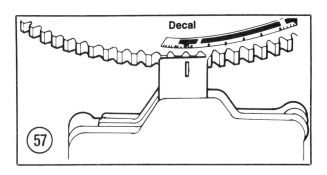

Decal

(57)

against the belt until the feeler gauge bends without moving the belt further. This should provide the desired 3/16-1/4 in. belt deflection. See **Figure 46**.

15. Check the length of the control rod between the tower shaft swivel and distributor. If it is not between 3/8-1/2 in., adjust the nylon swivels as required. See **Figure 49**.

16. Perform *Static Timing* or *Dynamic Timing* as described in this chapter.

Static Timing

1. Disconnect the white/black stripe wire from the distributor housing terminal.

2. Move the tower shaft to the wide-open throttle position.

3. Rotate the flywheel clockwise to align the timing decal 32 degree mark with the timing pointer index mark.

4. Connect a voltmeter between the distributor blue wire and white/black wire terminals as shown in **Figure 58**.

5. Connect the negative battery lead and turn the ignition switch to the ON or RUN position without starting the engine.

6. Loosen the distributor control rod locknut (**Figure 49**). Rotate the control rod until the voltmeter needle moves from 0 volts to 12 volts. Hold control rod in this position and tighten locknut.

7. Lightly depress distributor drive belt several times while watching the meter scale. The needle should deflect each time the belt is depressed. If it does not, repeat Step 6.

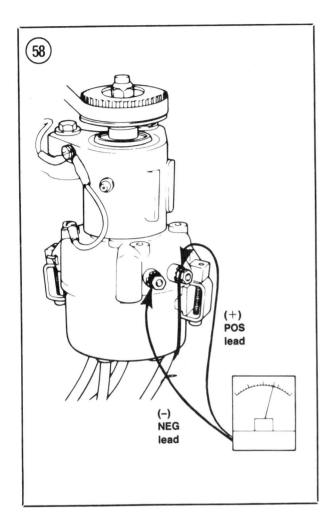

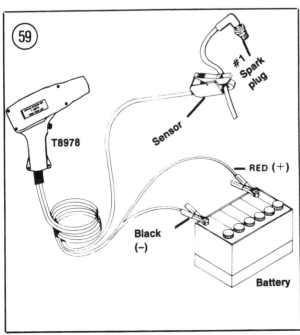

Dynamic Timing

1. Reconnect the negative battery lead and reinstall the spark plugs.

2. Connect a timing light as shown in **Figure 59**.

3. With the engine in a test tank or on the boat in the water, start the engine and accelerate to wide-open throttle.

4. Point the timing light at the flywheel and timing pointer and check timing mark alignment. If the 32 degree mark on the decal does not align with the timing pointer index mark, retard the engine and shut it off.

5. Loosen the control rod locknut (**Figure 49**) and adjust the control rod clockwise to retard or counterclockwise to advance the timing.

6. Restart the engine and repeat Step 4. If timing is correct, shut the engine off and tighten the control rod locknut. If timing is still incorrect, repeat the procedure as required until the marks align, then tighten the locknut.

Carburetor Adjustment, Throttle and Choke Shutter Synchronization

Adjust the carburetors and synchronize the throttle and choke shutters as described under *70-135 hp (Distributor Breaker Point Ignition)* in this chapter.

55, 125 AND 140 HP (PRESTOLITE CD IGNITION)

Engine Timing

The engine should be in a test tank or on the boat in the water for this procedure.

1. Remove the engine cover.

2. Connect a timing light according to manufacturer's instructions.

3. Move the throttle to the full retard position and crank the engine while pointing the timing light at the flywheel and carburetor flange adapter timing marks. The TDC mark

on the flywheel should align with the adapter timing mark ±2°.

> *WARNING*
> *Do not attempt to adjust the timing link with the engine running. It is close to the flywheel ring gear and serious personal injury can result.*

4. If the timing marks do not align as specified in Step 3, loosen the timing link locknut (**Figure 60**) and turn the adjusting screw clockwise to retard or counterclockwise to advance the timing. Tighten the locknut and repeat Step 3.

> *NOTE*
> *All 1978-1981 140 hp models were factory-timed at 32° BTDC. Due to variations in fuel quality and octane rating, Chrysler Marine changed the timing to 30° BTDC on 1982 models and recommends that the timing on all previous 140 hp models be reset from 32° to 30°.*

5. When the timing marks align in Step 3, start the engine and run at 4,500-5,500 rpm. Check timing mark alignment with the timing light. The 32 degree (55 and 125 hp) or 30 degree (140 hp) mark should align with the timing pointer index mark.

6. If the timing marks do not align as specified in Step 5, shut the engine OFF. Loosen the timing link locknut (**Figure 60**) and turn the adjusting screw clockwise to

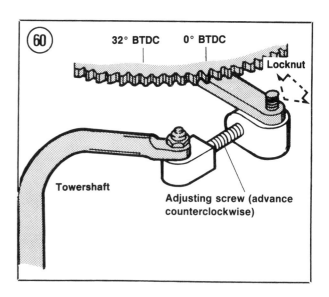

retard or counterclockwise to advance the timing.

7. Repeat Step 5 and Step 6 as required. When the timing is correct, shut the engine off and tighten the locknut.

Carburetor Adjustment, Throttle and Choke Shutter Synchronization

Adjust the carburetors and synchronize the throttle and choke shutters as described under *Chrysler outboards 55 hp (1977-1980), 60 hp (1974-1976), 65 hp (1977-1978) and 105, 120 and 135 hp (1975-1977) with Magnapower II Ignition* in this chapter.

Table 1 MANUAL ENGINE SERIAL NUMBERS

U.S.A. Model No.	Serial No.	Canada Model No.	Serial No.
352HA	3001	352BA	3027
353HA	3001	353BA	3022
452HA	3001	452BA	3006
453HA	3001	453BA	3006

5

Chapter Six

Fuel System

This chapter contains removal, overhaul, installation and adjustment procedures for fuel pumps, carburetors, reed valves, fuel tanks and connecting lines used with all outboards covered in this book. **Table 1** and **Table 2** are at the end of the chapter.

FUEL PUMP

All outboards equipped with an integral fuel tank use a gravity flow fuel system and require no fuel pump. The diaphragm-type fuel pump used on models with a remote fuel tank is operated by crankcase pressure. Since this type of fuel pump cannot create sufficient pressure to draw fuel from the tank during cranking, fuel is transferred to the carburetor for starting by operating the primer bulb installed in the fuel line.

A single-stage diaphragm displacement pump is used on 4.4-20 hp models. Pressure pulsations created by movement of the pistons reach the fuel pump through a passageway between the crankcase and pump. Changes in the crankcase pressure of the No. 1 cylinder cause the pump diaphragm to flex back and forth, transmitting the pressure to an inlet and outlet reed mounted on a plate inside the pump.

Upward piston motion creates a low pressure on the pump diaphragm. This low pressure opens the inlet check valve in the pump, drawing fuel from the line into the pump. At the same time, the low pressure draws the air-fuel mixture from the carburetor into the crankcase.

Downward piston motion creates a high pressure on the pump diaphragm. This pressure closes the inlet check valve and opens the outlet check valve, forcing the fuel into the carburetor and drawing the air-fuel mixture from the crankcase into the cylinder for combustion. **Figure 1** shows the operational sequence of a typical single stage fuel pump.

A 2-stage diaphragm displacement pump (designated Type A or Type B according to design) is used on all other engines. Pump operation is essentially the same. Vacuum and pressure pulsations from one cylinder bring fuel into the pump while similar pulsations from another cylinder deliver the fuel to the carburetors.

On 140 hp engines equipped with 2 fuel pumps, the lower pump is operated by the No. 3 and No. 4 cylinders. It delivers fuel to the upper pump (operated by the No. 1 and No. 2 cylinders) where it is sent to the carburetors.

Figure 2 shows the operational sequence of a typical 2-stage fuel pump used on a 3-cylinder engine.

The fuel pumps used on Chrysler outboards are self-contained units mounted on the power head. Pump design is extremely simple and reliable in operation. Diaphragm failures are the most common problem, although the use of dirty or improper fuel-oil mixtures can cause check valve problems.

Removal/Installation (Single-stage Pump)

1. Disconnect the fuel line at the support plate quick-disconnect fitting.
2. Remove the screw holding the stator ground wire to the pump cover, if so equipped.

6

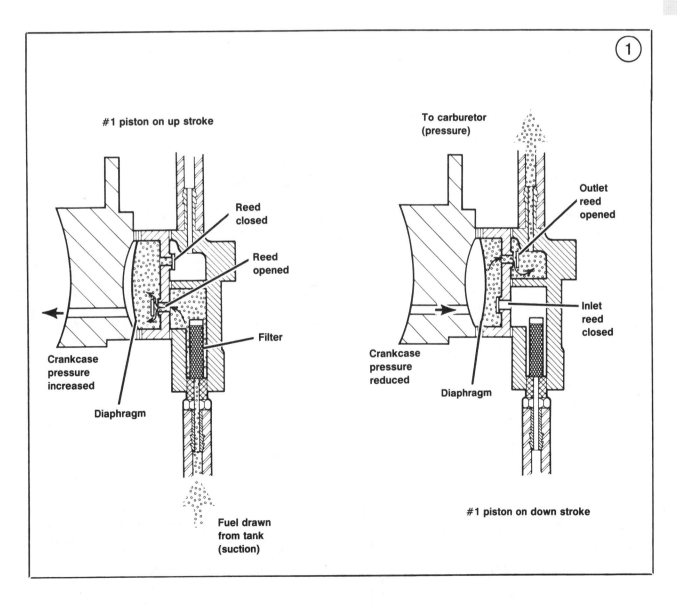

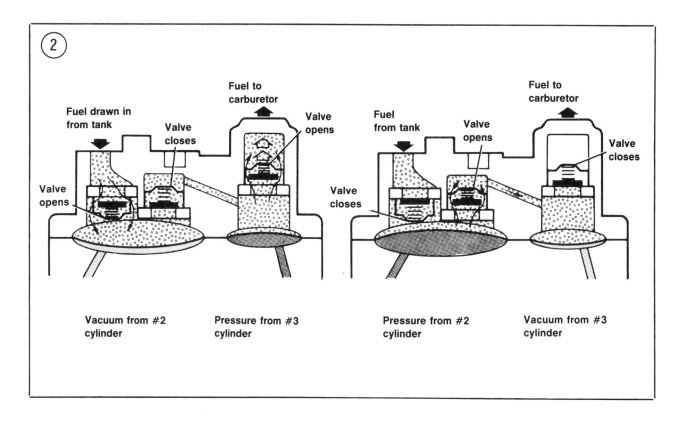

Fuel drawn in from tank — Valve closes — Fuel to carburetor — Valve opens — Valve opens — Vacuum from #2 cylinder — Pressure from #3 cylinder

Fuel from tank — Valve opens — Fuel to carburetor — Valve closes — Valve closes — Pressure from #2 cylinder — Vacuum from #3 cylinder

3. 9.9-15 hp electric start with CD ignition:
   a. Disconnect the negative battery lead.
   b. Remove the coils (A, **Figure 3**).
   c. Remove the starter mounting screws (B, **Figure 3**). Place starter on support plate.
   d. Remove starter support from power head support bracket.
   e. If a ratchet wrench with screwdriver bit is not available for use in Step 4, remove the starter switch and bracket.

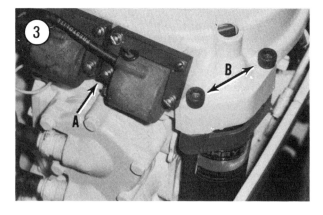

> *NOTE*
> *An offset screwdriver may be required in Step 4 to remove the pump cover screws on some models.*

4. Remove the 3 screws holding the pump cover to the power head or transfer port cover (**Figure 4**).

5. Compress the fuel line fitting clamp with hose clamp pliers and slide the clamp off the fitting. Repeat this step to disconnect the remaining fuel line from the pump cover.

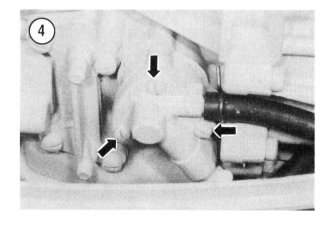

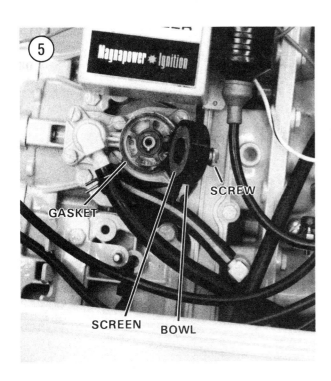

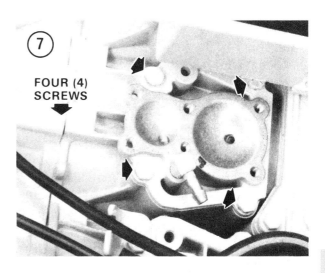

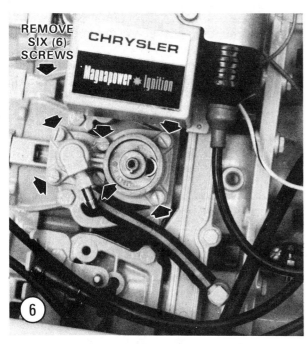

6. Remove the reed plate and pump diaphragm from the power head.

7. Remove the filter from the pump cover.

8. Clean all gasket residue from the engine mounting pad. Work carefully to avoid gouging or damaging the mounting surface.

9. Installation is the reverse of removal. Sandwich the diaphragm between new gaskets. Make sure diaphragm is not wrinkled and extends beyond the gasket on all sides.

Removal/Installation
(Type A 2-stage Pump)

1. Disconnect the fuel line at the support plate quick-disconnect fitting.

2. Unscrew sediment bowl screw. Remove sediment bowl and filter screen. See **Figure 5**.

3. Compress the fuel line fitting clamp with hose clamp pliers and slide the clamp off the fitting. Repeat this step to disconnect the remaining fuel line from the pump cover.

4. Remove the fuel pump body screws and lockwashers. See **Figure 6** (typical). Remove the pump body.

5. Remove the fuel line and impulse hose fitting from the fuel pump cover.

6. Remove the fuel pump cover screws. See **Figure 7** (typical). Remove the cover and gasket.

7. Repeat Steps 2-6 to remove the second pump on 140 hp engines.

8. Clean all gasket residue from the engine mounting pad. Work carefully to avoid gouging or damaging the mounting surface.

9. Installation is the reverse of removal. Use new gaskets. Apply EC-750 sealant to impulse hose and fuel line fittings and tighten securely. Install filter screen in sediment bowl with turned edge facing engine (**Figure 8**). Install new pump body gasket with slot over pump body key (**Figure 9**). Position sediment bowl with hose fitting as shown in **Figure 10**.

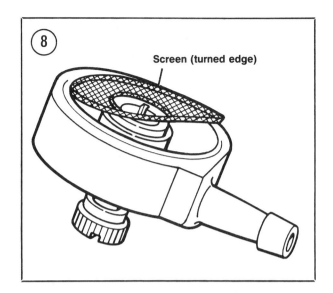

Removal/Installation (Type B 2-stage Pump)

Refer to **Figure 11** for this procedure.

1. Compress the fuel inlet line fitting clamp with hose clamp pliers and slide the clamp off the fitting. Repeat this step to disconnect the impulse hose from the pump cover.

2. Remove the 3 screws holding the fuel pump bracket to the power head.

3. Remove the 2 fittings from the pump cover.

4. Remove the screws holding the pump cover to the pump body. Remove the cover, gasket and diaphragm.

5. Remove the screws holding the pump body to the bracket. Remove the body.

6. Loosen the knurled nut on the pump wire bail, swing bail from under sediment bowl and remove bowl, gasket and filter screen.

7. Installation is the reverse of removal. Use new gaskets. Apply EC-750 sealant to impulse hose and fuel line fittings and tighten securely.

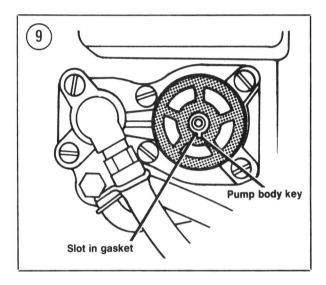

Disassembly/Assembly (Single-stage Pump)

The fuel pump is completely disassembled during removal. Proceed to *Cleaning and Inspection (Single-stage Pump)* in this chapter.

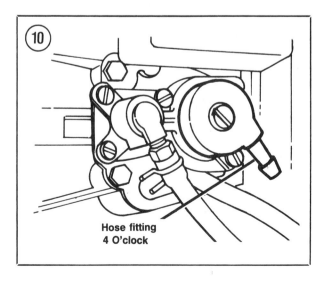

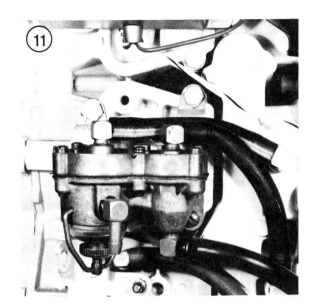

Disassembly/Assembly (2-stage Pumps)

Figure 12 shows the Type A pump; the Type B pump body is the same design.

1. Remove the 2 screws holding the middle valve to the pump body. See **Figure 13**. Remove the valve and gasket from the body.

NOTE
To make a hooked tool for use in Step 2, use a length of 16 gauge wire and a pair of needlenose pliers.

2. Remove the pressed-in second-stage valve with a suitable hooked tool.

6

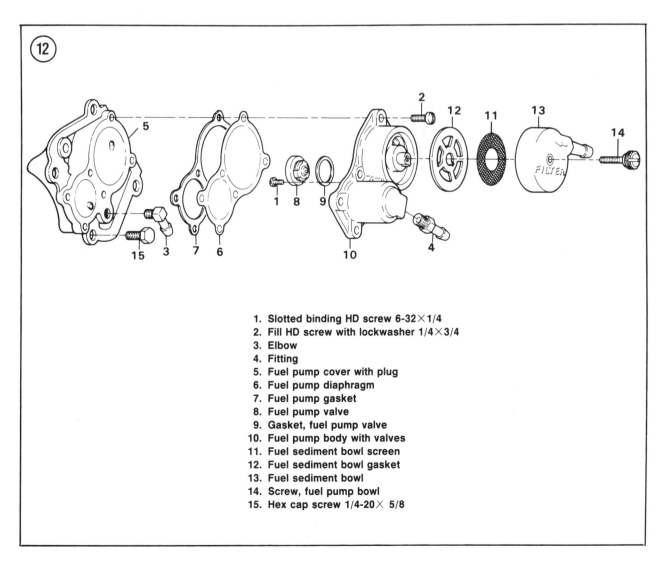

1. Slotted binding HD screw 6-32×1/4
2. Fill HD screw with lockwasher 1/4×3/4
3. Elbow
4. Fitting
5. Fuel pump cover with plug
6. Fuel pump diaphragm
7. Fuel pump gasket
8. Fuel pump valve
9. Gasket, fuel pump valve
10. Fuel pump body with valves
11. Fuel sediment bowl screen
12. Fuel sediment bowl gasket
13. Fuel sediment bowl
14. Screw, fuel pump bowl
15. Hex cap screw 1/4-20× 5/8

3. Remove the pressed-in first-stage valve by inserting a pin punch in the body hole behind the valve and tapping it lightly with a hammer.

4. Install a new center valve gasket in the pump body, then insert the valve spring end first (**Figure 14**). Install and tighten retaining screws.

5. Press first- and second-stage valves in pump body until they bottom out. Make sure valves are positioned as shown in **Figure 13**.

Cleaning and Inspection
(Single-stage Pump)

1. Clean all metal parts in solvent and blow dry with compressed air.

2. Check filter screen for damage. Replace as required.

3. Hold diaphragm up to a strong light source and check for pin holes, breaks or excessive stretching. Replace if any defects are found.

4. Check fuel pump reeds. If there are signs of warpage, tension or curling of the reed ends (**Figure 15**), install new reeds.

Cleaning and Inspection
(2-stage Pump)

1. Clean all metal parts (and glass sediment bowl on Type B pumps) in solvent. Dry housing with compressed air. Let check valves air dry.

2. Check filter screen for damage. Replace as required.

3. Hold diaphragm up to a strong light source and check for pin holes, breaks or excessive stretching. Replace if any defects are found.

4. Inspect check valves for warpage and spring tension. Replace any valve that is slightly warped, has weak tension or broken springs.

5. Check housing condition. Make sure the valve seats provide a flat contact area for the valve disc. Replace housing if cracks or rough gasket mating surfaces are found.

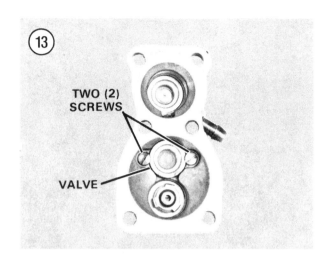

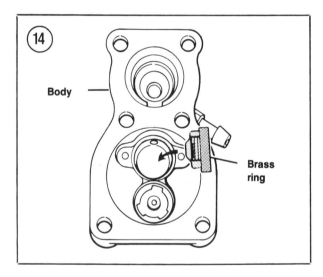

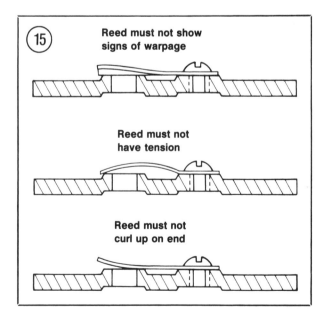

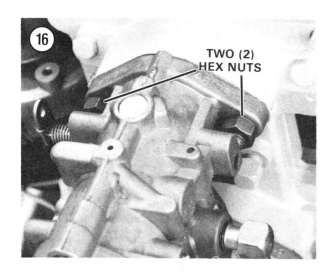

CARBURETORS

Chrysler outboards use a variety of Tillotson and Walbro carburetors. All operate essentially the same, but housing shape and design vary slightly according to engine size. Some Tillotson carburetors do not use an idle tube.

When removing and installing a carburetor, make sure that the mounting nuts (**Figure 16**) are securely tightened. A carburetor that is loose will cause a lean-running condition.

Before removing and disassembling any carburetor, be sure you have the correct overhaul kit, the proper tools and a sufficient quantity of fresh cleaning solvent. Chrysler/U.S. Marine dealers offer a gasket kit and a repair kit (which includes the gasket kit) for each carburetor.

CARBURETOR REMOVAL/INSTALLATION

3.5 hp (1966-1969)

1. Disconnect the fuel line from the shut-off valve. Plug the line to prevent fuel from draining out of the tank.
2. If equipped with a Tillotson carburetor, remove the idle screw adjusting knob.
3. Remove the 2 screws holding the control panel to the carburetor, if so equipped.
4. Remove the 2 hex nuts holding the carburetor to the reed plate. See **Figure 17**.
5. Loosen the swivel screw on the throttle bellcrank and remove the brass throttle link.
6. Remove the carburetor and gasket.
7. Installation is the reverse of removal. Use a new gasket. Adjust the carburetor (Chapter Five).

3.6-4 hp (1970-1978) and 4.9-5 hp

1. Remove the front and rear engine cover screws (1970) or rear cover screw (1971-on).
2. Remove the fuel tank filler cover, chain and anchor as an assembly. Remove the fuel tank filler neck grommet.
3. Lift the engine cover part way off and tilt forward. Tie a knot in the starter rewind rope inside the engine cover eyelet and rewind starter. See **Figure 18**.

6

4. Pry the starter rope retainer from the starter handle. Unthread rope from retainer and handle.

5. Remove the engine cover.

6. Remove the fuel tank as described in this chapter.

7. Remove the choke rod retaining clip. Disconnect choke rod from carburetor arm (**Figure 19**).

8. Move throttle control handle to full retard position. Remove 2 hex nuts holding carburetor to adapter plate. Remove the carburetor and gasket.

9. Slide fuel line clamp from fuel fitting with hose clamp pliers and pull carburetor free of fuel line.

10. Installation is the reverse of removal. Use a new gasket. Adjust the carburetor (Chapter Five).

3.5 and 4 hp
(1979-on)

1. Drain and remove the fuel tank as described in this chapter.

2. Slide fuel line clamp from fuel fitting with hose clamp pliers and pull fuel line from carburetor.

3. Pull knob from fuel shut-off valve. Remove valve from carburetor with an open-end wrench.

4. Remove the choke rod retaining clip and washer. Loosen choke link-to-knob screw and disconnect link from choke shaft.

5. Remove the 2 hex nuts holding the carburetor to the adapter plate. Remove the carburetor and gasket.

6. Installation is the reverse of removal. Use a new gasket. Adjust the carburetor (Chapter Five).

4.4-8 hp (1968-1979)

1. Remove the flywheel. See Chapter Eight.

2. Remove choke rod retaining clip. Disconnect rod from carburetor arm.

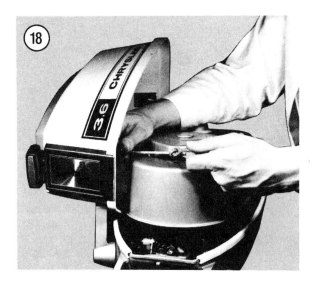

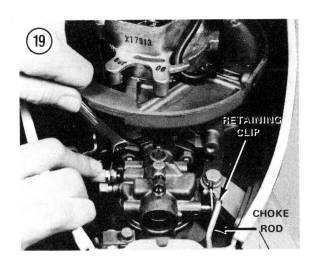

3. Loosen the hex screw holding the throttle cam follower arm to follower. See **Figure 20**. Slide follower out and disconnect the arm from the carburetor.

4. Remove 2 hex nuts holding carburetor to adapter plate. Remove the carburetor and gasket.

5. Slide fuel line clamp from fuel fitting with hose clamp pliers and pull carburetor free of fuel line.

6. Installation is the reverse of removal. Use a new gasket. Route MD series carburetor fuel line under the shift linkage before connecting line to carburetor. Install brass

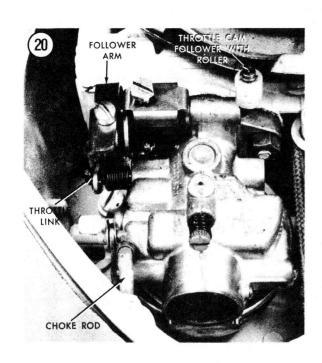

throttle link in inner hole (CO series carburetor) or outer hole (MD series carburetor) of follower arm with link ends facing carburetor. Adjust the carburetor (Chapter Five).

6-7.5 hp (1980-on)

1. Remove the choke rod retaining clip and washer.
2. Slide fuel line clamp from fuel pump fitting with hose clamp pliers and disconnect line at pump.
3. Remove the 2 hex nuts holding the carburetor to the intake manifold.
4. Slide carburetor forward on manifold studs and disconnect throttle shaft link.
5. Remove the carburetor and gasket.
6. Installation is the reverse of removal. Use a new gasket. Adjust the carburetor (Chapter Five).

9.2 hp

1. Slide fuel line clamp from fuel fitting with hose clamp pliers and pull fuel line from carburetor.

2. Remove the idle adjustment arm from the carburetor idle adjustment needle.
3. Disconnect the choke rod at the choke shaft.
4. Remove 2 hex nuts holding carburetor to adapter. Remove the carburetor and gasket.
5. Installation is the reverse of removal. Use a new gasket. Adjust the carburetor (Chapter Five).

9.9-15 hp
(1972-on)

1. Remove the flywheel. See Chapter Eight.
2. 9.9-15 hp electric start with CD ignition:
   a. Disconnect the negative battery lead.
   b. Remove the coils (A, **Figure 3**).
   c. Remove the starter mounting screws (B, **Figure 3**). Place starter on support plate.
   d. Remove starter support from power head support bracket.
   e. Remove the starter or neutral interlock switch as required.
3. Remove the choke rod retaining clip. Disconnect choke rod from carburetor arm.
4. Loosen the hex screw holding the throttle cam follower arm to follower. See **Figure 20**. Slide follower out and disconnect the arm from the carburetor.
5. Remove 2 hex nuts holding carburetor to adapter plate. Remove the carburetor and gasket.
6. Slide fuel line clamp from fuel fitting with hose clamp pliers and pull carburetor free of fuel line.
7. Installation is the reverse of removal. Use a new gasket. Adjust the carburetor (Chapter Five).

20 hp
(1966-1979)

1. Disconnect the choke rod at the choke shaft.

6

2. Remove the screw holding the throttle cam follower retainer to the cylinder block. See **Figure 21**.

3. Disconnect the throttle link at the throttle shaft.

4. Remove the starter spool. See Chapter Ten.

5. Slide fuel line clamp from fuel fitting with hose clamp pliers and pull fuel line from carburetor.

6. Remove 2 hex nuts holding carburetor to adapter plate. Remove the carburetor and gasket.

7. Installation is the reverse of removal. Use a new gasket. Adjust the carburetor (Chapter Five).

20-55 hp
(Single Carburetor)

1. On manual start models, remove the choke rod retaining ring and washer. Disconnect choke rod from choke lever. See **Figure 22**. If equipped with a choke cable instead of a rod, loosen the cable swivel screw and disconnect cable from swivel. See **Figure 23**.

2. On electric start models, loosen the choke solenoid rod screw and disconnect the rod from the choke lever. See **Figure 24**.

3. Slide fuel line clamp from fuel fitting with hose clamp pliers and pull fuel line from carburetor.

4. Remove 2 hex nuts holding carburetor to adapter plate. Remove the carburetor and gasket.

5. Installation is the reverse of removal. Use a new gasket. Adjust the carburetor (Chapter Five).

55-65 hp
(Dual Carburetor)

1. Remove the air silencer and gaskets.

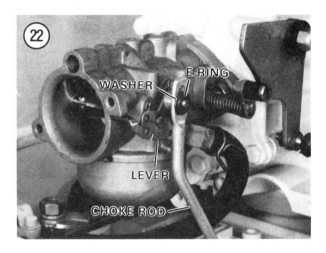

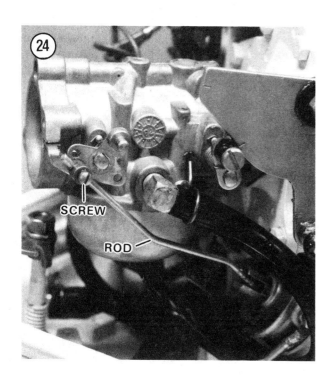

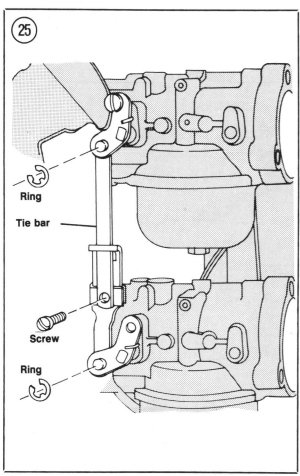

2. Slide fuel line clamp from bottom carburetor fuel fitting with hose clamp pliers and pull fuel line from carburetor.

3. Disconnect throttle link at tower shaft.

4. Remove throttle tie bar retaining clips (**Figure 25**).

5. Remove and discard the choke link cotter pins. Disconnect choke rod and choke link.

6. Remove the 2 hex nuts holding each carburetor to adapter plate. Remove the carburetors and gaskets.

7. Installation is the reverse of removal. Use new gaskets. Adjust the carburetors (Chapter Five).

70-85 hp

Top carburetor

1. Remove carburetor intake cover and gaskets.

2. Remove retaining rings holding throttle tie bar to top and middle carburetor arms. Disconnect tie bar. See **Figure 26**.

3. Loosen all 3 choke arm swivel screws. Let choke link slide down.

4. Remove and discard cotter pin holding choke rod and swivel to top lever. See **Figure 27**.

5. Remove washer, choke rod and 2 O-rings from swivel, then remove the swivel.

6. Slide fuel line clamp from carburetor fuel fitting with hose clamp pliers and pull fuel line from carburetor.

7. Remove the 2 hex nuts holding the carburetor to the adapter flange. Remove the carburetor and gasket.

8. Installation is the reverse of removal. Use new gaskets and cotter pin. If fuel line fitting was removed from carburetor, coat threads with EC-750 sealant and reinstall with barb at 6 o'clock position. Adjust the carburetor (Chapter Five).

6

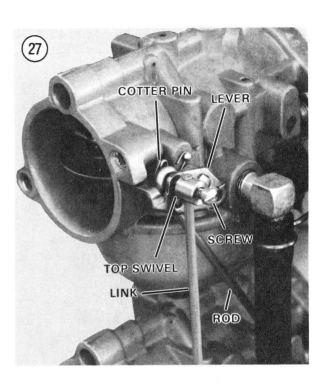

8. Installation is the reverse of removal. Use new gaskets and cotter pin. Adjust the carburetor (Chapter Five).

Middle carburetor

1. Remove carburetor intake cover and gaskets.

2. Remove retaining rings holding throttle tie bar to top and middle carburetor arms. Disconnect tie bar. See **Figure 26**.

3. Loosen all 3 choke arm swivel screws. Remove choke link.

4. Remove and discard cotter pin holding choke rod and swivel to middle lever. See **Figure 28**.

5. Remove washer, 2 O-rings and swivel from middle lever.

6. Slide fuel line clamp from each barb on fuel inlet fitting (**Figure 29**) and disconnect the hoses. Remove the barb.

7. Remove the 2 hex nuts holding the carburetor to the adapter flange. Remove the carburetor and gasket.

Bottom carburetor

1. Remove carburetor intake cover and gaskets.

2. Disconnect throttle link at tower shaft ball-joint.

3. Remove retaining rings holding throttle tie bar to middle and bottom carburetor arms. Disconnect tie bar. See **Figure 29**.

4. Loosen all 3 choke arm swivel screws. Pull choke link up and out of the way.

5. Remove and discard cotter pin holding choke rod and swivel to bottom lever.

6. Remove washer, 2 O-rings and swivel from bottom lever.

7. Slide fuel line clamp from carburetor fuel inlet fitting and disconnect the hose. Remove the fitting.

8. Remove the 2 hex nuts holding the carburetor to the adapter flange. Remove the carburetor and gasket.

9. Installation is the reverse of removal. Use new gaskets and cotter pin. Adjust the carburetor (Chapter Five).

90-140 hp

1. Remove carburetor cover(s) and gaskets.
2. Slide fuel line clamp from bottom carburetor fuel inlet fitting and disconnect the hose.

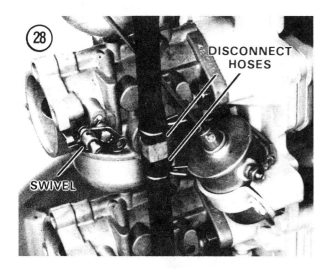

3. Disconnect throttle link at tower shaft.
4. Remove retaining ring holding each throttle arm to the throttle tie bar. See **Figure 30**.
5. Loosen each choke swivel screw and remove the choke link.
6. Remove and discard the cotter pin from each choke swivel. Disconnect the choke rod and remove the swivels, washers and O-rings.
7. Remove the 2 hex nuts holding each carburetor to the adapter flange. Remove the carburetors and gaskets.
8. Installation is the reverse of removal. Use new gaskets and cotter pins. Adjust the carburetors (Chapter Five).

6

CARBURETOR DISASSEMBLY/ASSEMBLY

Work slowly and carefully, follow the disassembly procedures and refer to the exploded drawing of your carburetor when necessary. When referring to the exploded

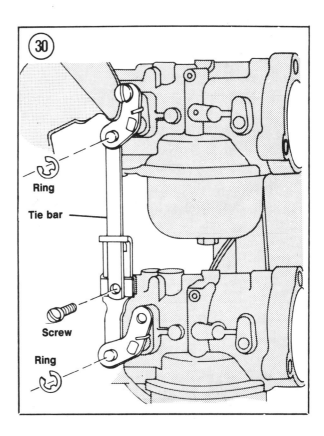

drawing, note that not all carburetors will use all of the components shown. Do not apply excessive force at any time.

It is not necessary to disassemble the carburetor linkage or remove the throttle cam or other external components. Remove the throttle or choke plate only if it is damaged or binds.

Refer to **Figure 31**, **Figure 32** or **Figure 33** for this procedure.

1. Turn idle speed adjustment needle clockwise, counting the number of turns required to lightly seat it. Back needle out and remove needle and spring from the carburetor. See **Figure 34**.

2A. MD series carburetor—Remove the 4 screws holding the fuel bowl to the carburetor body. Remove the fuel bowl and gasket. Discard the gasket.

2B. All others—Remove the hex head screw at the bottom center of the fuel bowl (**Figure 35**). Remove the fuel bowl, 2 bowl screw gaskets and fuel bowl gasket. Discard the gaskets.

3A. Tillotson carburetors—Remove the main fuel jet (**Figure 36**).

3B. Walbro carburetors—Remove the main fuel nozzle and compression spring, if so equipped. See **Figure 37** or **Figure 38**.

4A. MD series carburetors—Unscrew float lever pin from fuel bowl.

4B. All others—One end of float hinge pin is knurled. Push the smooth end of the pin through the float hinge bores with a small punch and remove with needlenose pliers.

5A. MD series carburetors—Remove fuel bowl plug screw and slide inlet needle from seat. Remove seat and gasket from fuel bowl with a wide-blade screwdriver.

5B. Lift float straight up and remove from the carburetor. The inlet needle will come out with the float on Walbro carburetors (**Figure 39**). With Tillotson carburetors, it will remain in the valve seat (**Figure 36**).

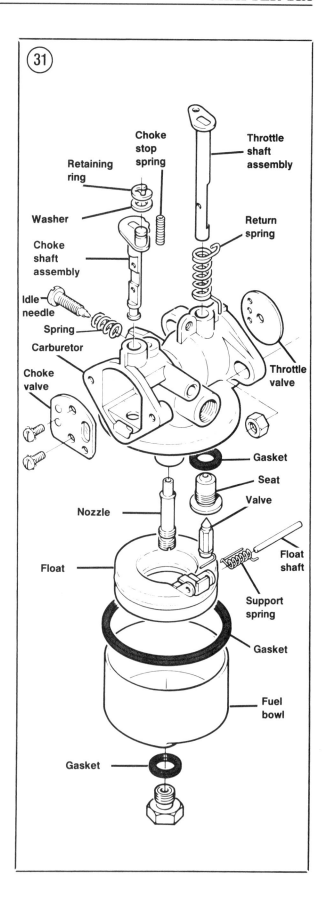

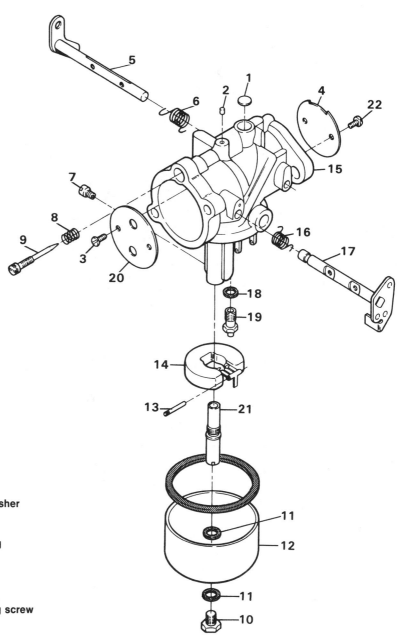

1. Body channel welch plug
2. Body channel cup plug
3. Shutter screw with lockwasher
4. Throttle shutter
5. Throttle shaft and lever
6. Throttle shaft return spring
7. Main fuel jet
8. Idle screw spring
9. Idle screw
10. Fuel bowl retaining screw
11. Gasket, fuel bowl retaining screw
12. Fuel bowl
13. Float lever pin
14. Float
15. Carburetor
16. Choke shaft return spring
17. Choke shaft and lever
18. Inlet seat gasket
19. Inlet needle, seat and gasket
20. Choke shutter
21. Main nozzle
22. Throttle shutter screw and lockwasher

6

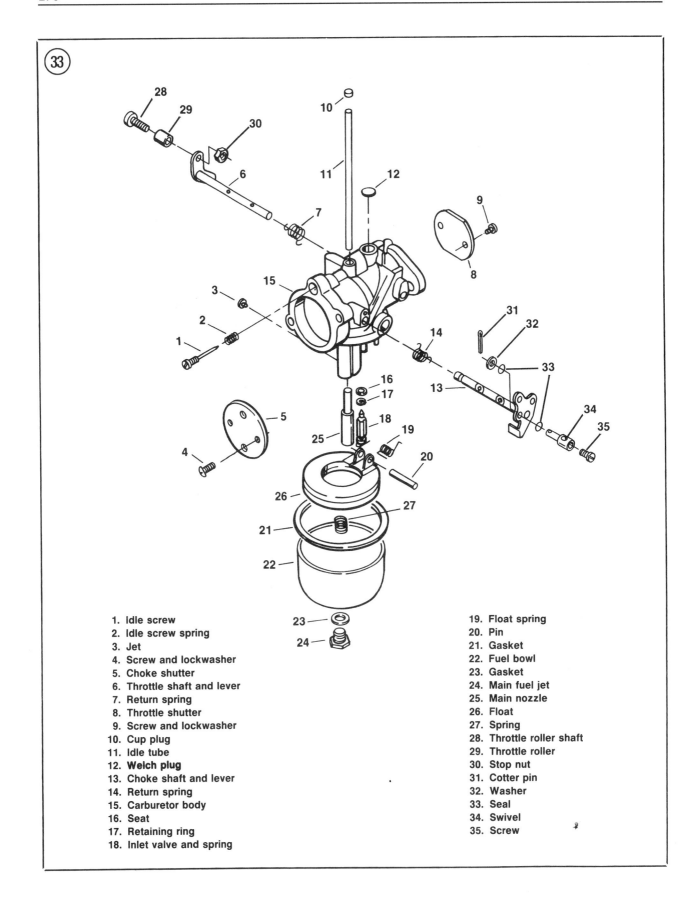

1. Idle screw
2. Idle screw spring
3. Jet
4. Screw and lockwasher
5. Choke shutter
6. Throttle shaft and lever
7. Return spring
8. Throttle shutter
9. Screw and lockwasher
10. Cup plug
11. Idle tube
12. **Welch plug**
13. Choke shaft and lever
14. Return spring
15. Carburetor body
16. Seat
17. Retaining ring
18. Inlet valve and spring

19. Float spring
20. Pin
21. Gasket
22. Fuel bowl
23. Gasket
24. Main fuel jet
25. Main nozzle
26. Float
27. Spring
28. Throttle roller shaft
29. Throttle roller
30. Stop nut
31. Cotter pin
32. Washer
33. Seal
34. Swivel
35. Screw

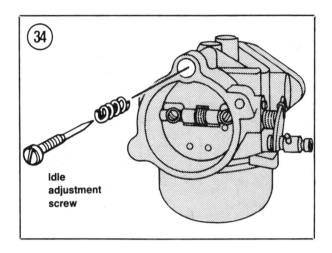

(34)

Idle
adjustment
screw

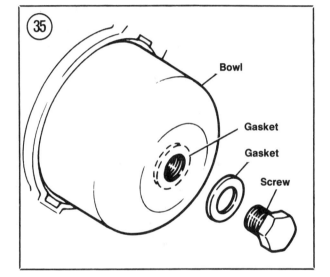

(35)

Bowl

Gasket

Gasket

Screw

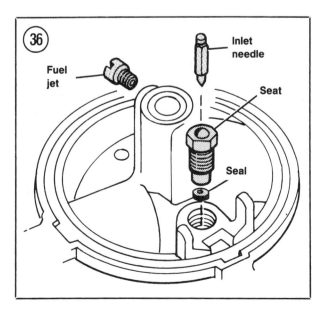

(36)

Fuel
jet

Inlet
needle

Seat

Seal

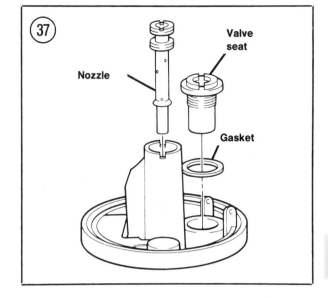

(37)

Nozzle

Valve
seat

Gasket

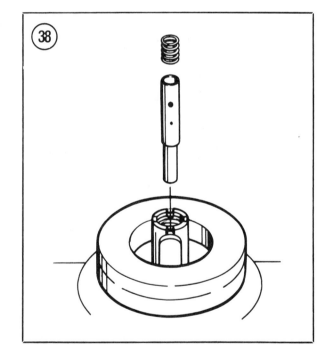

(38)

6

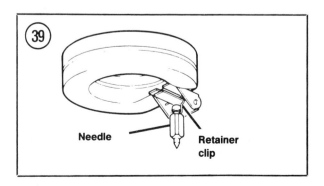

(39)

Needle

Retainer
clip

6A. Tillotson carburetors—Remove the inlet needle valve seat and gasket. See **Figure 36**.

> *NOTE*
> *A suitable hooked tool for use in Step 6B can be made from a length of wire with a pair of needlenose pliers.*

6B. Walbro carburetors—Remove the inlet seat retaining ring with a small O-ring remover. Remove the seat with the same tool. See **Figure 40**.

7. Assembly is the reverse of disassembly, plus the following:

 a. Remove any loose gasket fibers or stamping crumbs adhering to the new gaskets.

 b. On Walbro carburetors, position float spring as shown in **Figure 41** before installing float. Install float and push smooth end of hinge pin partially through float and one hinge, then move end of spring behind float tab and push the pin through the remaining hinge.

 c. On MD series carburetors, install float in bowl and align slot in float tab with inlet needle notch. Coat float lever pin threads with Gasoila and tighten pin securely.

 d. Adjust the float as described in this chapter.

 e. Tighten fuel bowl screw(s) to 20-25 in.-lb.

 f. Lightly seat needle valve. Back low-speed needle out one full turn.

 g. Reinstall on engine and adjust carburetor (Chapter Five).

CARBURETOR CLEANING AND INSPECTION

Wipe the carburetor casting and linkage with a cloth moistened in solvent to remove any contamination and operating film. Clean the carburetor castings with fresh gasoline or an aerosol type solvent and a brush. Do not

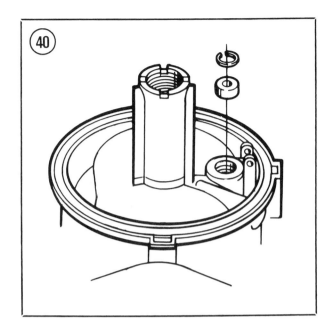

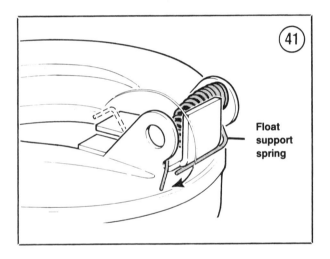

Float support spring

submerge them in a hot tank or carburetor cleaner. A sealing compound is used around the metering tubes and on the casting to eliminate porosity problems. A hot tank or submersion in carburetor cleaner will remove this sealing compound.

> *WARNING*
> *Gasoline is recommended as a suitable cleaning agent, as carburetor cleaner may remove the sealing compound applied to the casting to prevent porosity. Be extremely careful when using gasoline as it is a very real fire hazard.*

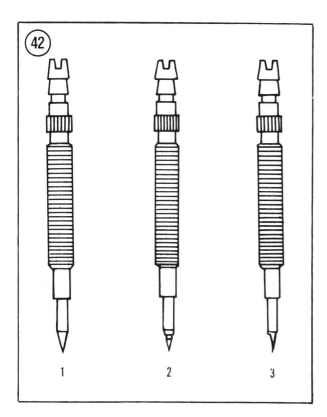

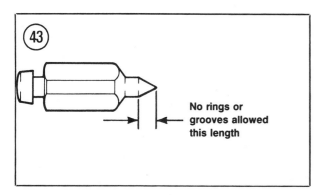

needle valve contact areas. Replace as required.

Check the idle needle tip for grooving, nicks or scratches. **Figure 42** shows a good valve tip (1), a valve tip damaged from excessive pressure when seating (2) and one with wear on one side caused by vibration resulting from the use of a damaged propeller (3).

Check the inlet needle tip (seat surface) for grooving, nicks or scratches. See **Figure 43**. Replace idle needle if tip is damaged.

Check the throttle and choke shafts for excessive wear or play. The throttle and choke valves must move freely without binding. Replace carburetor if any of these defects are noted.

Clean all gasket residue from mating surfaces and remove any nicks, scratches or slight distortion with a surface plate and emery cloth.

FLOAT ADJUSTMENT

Bend the float adjustment arm or tang carefully when adjustment is required—do *not* press down on the float. Downward pressure on the float will press the inlet needle tip into its seat and can damage the tip surface.

MD Series Carburetors

Invert fuel bowl. If top of float is not parallel with top of fuel bowl, bend the metal tab which attaches to the inlet needle as required.

OM Series Carburetors

Invert the carburetor body. The float should be perfectly level. Measure the clearance between the edge of the body casting and the float. If it is not 13/32 in. ± 1/64 in., bend the float tang as required. See A, **Figure 44**.

Spray the gasoline or cleaner on the casting and scrub off any gum or varnish with a small bristle brush. Spray the cleaner through the casting metering passages. Never clean passages with a wire or drill as you may enlarge the passage and change the carburetor calibration.

Blow castings dry with low-pressure (25 psi or less) compressed air. The use of higher pressures can damage the sealing compound.

Check the float for fuel absorption. Check the float arm for wear in the hinge pin and

Once the float level is correct, hold the carburetor upright and check the float drop. If it is not 13/32 in. ± 1/32 in., bend the short curved arm on the float as required. See B, **Figure 44**.

CO, WB and TC
Series Carburetors

Invert the carburetor body. The float should be perfectly level. Measure the distance between the edge of the body casting and the float. See **Figure 45**. If it is not 13/32 in. ± 1/64 in., bend the float tang as required. See A, **Figure 44**.

3.5-7.5 hp (1979-1984),
9.9-15 hp (1981-1984)
Carburetors

Invert the carburetor body. Weight of float will close the needle valve. Measure the distance between the float and casting (**Figure 46**). If it is not 1/8 in., bend tab A as required.

Hold the carburetor upright and measure the distance between the float and casting

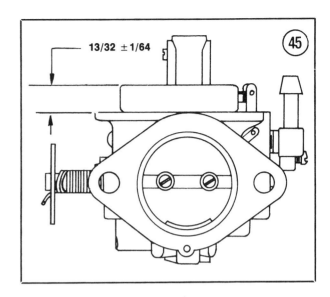

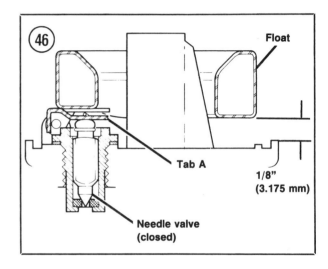

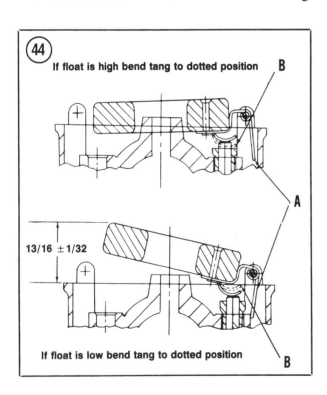

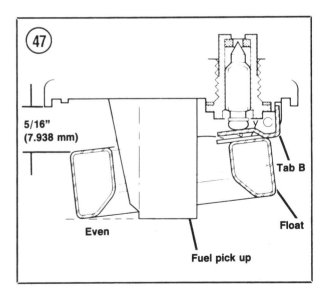

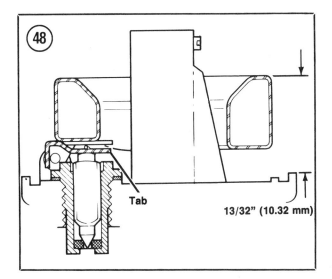

48

Tab

13/32" (10.32 mm)

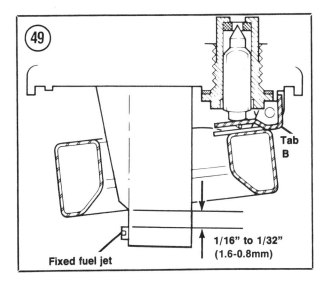

49

Tab B

Fixed fuel jet

1/16" to 1/32"
(1.6-0.8mm)

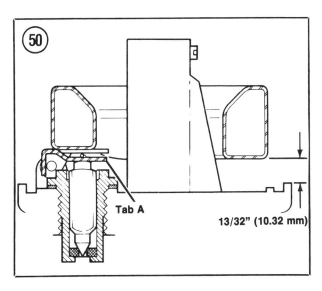

50

Tab A

13/32" (10.32 mm)

6

(**Figure 47**). If it is not 5/16 in., bend tab B as required.

9.9-15 hp (1979-1981)
Carburetors

Invert the carburetor body. Weight of float will close the needle valve. Measure the distance between the float and casting (**Figure 48**). If it is not 13/32 in., bend tab as required.

Hold the carburetor upright and measure the distance between the bottom of the float and the top of the fixed fuel jet (**Figure 49**). If it is not 1/16-1/32 in., bend tab as required.

20-30 hp (1982),
45-50 hp (1982-1984)
Carburetors

Invert the carburetor body. Weight of float will close the needle valve. Measure the distance between the float and casting (**Figure 50**). If it is not 13/32 in., bend tab as required.

Hold the carburetor upright and measure the distance between the bottom of the float and the top of the fixed fuel jet (**Figure 51**). If it is not 1/16-1/32 in., bend tab as required.

25, 35 and 55 hp (1983-1984),
90-140 hp (1983-1984 Walbro)
Carburetors

Invert the carburetor body. Weight of float will close the needle valve. Measure the

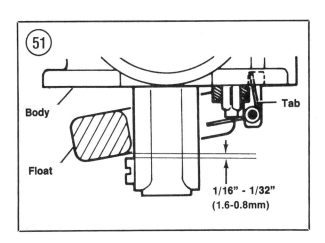

51

Body

Float

Tab

1/16" - 1/32"
(1.6-0.8mm)

distance between the float and casting (**Figure 52**). If it is not 0.150-0.190 in., bend tab 1 (**Figure 53**) as required.

Hold the carburetor upright and measure the distance between the casting and float as shown in **Figure 53**. If it is not 1.00-1.08 in., bend tab 2 as required.

55-140 hp (1979-1982), 75-140 hp (1983-on Tillotson) Carburetors

Invert the carburetor body. Weight of float will close the needle valve. Measure the distance between the float and casting (**Figure 54**). If it is not 13/32 in. ± 1/64 in., bend tab 1 (**Figure 55**) as required.

Hold the carburetor upright and measure the distance between the casting and float as shown in **Figure 55**. If it is not 1/16-1/32 in., bend tab 2 as required.

REED VALVE ASSEMBLY

The reed valve assembly is mounted on the rear of the carburetor adapter plate or intake

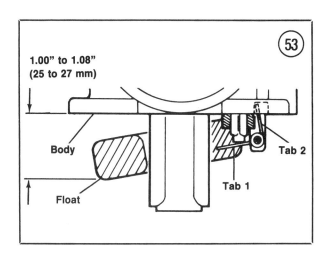

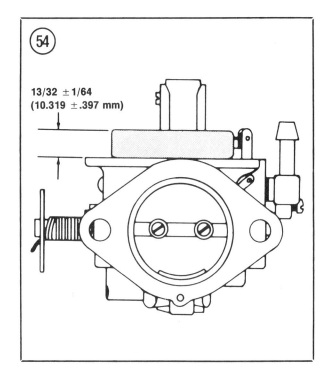

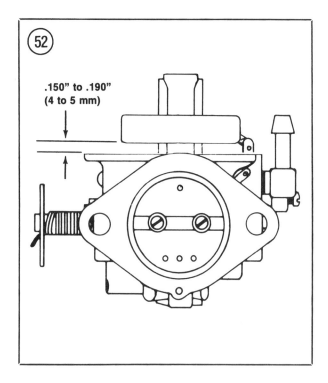

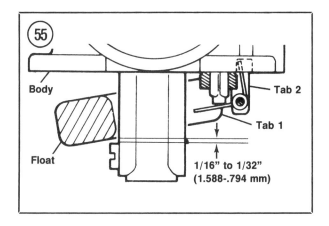

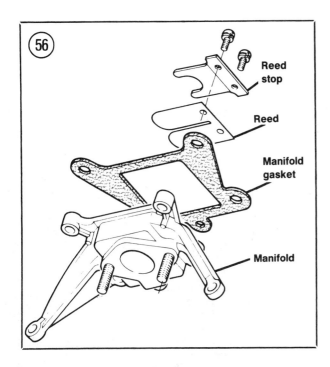

manifold. Reed valves control the passage of air-fuel mixture into the crankcase by opening and closing as crankcase pressure changes. When crankcase pressure is high, the reeds maintain contact with the reed plate to which they are attached. As crankcase pressure drops on the compression stroke, the reeds move away from the plate and allow air-fuel mixture to pass. Reed travel is limited by the reed stop. As crankcase pressure increases, the reeds return to the reed plate.

Figure 56 (1-cylinder), **Figure 57** (2-cylinder) and **Figure 58** (3-cylinder) show typical reed plate/intake manifold configurations. Note that larger engines have a deflector plate or reed plate adapter between the intake manifold/carburetor adapter and the reed valve assemblies.

6

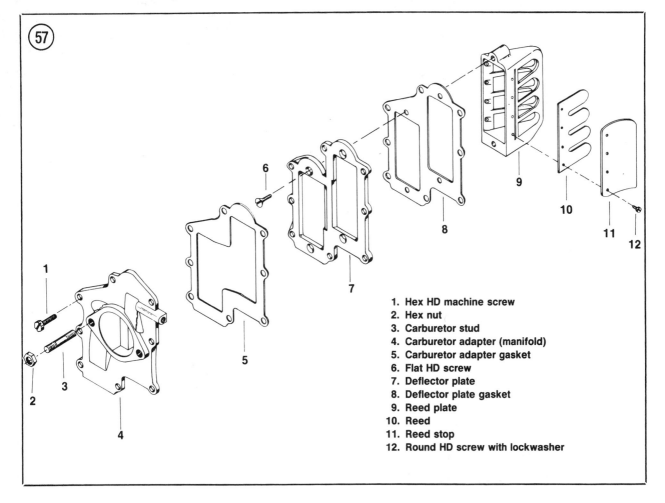

1. Hex HD machine screw
2. Hex nut
3. Carburetor stud
4. Carburetor adapter (manifold)
5. Carburetor adapter gasket
6. Flat HD screw
7. Deflector plate
8. Deflector plate gasket
9. Reed plate
10. Reed
11. Reed stop
12. Round HD screw with lockwasher

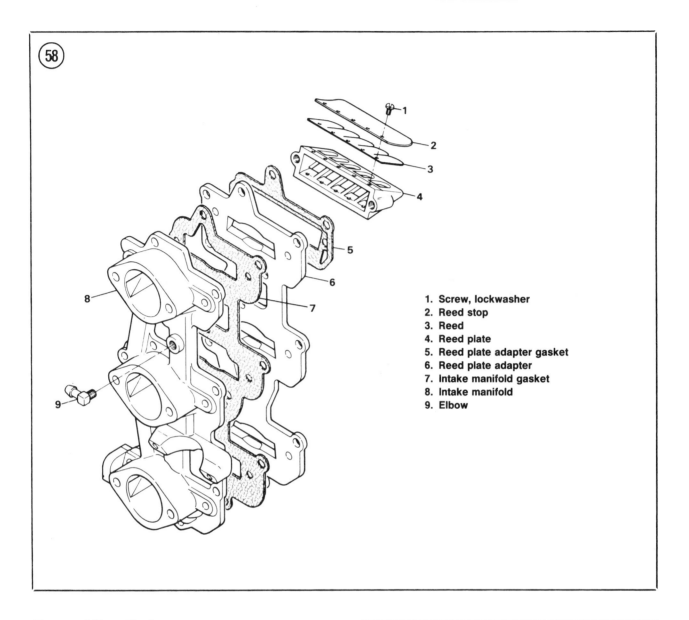

1. Screw, lockwasher
2. Reed stop
3. Reed
4. Reed plate
5. Reed plate adapter gasket
6. Reed plate adapter
7. Intake manifold gasket
8. Intake manifold
9. Elbow

Removal/Installation

1. Remove the carburetor as described in this chapter.

2. Disconnect any hoses connected to the intake manifold or carburetor adapter.

3. On 3- and 4-cylinder models, it may be necessary to remove the throttle cam from the intake manifold.

4. Remove the screws holding the intake manifold or carburetor adapter to the crankcase cover.

5. Remove the intake manifold or carburetor adapter and gasket. Discard the gasket.

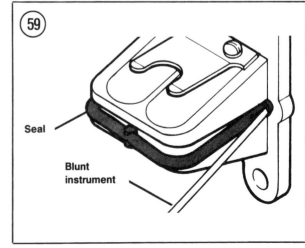

Seal

Blunt
instrument

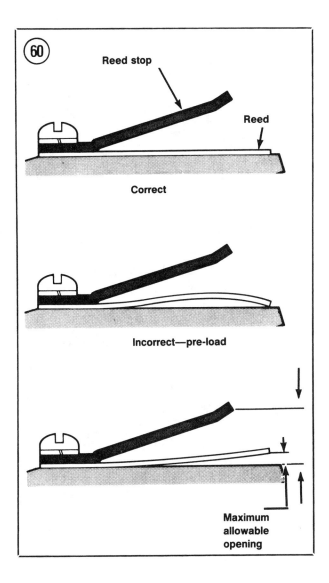

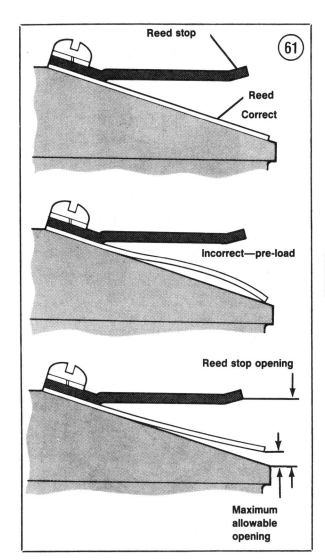

6. If a deflector plate is used, separate it from the intake manifold/carburetor adapter and remove the screws holding the reed valve assemblies.

7. Clean all mating surfaces of gasket or sealant residue.

8. Installation is the reverse of removal, plus the following:

   a. Run a thin bead of RTV sealant around the reed plate opening on the power head.

   b. If reed plate uses a seal, remove and discard the old seal. Lubricate reed plate groove with engine oil and install a new seal, pushing its ends through the reed plate hole with a blunt instrument. See **Figure 59**. Seal should extend no more than 0.06 in. above face of plate after installation.

   c. Use new gaskets.

   d. Tighten intake manifold/carburetor adapter screws to specifications. See **Table 1**.

Disassembly, Inspection and Assembly

Refer to **Figure 60** (1-cylinder), **Figure 61** (2-cylinder) or **Figure 62** (3- and 4-cylinder) for this procedure.

1. Check reeds for cracking or other damage. Replace if any defects are noted.

2. Reeds should lie flat on the reed plate with no preload. To check flatness, gently push each reed petal out. Constant resistance should be felt with no noise.

3. If reeds do not lie perfectly flat on the reed plate, measure the amount they are open and compare to maximum allowable reed opening specifications (**Table 2**).

4. Measure the distance between the reed stop and reed plate. Compare to specifications (**Table 2**).

5. Replace any reeds that do not meet the specifications in Step 3 or Step 4.

Reed Replacement

NOTE
Some late-model reed valve assemblies use spacers between the reed stop and reed.

1. Remove the screws, spacers (if used) and lockwashers holding the reed stop and reeds to the reed plate.

2. Remove the reed stop and reeds.

3. Place a new reed on the reed plate and check for flatness.

4. Locate reed over reed plate openings. There should be a minimum overlap of 0.030 in. (3.5-15 hp) or 0.040 in. (20-140 hp) over the reed plate opening. See **Figure 63** (3.5-15 hp) or **Figure 64** (20-140 hp).

5. Install reed stop with screws, spacers (if used) and lockwashers.

6. Check reed stop tension and opening. See *Disassembly, Cleaning and Assembly* in this chapter.

CHOKE SOLENOID

A choke solenoid is used on 20-140 hp engines. Test procedures are given in Chapter Three.

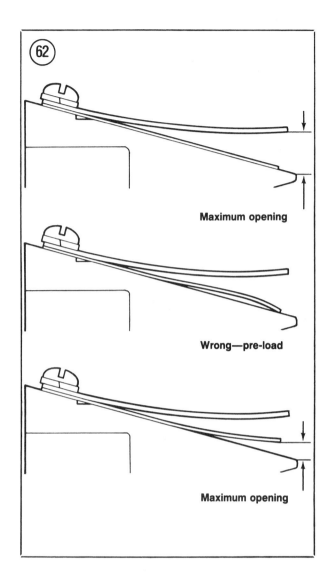

Maximum opening

Wrong—pre-load

Maximum opening

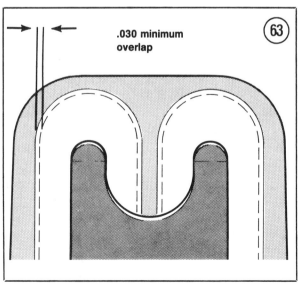

.030 minimum overlap

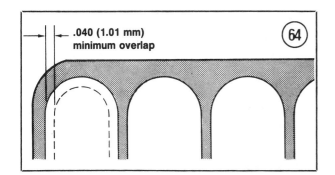

.040 (1.01 mm) minimum overlap

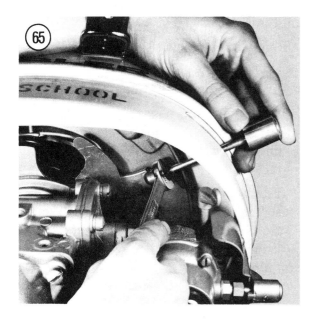

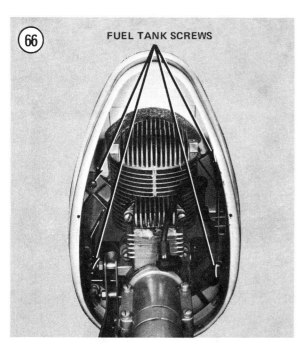

FUEL TANK SCREWS

Removal/Installation

1. Disconnect the solenoid terminal post wire. Note the position of the terminal post for reinstallation reference.
2. Loosen the fastener(s) holding the solenoid clamp.
3. Remove solenoid from clamp and disconnect it from the plunger.
4. Remove the strap holding the wire to the solenoid body.
5. Installation is the reverse of removal. Align solenoid with terminal post in the same relative position noted in Step 1.

6

INTEGRAL FUEL TANK

All 3.5, 3.6, 4, 4.9 and 5 hp models are fitted with an integral fuel tank.

Removal/Installation
(1966-1969 3.5 hp)

1. Loosen the magneto control lever jam nut. See **Figure 65**.
2. Unscrew the control lever and remove it from the engine.
3. Disconnect the fuel line at the fuel shut-off valve (port side of carburetor). Plug the end of the line to prevent leakage.
4. Remove the 4 screws holding the fuel tank to the support plate (**Figure 66**).
5. Lift tank from engine and remove the 3 screws holding the starter to the tank. Separate the tank and starter.
6. Installation is the reverse of removal. Loosely install starter to fuel tank, pull starter handle to engage the friction shoe plates against the starter cup, then tighten the starter screws.

**Removal/Installation
(3.6 hp, 1976-1978 4 hp,
4.9 and 5 hp)**

1. Remove the engine cover.
2. Remove the 4 nuts and lockwashers holding the starter to the fuel tank.
3. Disconnect the fuel line at the fuel valve center fitting. Plug the end of the line to prevent leakage.
4. Remove the spark plug. See Chapter Four.
5. Remove the 2 screws holding the front of the tank to the bearing cage. See **Figure 67**.

NOTE
Late models use only one bracket on the port side.

6. Remove the 2 screws holding the rear of the tank to the brackets. See **Figure 68**.
7. Remove tank from engine.
8. Installation is the reverse of removal. Loosely install starter to fuel tank, pull starter handle to engage the friction shoe plates against the starter cup, then tighten the starter mounting nuts.

**Removal/Installation
(1980-on 3.5 hp and 1979-on 4 hp)**

1. Remove the engine cover.
2. Slide fuel line clamp from carburetor fuel fitting with hose clamp pliers and pull fuel line free of carburetor. Plug the line to prevent leakage.
3. Remove the 2 front and 2 rear bracket bolts and star washers.
4. Remove the bracket screw on each side of the tank.
5. Remove the fuel tank with starter attached.
6. Remove the 4 nuts and lockwashers holding the starter to the fuel tank. Separate the tank and starter.
7. Installation is the reverse of removal. Loosely install starter to fuel tank, pull starter handle to engage the friction shoe plates against the starter cup, then tighten the starter mounting nuts.

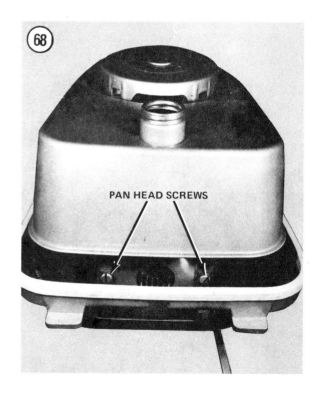

PORTABLE FUEL TANK

Figure 69 shows the components of the portable fuel tank, including the fuel cap gauge sender and primer bulb assembly. The fuel tank adapter threads into the top of the tank and acts as the fuel pick-up. A screen clamped to the end of the adapter prevents fuel sediment from reaching the fuel pump.

When some oils are mixed with gasoline and stored in a warm place, a bacterial substance will form. This clear substance covers the fuel pickup, restricting flow through the fuel system. Bacterial formation can be prevented by using a good quality fuel conditioner on a regular basis. If present, it can be removed with a good marine engine cleaner.

To remove any dirt or water that may have entered the tank during refilling and to prevent the build-up of gum and varnish, clean the inside of the tank once each season by flushing with clean lead-free gasoline or kerosene.

Check the inside and outside of the tank for signs of rust, leakage or corrosion. Replace as required. Do not attempt to patch the tank with automotive fuel tank repair materials. Portable marine fuel tanks are subject to much greater pressure and vacuum conditions.

6

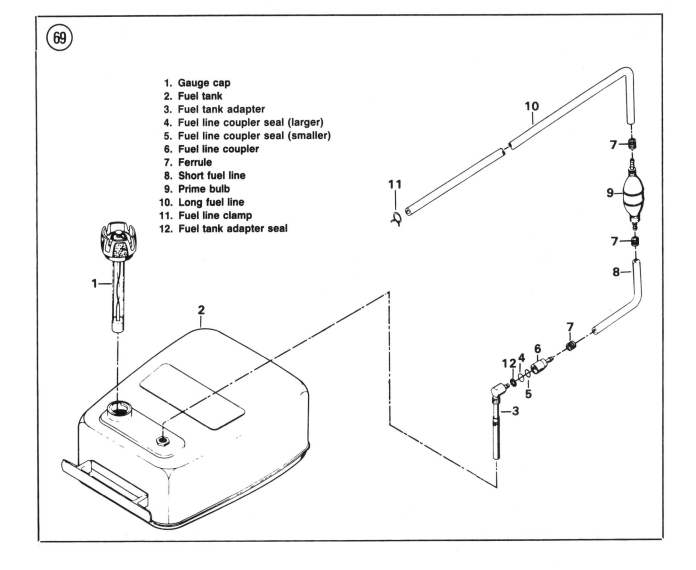

(69)

1. Gauge cap
2. Fuel tank
3. Fuel tank adapter
4. Fuel line coupler seal (larger)
5. Fuel line coupler seal (smaller)
6. Fuel line coupler
7. Ferrule
8. Short fuel line
9. Prime bulb
10. Long fuel line
11. Fuel line clamp
12. Fuel tank adapter seal

To check the fuel tank adapter screen for possible restrictions, remove the adapter and inspect the hose and screen for damage. Remove the hose clamp which holds the screen in place. Remove the screen and clean with solvent, then blow low-pressure compressed air through the plastic collar. See **Figure 70**.

FUEL LINE AND PRIMER BULB

When priming the engine, the primer bulb should gradually become firm. If it does not become firm or if it stays firm even when disconnected, the check valve inside the primer bulb is malfunctioning.

The line should be checked periodically for cracks, breaks, restrictions and chafing. The bulb should be checked periodically for proper operation. Make sure all fuel line connections are tight and securely clamped.

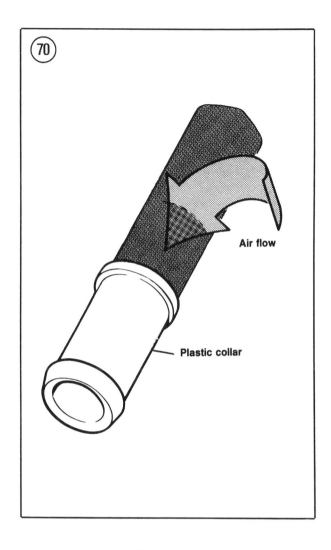

Air flow

Plastic collar

FUEL RECIRCULATION SYSTEM

Larger Chrysler outboard models may be equipped with a fuel recirculation or puddle drain system located on the starboard side of the crankcase housing. This consists of a cover containing directional reed valves and screens which connect each cylinder in the crankcase. See **Figure 71**.

Since the cylinders in an outboard engine do not burn all of the fuel sent to their combustion area, this system provides a method of collecting the excess fuel/oil mixture and routing it to the combustion area of another cylinder. When the system works correctly, it promotes better fuel economy by burning more of the available fuel/oil mixture and also minimizes the pollution caused by fuel/oil in the exhaust.

Correct puddle valve operation is important to good engine operation. If the system fails, crankcase pressure problems will result, causing the following symptoms:

a. Poor engine performance at low rpm

b. Excessive exhaust smoke

c. Stalling or popping at idle

d. Poor acceleration

To service the puddle drain system, use the following procedures.

1. Disconnect and remove the necessary electrical components positioned in front of the puddle drain cover.

2. Remove the cover, plate and gasket assembly. See **Figure 71**.

3. Carefully separate the cover from the plate and discard the gasket.

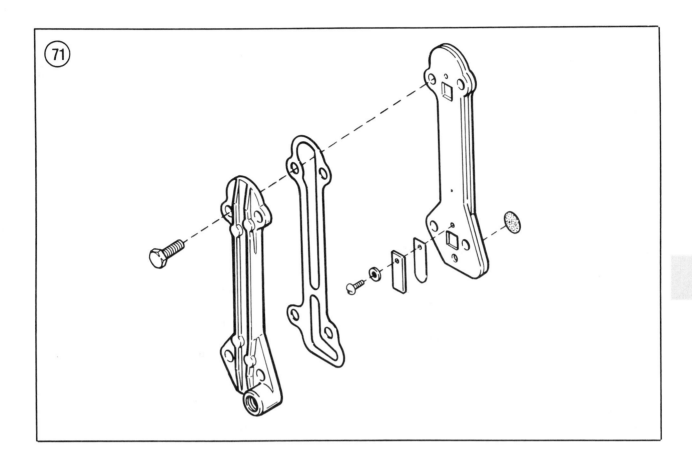

6

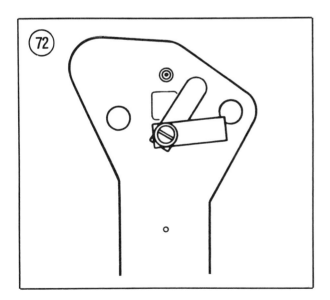

4. Remove the reeds and reed stops (**Figure 72**). Check the reeds for flatness and damage. Replace as required.

5. Check and clean the screens.

6. Check condition of all hoses connected to the puddle drain system.

7. Clean all parts in solvent and blow dry with low-pressure compressed air.

8. Reverse Steps 1-4 to complete installation. Use new gaskets.

Tables are on the following page.

Table 1 STANDARD TORQUE VALUES

Screw or nut size	in.-lb.
6-32	9
8-32	20
10-24	30
10-32	35
12-24	45
1/4-20	70
5/16-18	160
3/8-16	270

Table 2 REED VALVE SPECIFICATIONS

Engine	Maximum allowable reed opening	Maximum distance between stop and plate
3.5-4 hp	0.006 in.	0.250 in. $\pm$0.10 in.
4.4-15 hp	0.012 in.	0.210 in. $\pm$0.10 in.
20 hp (1966-1978)	0.005 in.	0.265 in. $\pm$0.10 in.
85-140 hp		
1979-1984	0.010 in.	0.310 in. $\pm$0.10 in.
All others	0.010 in.	0.280 in. $\pm$0.10 in.

Chapter Seven

Ignition and Electrical Systems

This chapter provides service procedures for the battery, starter motor on electric start models and each ignition system used on Chrysler outboard motors during the years covered by this manual. Wiring diagrams are included at the end of the book. **Tables 1-4** are at the end of the chapter.

BATTERY

Since batteries used in marine applications endure far more rigorous treatment than those used in an automotive charging system, they are constructed differently. Marine batteries have a thicker exterior case to cushion the plates inside during tight turns and rough weather. Thicker plates are also used, with each one individually fastened within the case to prevent premature failure. Spill-proof caps on the battery cells prevent electrolyte from spilling into the bilges.

Automotive batteries are not designed to be run down and recharged repeatedly. For this reason, they should *only* be used in an emergency situation when a suitable marine battery is not available.

Chrysler recommends that any battery used to crank an outboard motor have a minimum rating of 70 amp hours.

> *CAUTION*
> *Sealed or maintenance-free batteries are **not** recommended for use with the unregulated charging systems used on Chrysler and Force outboards. Excessive charging during continued high-speed operation will cause the electrolyte to boil, resulting in its loss. Since water cannot be added to such batteries, such overcharging will ruin the battery.*

Separate batteries may be used to provide power for any accessories such as lighting, fish finders, depth finder, etc. To determine the required capacity of such batteries, calculate the average discharge rate of the accessories and refer to **Table 1**.

Batteries may be wired in parallel to double the ampere hour capacity while maintaining a 12-volt system. See **Figure 1**. For accessories which require 24 volts, batteries may be wired in series (**Figure 2**) but only accessories specifically requiring 24 volts should be connected into the system. Whether wired in parallel or in series, charge the batteries individually.

Battery Installation in Aluminum Boats

If a battery is not properly secured and grounded when installed in an aluminum boat, it may contact the hull and short to ground. This will burn out remote control cables, tiller handle cables or wiring harnesses.

The following preventive steps should be taken when installing a battery in a metal boat.

1. Choose a location as far as practical from the fuel tank while providing access for maintenance.

2. Install the battery in a plastic battery box with cover and tie-down strap (**Figure 3**).

3. If a covered container is not used, cover the positive battery terminal with a non-conductive shield or boot (**Figure 4**).

4. Make sure the battery is secured inside the battery box and that the box is fastened in position with the tie-down strap.

Care and Inspection

1. Remove the battery container cover (**Figure 3**) or hold-down (**Figure 4**).

2. Disconnect the negative battery cable. Disconnect the positive battery cable.

> *NOTE*
> *Some batteries have a built-in carry strap (**Figure 5**) for use in Step 3.*

3. Attach a battery carry strap to the terminal posts. Remove the battery from the battery tray or container.

4. Check the entire battery case for cracks.

5. Inspect the battery tray or container for corrosion and clean if necessary with a solution of baking soda and water.

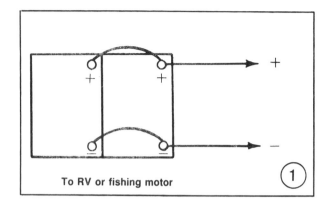

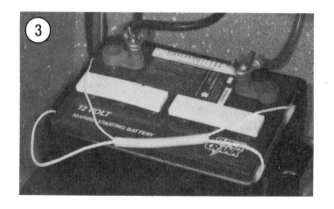

To RV or fishing motor

To RV or fishing motor

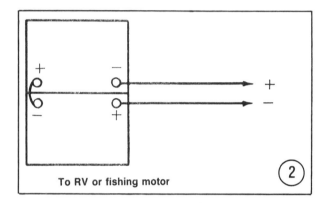

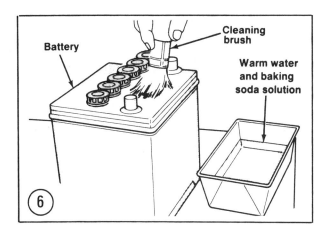

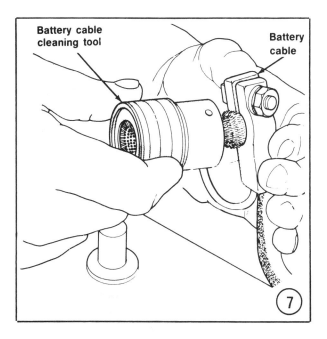

7

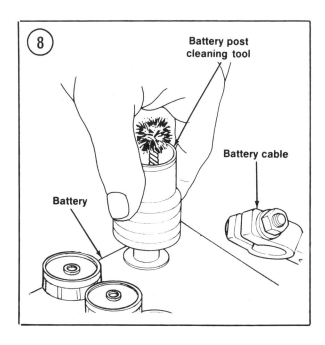

NOTE
Keep cleaning solution out of the battery cells in Step 6 or the electrolyte will be seriously weakened.

6. Clean the top of the battery with a stiff bristle brush using the baking soda and water solution (**Figure 6**). Rinse the battery case with clear water and wipe dry with a clean cloth or paper towel.

7. Position the battery in the battery tray or container.

8. Clean the battery cable clamps with a stiff wire brush or one of the many tools made for this purpose (**Figure 7**). The same tool is used for cleaning the battery posts. See **Figure 8**.

9. Reconnect the positive battery cable, then the negative cable.

CAUTION
Be sure the battery cables are connected to their proper terminals. Connecting the battery backwards will reverse the polarity and damage the rectifier.

10. Tighten the battery connections and coat with a petroleum jelly such as Vaseline or a light mineral grease.

11. Remove the filler caps and check the electrolyte level. Add distilled water, if necessary, to bring the level up to 3/16 in. above the plates in the battery case. See **Figure 9**.

Testing

Hydrometer testing is the best way to check battery condition. Use a hydrometer with numbered graduations from 1.100-1.300 rather than one with just color-coded bands. To use the hydrometer, squeeze the rubber ball, insert the tip in a cell and release the ball (**Figure 10**).

Draw enough electrolyte to float the weighted float inside the hydrometer. When using a temperature-compensated hydrometer, release the electrolyte and repeat this process several times to make sure the thermometer has adjusted to the electrolyte temperature before taking the reading.

Hold the hydrometer vertically and note the number in line with the surface of the electrolyte (**Figure 11**). This is the specific gravity for the cell. Return the electrolyte to the cell from which it came.

The specific gravity of the electrolyte in each battery cell is an excellent indicator of that cell's condition. A fully charged cell will read 1.260 or more at 80° F (25° C). A cell

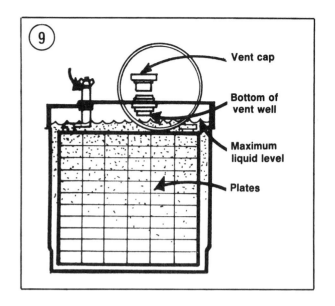

9

Vent cap

Bottom of vent well

Maximum liquid level

Plates

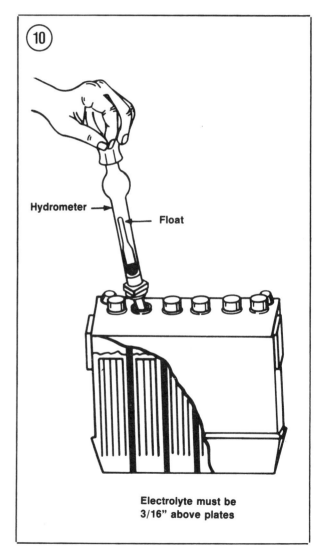

10

Hydrometer

Float

Electrolyte must be 3/16" above plates

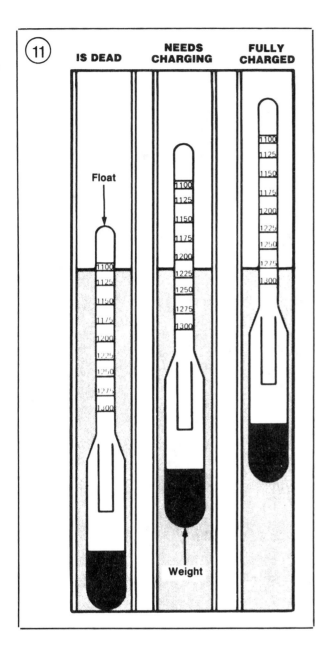

that is 75 percent charged will read from 1.220-1.230 while one with a 50 percent charge reads from 1.170-1.180. If the cell tests below 1.200, the battery must be recharged and one that reads 1.140 or below is dead. Charging is also necessary if the specific gravity varies more than 0.050 from cell to cell.

NOTE
If a temperature-compensated hydrometer is not used, add 0.004 to the specific gravity reading for every 10° above 80° F (25° C). For every 10° below 80° F (25° C), subtract 0.004.

Battery Storage

Wet cell batteries slowly discharge when stored. They discharge faster when warm than when cold. See **Table 2**. Before storing a battery for the season, clean the case with a solution of baking soda and water. Rinse with clear water and wipe dry. The battery should be fully charged (no change in specific gravity when 3 readings are taken 1 hour apart) and then stored in as cool and dry a place as possible.

Charging

A good state of charge should be maintained in batteries used for starting. Check the battery with a voltmeter as shown in **Figure 12**. Any battery that cannot deliver

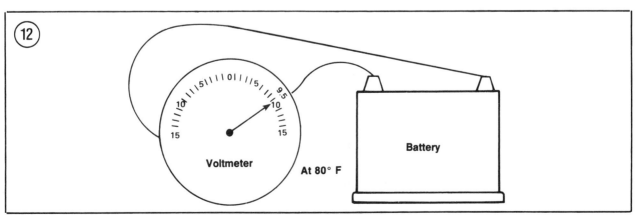

at least 9.6 volts under a starting load should be recharged. If recharging does not bring it up to strength or if it does not hold the charge, replace the battery.

The battery does not have to be removed from the boat for charging, but it is a recommended safety procedure since a charging battery gives off highly explosive hydrogen gas. In many boats, the area around the battery is not well ventilated and the gas may remain in the area for hours after the charging process has been completed. Sparks or flames occuring near the battery can cause it to explode, spraying battery acid over a wide area.

For this reason, it is important that you observe the following precautions:

a. Do not smoke around batteries that are charging or have been recently charged.

b. Do not break a live circuit at the battery terminals and cause an electrical arc that can ignite the hydrogen gas.

Disconnect the negative battery cable first, then the positive cable. Make sure the electrolyte is fully topped up.

Connect the charger to the battery—negative to negative, positive to positive. If the charger output is variable, select a 4 amp setting. Set the voltage regulator to 12 volts and plug the charger in. If the battery is severely discharged, allow it to charge for at least 8 hours. Batteries that are not as badly discharged require less charging time. **Table 3** gives approximate charge rates for batteries used primarily for cranking. Check the charging progress with the hydrometer.

Jump Starting

If the battery becomes severely discharged, it is possible to start and run an engine by jump starting it from another battery. If the proper procedure is not followed, however,

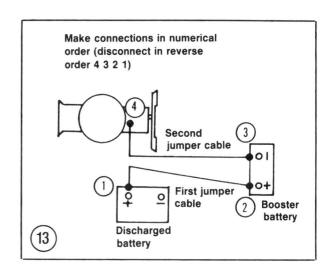

Make connections in numerical order (disconnect in reverse order 4 3 2 1)

Second jumper cable

First jumper cable

Discharged battery

Booster battery

13

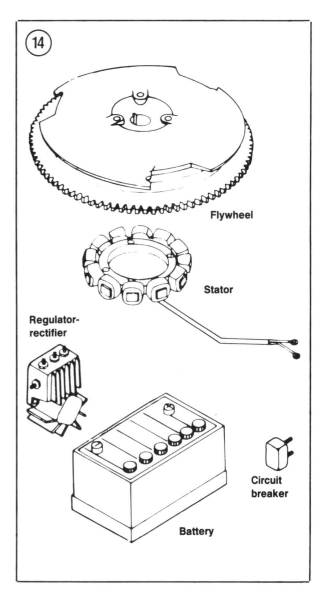

14

Flywheel

Stator

Regulator-rectifier

Battery

Circuit breaker

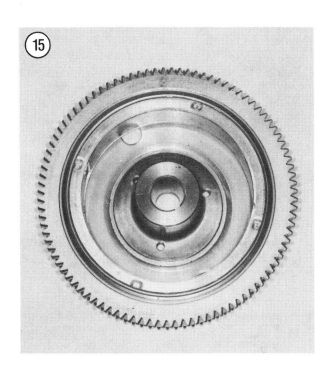

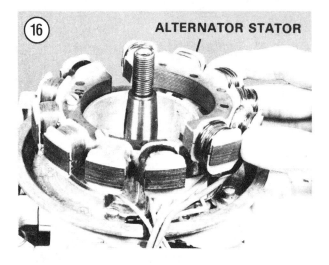

ALTERNATOR STATOR

jump starting can be dangerous. Check the electrolyte level before jump starting any battery. If it is not visible or if it appears to be frozen, do not attempt to jump start the battery.

WARNING
Use extreme caution when connecting a booster battery to one that is discharged to avoid personal injury or damage to the system.

1. Connect the jumper cables in the order and sequence shown in **Figure 13**.

WARNING
An electrical arc may occur when the final connection is made. This could cause an explosion if it occurs near the battery. For this reason, the final connection should be made to a good ground away from the battery and not to the battery itself.

2. Check that all jumper cables are out of the way of moving engine parts.
3. Start the engine. Once it starts, run it at a moderate speed.

CAUTION
Running the engine at wide-open throttle may cause damage to the electrical system.

4. Remove the jumper cables in the exact reverse order shown in **Figure 13**. Remove the cables at point 4, then 3, 2 and 1.

ALTERNATOR CHARGING SYSTEM

An alternator charging system is standard on all electric start models and optional on 4-7.5 hp manual models equipped with an AC lighting system.

The alternator charging system on Chrysler outboards consists of the alternator stator coils, permanent magnets located within the flywheel, a rectifier (Prestolite) or regulator-rectifier (Motorola), the circuit breaker, battery and connecting wiring. **Figure 14** shows the Motorola system components.

Magneto breaker point and CD ignition systems both use a flywheel containing magnets (**Figure 15**). Manual start flywheels use 2 magnets for the ignition system. Electric start flywheels use an additional 2 magnets for the charging system. Rotation of the flywheel stator magnets past the stator coils (**Figure 16**) creates alternating current. This

7

current is sent to the rectifier (**Figure 17** or **Figure 18**, typical) where it is converted into direct current to charge the battery or power the accessories.

A malfunction in the battery charging system will result in an undercharged battery. Perform the following visual inspection to determine the cause of the problem. If the visual inspection proves satisfactory, test the stator coils and rectifier. See Chapter Three.

1. Make sure that the battery cables are connected properly. The red cable must be connected to the positive battery terminal. If polarity is reversed, check for a damaged rectifier.

2. Inspect the battery terminals for loose or corroded connections. Tighten or clean as required.

3. Inspect the physical condition of the battery. Look for bulges or cracks in the case, leaking electrolyte or corrosion build-up.

4. Carefully check the wiring between the stator coils and battery for signs of chafing, deterioration or other damage.

5. Check the circuit wiring for corroded, loose or disconnected connections. Clean, tighten or connect as required.

6. Determine if the electrical load on the battery from accessories is greater than the battery capacity.

**Stator Coil Replacement
(Electric Start Models)**

1. Disconnect the negative battery cable.
2. Remove the engine cover.
3. Remove the flywheel. See Chapter Eight.
4A. Magnapower II models—Disconnect the leads at the stator module.
4B. Prestolite—Disconnect the 2 green/yellow stator leads at the rectifier (**Figure 18**).
4C. Motorola—Disconnect the black and purple leads at the regulator-rectifier (**Figure 19**).

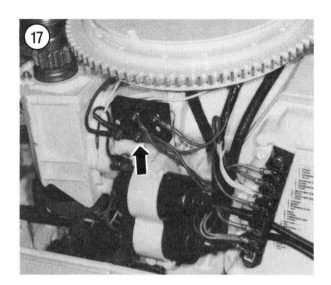

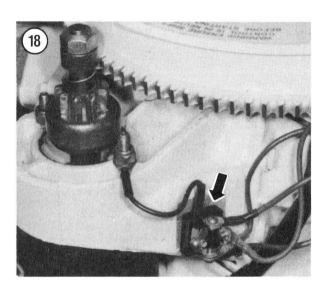

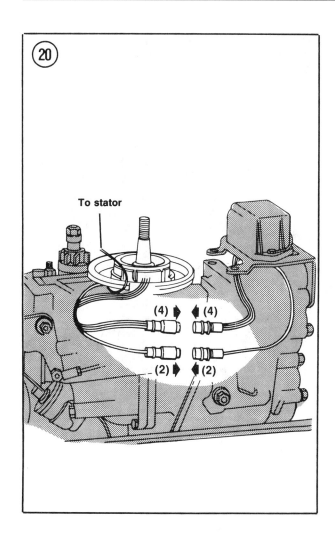

To stator

(4) ▶ ◀ (4)

(2) ▶ ◀ (2)

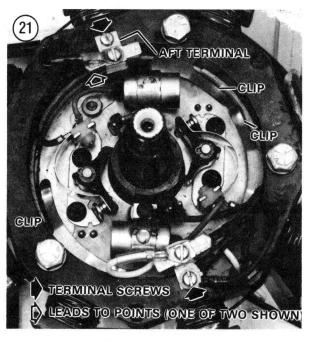

AFT TERMINAL

CLIP

CLIP

CLIP

TERMINAL SCREWS

LEADS TO POINTS (ONE OF TWO SHOWN)

5. Alternator CD ignition—Disconnect the stator connector plugs. See **Figure 20** (typical).

6. Battery ignition—Pull the breaker point leads off the stator spade terminals. Remove the terminal mounting screws and terminals from the stator. See **Figure 21**.

7. Remove the stator mounting screws. Remove stator from trigger housing or bearing cage.

8. Installation is the reverse of removal. Route wires so they do not contact or interfere with any moving components.

7

**Stator Coil Replacement
(Manual Models Converted to
Battery Charging System)**

1. Disconnect the negative battery cable.
2. Remove the engine cover.
3. Remove the flywheel. See Chapter Eight.
4. Remove the screws holding the alternator stator to the trigger stator.
5. Cut any straps holding stator leads to other wiring.
6. Unwrap the lighting coil cable splices. Cut stator leads from splices. Remove stator from power head.
7. Installation is the reverse of removal. Route stator leads so they do not contact or interfer with any moving components.

**Rectifier and
Regulator-rectifier
Removal/Installation
(Except Magnapower II Ignition)**

Refer to **Figures 17-19** and **Figure 22** for this procedure.

1. Disconnect the negative battery cable.
2. Remove the engine cover.

3. Label and disconnect all lead wires at the rectifier or regulator-rectifier.

> *CAUTION*
> *When removing a selenium rectifier (Figure 22), place the wrench on the large (inner) hex head of the fastener to prevent internal damage to the component.*

4. Remove the screw(s) holding the rectifier or regulator-rectifier to the power head or mounting bracket.
5. Remove the rectifier or regulator-rectifier.
6. Installation is the reverse of removal.

Magnapower II
Regulator-rectifier
Module Removal/Installation
(Except 55 and 65 hp)

1. Disconnect the negative battery cable.
2. Remove the engine cover.
3. Remove the flywheel. See Chapter Eight.
4. Remove the capacitor module. See **Figure 23**.
5. Remove the screws and lockwashers holding the regulator-rectifier leads to the alternator module. See **Figure 24**.
6. Disconnect the regulator-rectifier lead at the terminal block.
7. Remove the screws and lockwashers holding the regulator-rectifier module to the bearing cage (**Figure 25**). Remove the module.
8. Installation is the reverse of removal.

Magnapower II
Regulator-rectifier Module
Removal/Installation
(55 and 65 hp)

1. Disconnect the negative battery cable.
2. Remove the engine cover.
3. Remove the flywheel. See Chapter Eight.
4. Disconnect the leads at the regulator-rectifier.

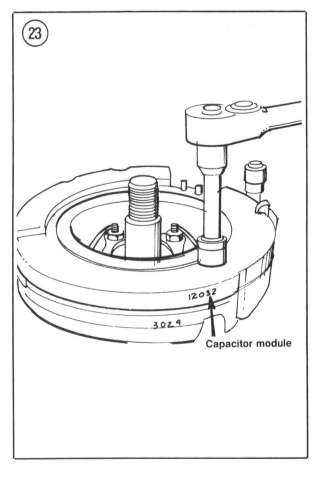

Capacitor module

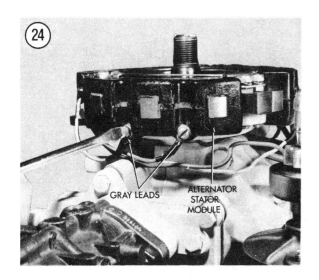

GRAY LEADS ALTERNATOR STATOR MODULE

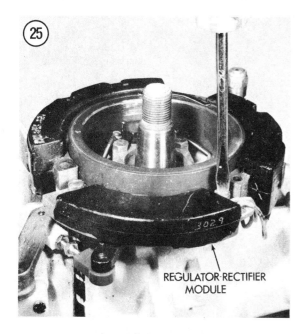

REGULATOR-RECTIFIER MODULE

5. Remove the 2 screws holding the regulator-rectifier to the power head (**Figure 26**). Remove the regulator-rectifier.

6. Installation is the reverse of removal.

Trip Regulator Replacement

A trip regulator (**Figure 27**) is used in conjunction with the rectifier on some early 3- and 4-cylinder models.

1. Disconnect the negative battery cable.

2. Remove the engine cover.

3. Remove the rectifier as described in this chapter.

4. Disconnect the 2 red lead wires at the trip regulator.

5. Remove the hex head screw holding the trip regulator to the exhaust cover. Remove the trip regulator.

6. Installation is the reverse of removal.

Circuit Breaker Replacement

The circuit breaker is generally located on or near the terminal board. See **Figure 28** for typical location.

7

AC
TRIP REGULATOR
AC +

1. Disconnect the negative battery cable.
2. Remove the engine cover.

> *NOTE*
> *Depending upon the circuit breaker location and positioning, it may be more efficient to reverse Step 3 and Step 4 on some models.*

3. Remove the 2 screws and washers holding the circuit breaker to the terminal board bracket or crankcase cover. See **Figure 29** (typical).
4. Remove the nuts and disconnect the leads from the circuit breaker terminals. See **Figure 29** (typical).
5. Installation is the reverse of removal.

GENERATOR CHARGING SYSTEM

Autolectric models use a starter-generator to provide charging current. The system contains a diode which acts as a rectifier, a voltage regulator, circuit breaker, battery and connecting wiring. Troubleshooting the system is discussed in Chapter Three. Component replacement is covered in this chapter under *Electric Starting System*.

ELECTRIC STARTING SYSTEM

Outboards covered in this manual may use a rope-operated mechanical (rewind) starting system or an electric (starter motor or starter-generator) starting system. The starting circuit consists of the battery, an ignition or starter switch, an interlock switch, the starter motor or starter-generator and connecting wiring. Smaller displacement engines use a heavy-duty starter switch; larger displacement engines incorporate a starter solenoid or relay in the circuit.

Starting system operation and trouble-shooting is described in Chapter Three.

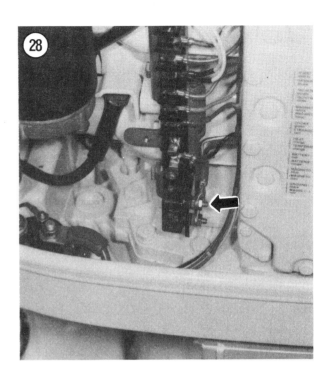

STARTER MOTOR SYSTEM

Marine starter motors are very similar in design and operation to those found on automotive engines. They use an inertia-type drive in which external spiral splines on the armature shaft mate with internal splines on the drive assembly.

The starter motor produces a very high torque but only for a brief period of time, due to heat buildup. Never operate the starter motor continuously for more than 15 seconds. Let the motor cool for at least 3 minutes before operating it again.

If the starter motor does not turn over, check the battery and all connecting wiring for loose or corroded connections. If this does not solve the problem, refer to Chapter Three. Except for brush replacement, service to the starter motor by amateur mechanics is limited to replacement with a new or rebuilt unit.

The starter bracket plate on certain models bolts to bosses on the crankcase cover. If a crankcase cover boss breaks, a steel starter bracket plate (part No. A191090) can be

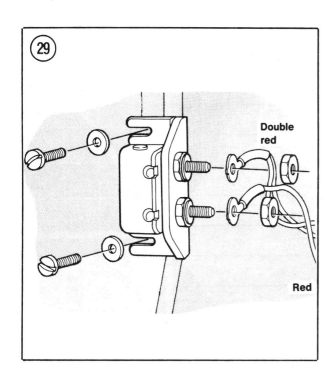

Double
red

Red

installed on the rear of the starter mounting area on the crankcase cover. This eliminates the necessity of replacing the crankcase cover. This bracket can be used on all electric start 1966-1975 35 hp, 1966-on 45 hp, 1966 50 hp and 1967-1976 55 hp models.

Starter Motor
Removal/Installation
(1968-1978 9.9-15 hp)

1. Disconnect the negative battery cable.
2. Remove the engine cover.
3. Remove the ignition coils as described in this chapter.
4. Remove terminal block from starter support to gain access to the forward port starter mounting screw.
5. Remove the 2 capscrews holding the starter to the starter support.
6. Remove the 4 starter support screws and lift rear of support assembly up enough to remove the starter.
7. Lower starter support after starter is removed and disconnect the battery lead from the starter terminal.

8. Installation is the reverse of removal. Position starter with terminal facing toward the rear of the engine.

Starter Motor
Removal/Installation
(9.9-15 hp 1979-on)

1. Disconnect the negative battery cable.
2. Remove the engine cover.
3. Disconnect the battery lead at starter motor terminal.
4. Remove the 2 capscrews holding the starter to the starter support housing. Lower the starter and remove from the engine.
5. Installation is the reverse of removal.

Starter Motor
Removal/Installation
(1973-1978 25 hp; 1973-1975 30 hp;
35 hp 1976-1978)

1. Disconnect the negative battery cable.
2. Remove the engine cover.
3. Disconnect the white/red and red-tipped leads from the starter relay upper terminal post.
4. Disconnect the yellow and black/yellow leads connected to the interlock switch on the exhaust port cover.
5. Disconnect the green choke solenoid lead and the black support plate double ground lead.
6. Remove the screw holding the starter clamp to the crankcase cover. See **Figure 30**.
7. Remove the upper and lower mounting screws holding the starter bracket to the crankcase cover. See **Figure 31**. Remove the starter bracket.
8. Disconnect the red lead from the starter relay lower terminal post.
9. Remove the 2 nuts and lockwashers from the through-bolts holding starter to bracket. See **Figure 30**.

7

10. Lower the starter from the bracket and remove the clamp.

11. Installation is the reverse of removal. Make sure to position the fuel line clip under the lower starter bracket port screw and the negative lead from the battery under the bracket starboard screw.

Starter Motor
Removal/Installation
(1979-1982 20 and 30 hp;
1982-on 25 hp; 1983-on 35 hp)

1. Disconnect the negative battery cable.
2. Remove the engine cover.
3. Remove the 3 screws holding the starter housing.
4. Remove the 5 screws holding the starter bracket. See **Figure 32**. Remove the starter.
5. Installation is the reverse of removal.

Starter Motor
Removal/Installation
(1966-1981 35 hp; 1966-on
45-50 hp; 1967-1978 55 hp)

1. Disconnect the negative battery cable.
2. Remove the engine cover.

3. If rectifier is mounted on the starter bracket, remove the mounting screws and move rectifier to one side.
4. Disconnect the starter-to-relay lead at the lower front of the starter.

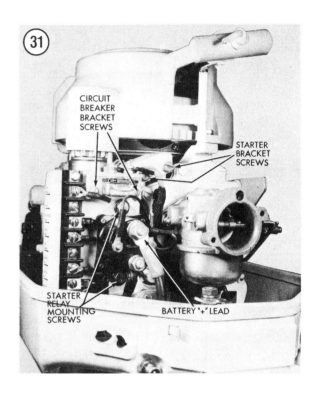

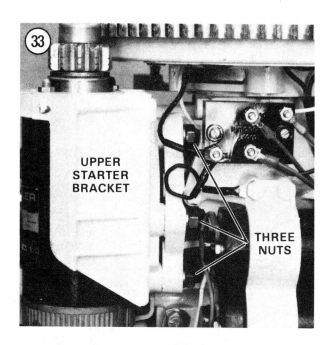

UPPER STARTER BRACKET

THREE NUTS

LOWER STARTER BRACKET

THRU BOLT
(1 OF 2 SHOWN)

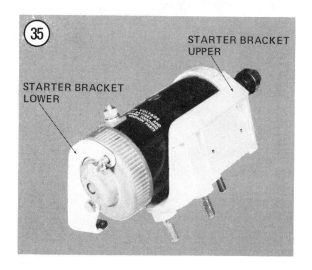

STARTER BRACKET UPPER

STARTER BRACKET LOWER

5. Remove the 3 nuts and lockwashers from the upper starter bracket studs. See **Figure 33**.

6. Remove the 2 screws holding the lower bracket to the crankcase (**Figure 34**).

7. If choke solenoid interferes with starter removal, remove starboard solenoid bracket screw. Loosen port bracket screw and reposition solenoid to starboard side.

8. Remove the starter and bracket assembly (**Figure 35**).

9. If starter is to be removed from bracket assembly, remove the 2 starter through-bolts, holding the commutator and drive end caps together to prevent the starter from becoming disassembled. Separate starter from brackets. If starter is not to be disassembled, reinstall the through-bolts.

10. If starter is removed from bracket assembly, reinstall by inserting through-bolts and lockwashers through the lower bracket, starter and upper bracket. Install through-bolt locknuts and tighten securely.

11. Install starter and bracket assembly and reverse Steps 1-7 to complete installation.

Starter Motor Removal/Installation (1979-on 55 hp; All 60 and 65 hp)

1. Disconnect the negative battery cable.

2. Remove the engine cover.

3. Disconnect the starter-to-relay lead at the starter terminal.

4. Remove the 3 capscrews and washers holding the starter to the power head. See **Figure 36**. Remove the starter.

5. Installation is the reverse of removal.

Starter Motor Removal/ Installation (70-140 hp)

Early models use one of two starter mounts, designated as Type A or Type B. Later models all use the Type B mount.

7

Type A mount

1. Disconnect the negative battery cable.
2. Remove the engine cover.
3. Remove the bolts holding the lower starter bracket to the power head.
4. Remove the through-bolt locknuts holding the lower starter bracket to the starter.
5. Disconnect the starter-to-relay lead at the starter terminal.
6. Thread the through-bolts from the power head flange and remove the starter motor and bracket assembly.
7. If starter is to be removed from bracket assembly, remove the 2 starter through-bolts (holding the commutator and drive end caps together) to prevent the starter from becoming disassembled. Separate starter from bracket. If starter is not to be disassembled, reinstall the through-bolts.
8. If starter is removed from bracket assembly, reinstall by inserting through-bolts and lockwashers through the lower bracket and into the starter motor. Install through-bolt locknuts and tighten securely.
9. Install starter and bracket assembly and reverse Steps 1-6 to complete installation.

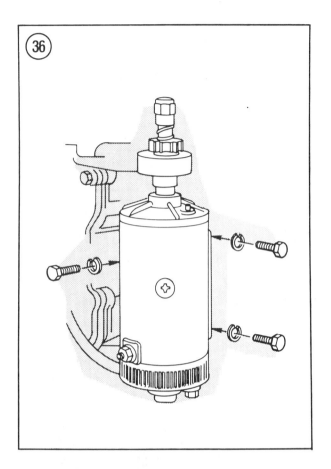

Type B mount

1. Disconnect the negative battery cable.
2. Remove the engine cover.
3. Remove the 2 nuts holding the starter to the upper starter bracket (A, **Figure 37**).
4. Remove the screw holding the lower starter bracket to the crankcase cover (B, **Figure 37**).
5. Disconnect the starter-to-relay lead at the starter terminal (C, **Figure 37**).
6. Lower the starter and remove from the engine.
7. Installation is the reverse of removal.

Brush Replacement

Chrysler outboards use a variety of starter motors, manufactured primarily by Bosch or Prestolite. Engines may use either a 2- or 4-brush starter design. See **Figure 38** (2-brush) and

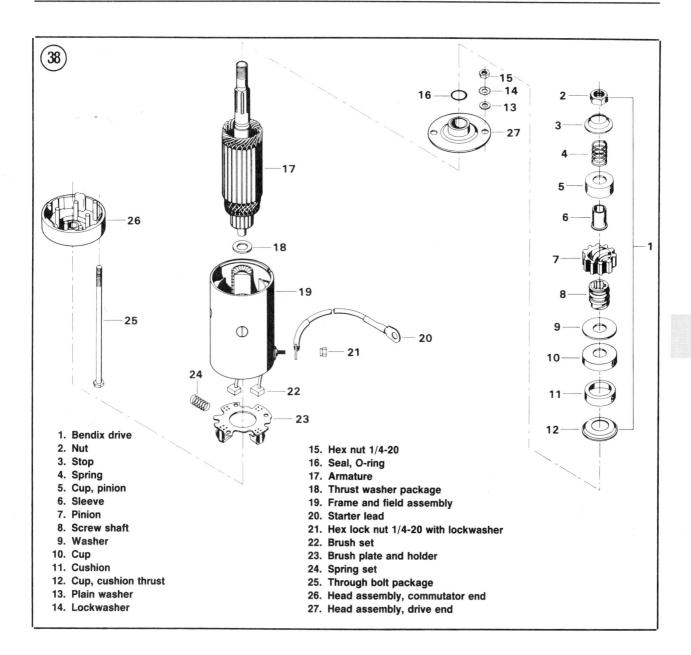

1. Bendix drive
2. Nut
3. Stop
4. Spring
5. Cup, pinion
6. Sleeve
7. Pinion
8. Screw shaft
9. Washer
10. Cup
11. Cushion
12. Cup, cushion thrust
13. Plain washer
14. Lockwasher
15. Hex nut 1/4-20
16. Seal, O-ring
17. Armature
18. Thrust washer package
19. Frame and field assembly
20. Starter lead
21. Hex lock nut 1/4-20 with lockwasher
22. Brush set
23. Brush plate and holder
24. Spring set
25. Through bolt package
26. Head assembly, commutator end
27. Head assembly, drive end

Figure 39 (4-brush) for typical starter components. Always replace brushes in complete sets.

2-brush starter

1. Remove the starter as described in this chapter.

2. Remove the 2 through-bolts from the starter.

3. Lightly tap on end of starter drive with a rubber mallet until the lower end cap breaks free of starter housing. Remove end cap, taking care not to lose the brush springs.

NOTE
*On starters with the brushes attached to the commutator end cap (**Figure 40**), the end cap and brushes are replaced as an assembly. If Step 4 inspection indicates that brushes require replacement, proceed with Step 9.*

4. Inspect the brushes in the end cap (**Figure 40**) or brush holder plate (**Figure 41**). Replace

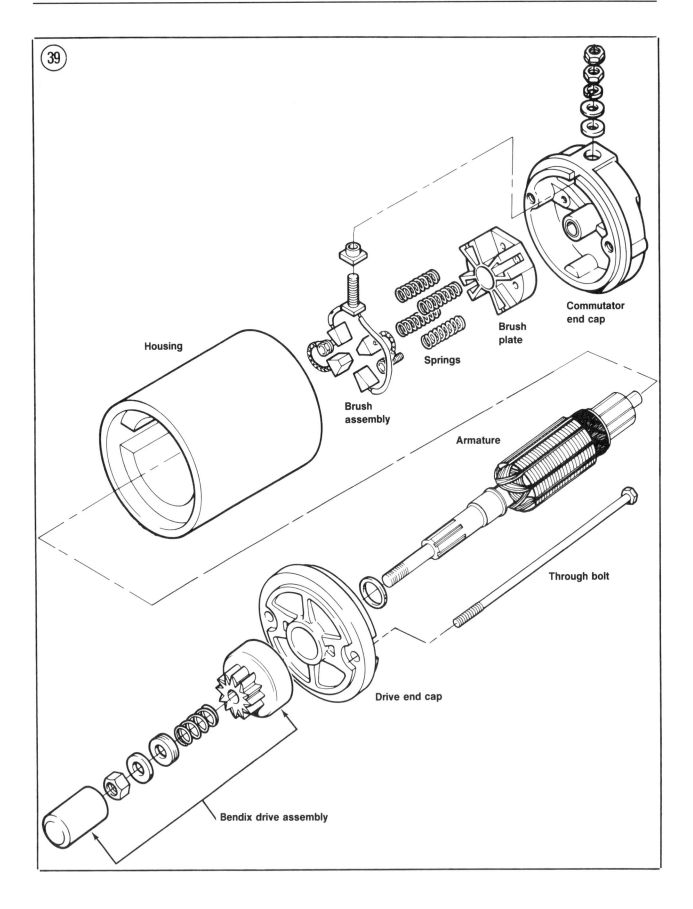

39

Housing

Brush
assembly

Springs

Brush
plate

Commutator
end cap

Armature

Through bolt

Drive end cap

Bendix drive assembly

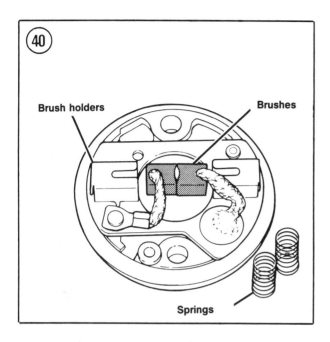

Brush holders

Brushes

Springs

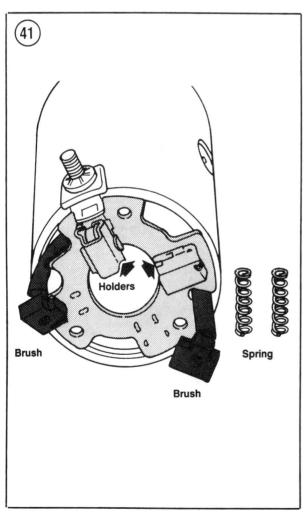

Holders

Brush

Brush

Spring

both brushes if either is pitted, oil-soaked or worn to 3/16 in. or less.

5. Remove the brush holder plate (**Figure 41**). Remove hex nut and washers from positive terminal. Remove positive terminal and brush from starter frame.

6. Remove the screw holding the ground brush to the brush holder plate.

7. Install a new positive terminal and brush assembly to the starter frame.

8. Install a new ground brush to the brush holder plate.

9. Fit springs and brushes into brush holders.

10. Press the brushes into the holders and use a narrow strip of flexible metal or plastic as shown in **Figure 42** to keep them in place.

11. Fit end cap in place, removing the temporary brush retainer as the brush holders slip over the commutator. Align end cap mark with center of positive terminal or end cap tab with starter frame notch, as appropriate.

12. Install through-bolts and tighten to specifications (**Table 4**).

7

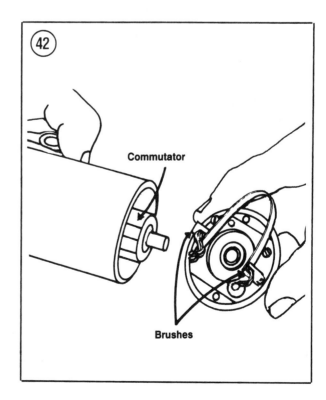

Commutator

Brushes

Bosch 4-brush starter

Fabricate a brush retainer tool from a putty knife with a $1 \times 1/2$ in. opening as shown in **Figure 43**. This tool is necessary to position the brushes properly and prevent damaging them when reassembling the starter end cap to the housing.

1. Remove the starter as described in this chapter.

2. Remove the 2 through-bolts and commutator end cap from the starter. Do not lose brush springs from end cap.

3. Inspect the brushes in the end cap. Replace all brushes if any are pitted, oil-soaked or worn to 1/4 in. or less.

4. If replacement is necessary, remove the screws holding the ground brushes. Remove the hex nut and washers from the positive terminal. Remove positive terminal and insulated brushes from the end cap. See **Figure 44**.

5. Install new insulated brush and terminal assembly in end cap, positioning the long brush lead in the cap slot. See **Figure 45**.

6. Install new ground brushes to holder. Tighten retaining screws snugly.

7. Fit springs and brushes into the holder. Retain the brushes with the brush holder tool and lower the starter frame in place, aligning the starter frame rib with the cap notch. See **Figure 46**.

8. Remove the putty knife and install the through-bolts. Tighten bolts to specifications (**Table 4**).

Prestolite 4-brush starter

Fabricate a retainer clip from flexible metal for each brush as shown in **Figure 47**. These clips are necessary to position the brushes properly and prevent damaging them when reassembling the starter end cap to the housing.

1. Remove the starter as described in this chapter.

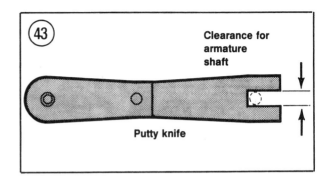

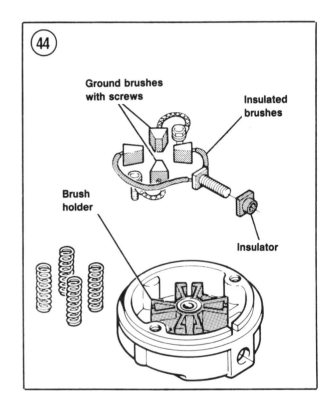

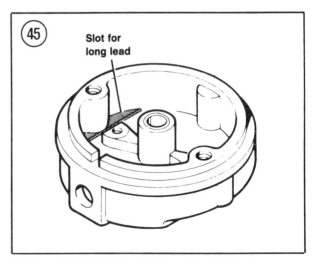

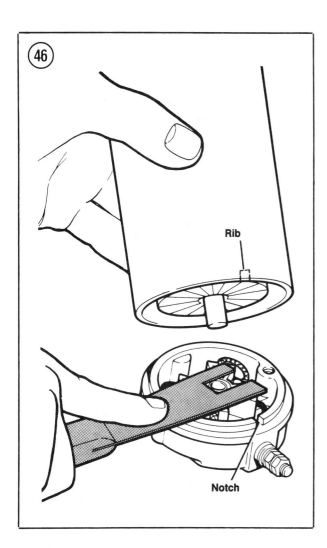

Rib

Notch

2. Remove the 2 through-bolts. Pull the commutator end cap from the starter frame.

3. Remove the washers from the end of the armature shaft, noting the quantity and sequence of installation for reassembly. Remove the armature from the starter frame.

4. Remove the brushes and springs from the brush plate. Remove the brush plate from the starter frame (**Figure 48**).

5. Inspect the brushes. Replace all if any are pitted, oil-soaked or worn to 1/4 in. or less.

6. If brush replacement is required, drive the positive terminal and insulated brush assembly from the starter frame.

7A. Early models—Cut the ground brush leads at the point where they are attached to the field coils. File or grind solder from ends of field coil leads.

7B. Late models—Remove the screw holding the ground brush assembly to the starter frame. Remove the ground brush assembly.

8. Install a new positive terminal and insulated brush assembly.

9A. Early models—Use rosin soldering flux and solder the ground brush leads to the back sides of the field coils, making sure they are in the right position to reach the brush holders.

7

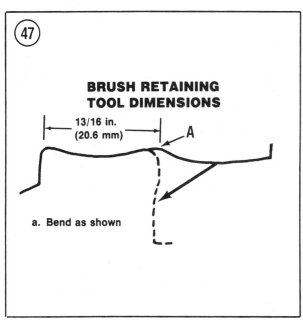

BRUSH RETAINING TOOL DIMENSIONS

13/16 in.
(20.6 mm)

A

a. Bend as shown

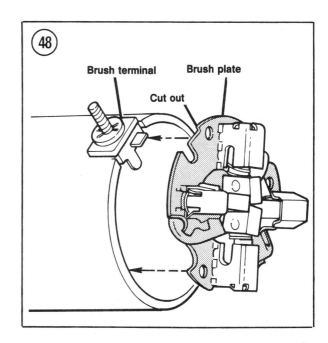

Brush terminal   Brush plate

Cut out

9B. Late models—Install a new ground brush assembly.

10. Install the brush plate, aligning the terminal tab with the plate slot. See **Figure 48**.

11. Install the brush springs and brushes in the plate and hold each in place with a fabricated clip.

12. Install armature in starter frame. When commutator fits between the brushes, remove the fabricated clips and let the brushes move into position.

13. Install end cap to starter frame. Align cap mark with positive terminal (**Figure 49**).

14. Install and tighten through-bolts to specifications (**Table 4**).

Starter Solenoid (Relay) Replacement

Two types of solenoids (relays) are used. See **Figure 50** and **Figure 51**.

1. Disconnect the negative battery cable.

2. Remove the engine cover.

3. Note the wire colors connected to each terminal for reinstallation reference.

4. Remove the nut from each terminal and disconnect the wire(s) from the terminals.

5. Remove the 2 screws holding the solenoid (relay) to the support plate. On 1971 and later models, there is a ground wire attached to the forward screw. Remove the solenoid (relay).

6. Installation is the reverse of removal. Be sure to reattach the correct wire(s) to each terminal as noted in Step 3.

Interlock Switch Replacement

The interlock switch is mounted in different locations according to engine size and model year. See **Figure 52** for typical locations.

1. Disconnect the negative battery cable.

2. Remove the engine cover.

3. Disconnect the lead wires at each end of the interlock switch.

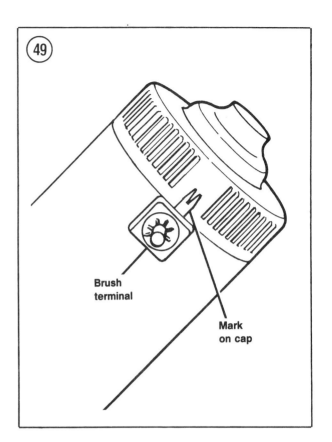

Brush terminal

Mark on cap

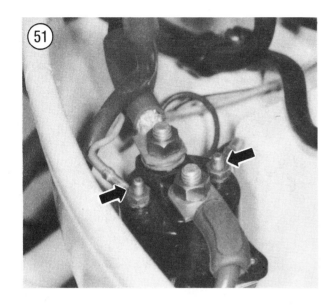

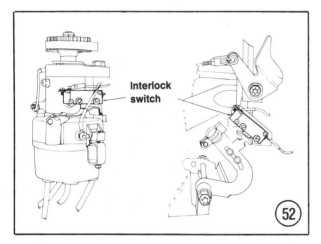

Interlock switch

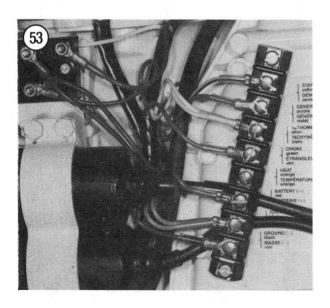

4. Remove the 2 mounting screws and washers. Remove the switch.

5. Installation is the reverse of removal.

Terminal Block Replacement

Refer to **Figure 53** (typical) for this procedure.

1. Disconnect the negative battery cable.

2. Remove the engine cover.

3. Disconnect each lead wire from the terminal block. If the wire connection decal on the power head is damaged, be sure to write down the color and location of each wire before disconnecting it.

4. Remove the 2 mounting screws. Remove the terminal block.

5. Installation is the reverse of removal.

AUTOLECTRIC STARTER-GENERATOR SYSTEM

The Autolectric system is used on 1966-1975 20 hp models and 1966-1975 9.2-15 hp engines with magneto breaker point ignitions.

Starter-generator Cover Removal/Installation

1. Disconnect the negative battery cable.

2. Remove the engine cover.

3. Remove the screws holding the starter-generator cover to the housing. Lift the cover up and off the housing. See **Figure 54**.

4. Disconnect the black ground lead from the light socket at the armature brush assembly.

5. Disconnect the brown lead from the light socket at the resistor terminal. Remove the cover.

6. If light socket requires replacement, pull it from the cover grommet. Remove grommet if necessary.

7

7. Reinstall grommet and light socket in cover, if removed. Route the leads as shown in **Figure 54** to prevent them from rubbing on the armature and causing a short circuit.

8. Route the brown light socket lead through the cover cutout and down toward the rear of the engine. Connect lead to resistor.

9. Connect the black light socket lead to the armature brush mounting screw.

10. Install the cover and tighten screws securely.

11. Reconnect the negative battery cable.

Brush Replacement

Refer to **Figure 55** (9.2-15 hp) or **Figure 56** (20 hp) for this procedure.

1. Disconnect the negative battery cable.

2. Remove the engine cover.

3. Remove the starter-generator cover as described in this chapter.

4. Check the brushes. If pitted, cracked, oil-soaked or worn to less than half their original length, replace both brushes as an assembly.

5. If brush replacement is required, remove the screws holding the brushes to the housing. Pull the brush springs back until the brushes clear the springs and remove them from their holders.

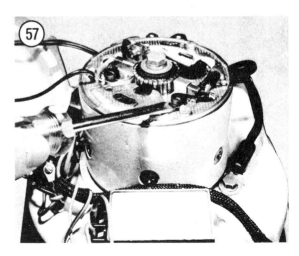

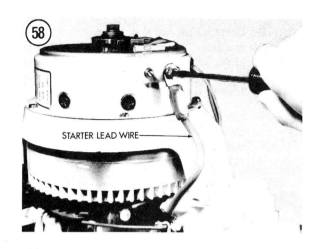

STARTER LEAD WIRE

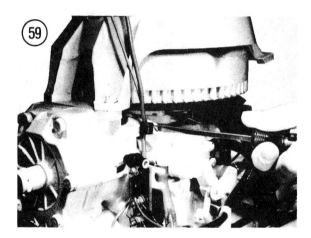

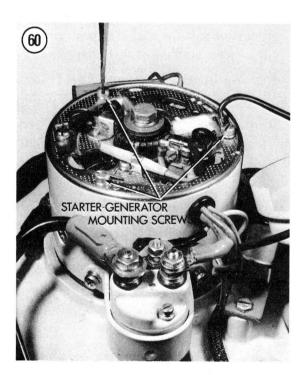

STARTER-GENERATOR MOUNTING SCREW

6. If brush spring replacement is necessary, pry the spring off its mounting stud as shown in **Figure 57**. Slide the new spring loop over the stud and pull spring back and over the rear of the brush.

7. Installation is the reverse of removal.

Starter-generator Housing Removal/Installation

1. Disconnect the negative battery cable.

2. Remove the engine cover.

3. Remove the starter-generator cover, brushes and springs as described in this chapter.

4. Disconnect the starter lead wires at the starter solenoid. If mounted to starter-generator housing, remove the solenoid as described in this chapter.

5. On 20 hp models, disconnect the starter lead wire at the housing terminal (**Figure 58**) and remove the clamp holding the starter-generator lead wires to the transfer port cover (**Figure 59**).

6. Disconnect the gray starter-generator lead from the terminal block on the voltage regulator.

7. Disconnect the red starter-generator lead from the diode terminal.

8. Remove the screws holding the starter-generator housing to its support. See **Figure 60** (9.2-15 hp) or **Figure 61** (20 hp). Remove the housing.

9. Installation is the reverse of removal, plus the following:

    a. 9.2-15 hp—Position housing on support with housing leads facing toward front

7

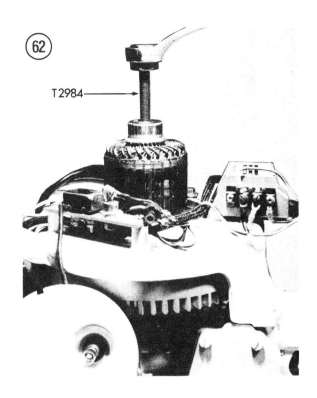

and slightly toward the port side of the engine.

  b. 20 hp—Position housing on support with starter lead wire terminal facing to rear port side of engine.

Armature Removal/Installation

1. Disconnect the negative battery cable.
2. Remove the engine cover.
3. Remove the starter-generator housing as described in this chapter.
4. Remove the hex bolt, plain washer and spring lockwasher holding the armature to the crankshaft.
5. Install armature puller part No. T 2984 (9.2-15 hp) or part No. T 8983 (20 hp) in armature threads. Tighten puller until armature is free from crankshaft. See **Figure 62** (9.2-15 hp shown).
6. Remove armature, armature key and spacer from crankshaft.
7. Installation is the reverse of removal. Tighten hex bolt to specifications (**Table 3**).

Starter-generator
Support and Bearing Removal
(9.2-15 hp)

1. Disconnect the negative battery cable.

2. Remove the engine cover.

3. Remove the starter-generator housing and armature as described in this chapter.

4. Remove the circuit breaker bracket from the starter-generator support.

5. Remove the voltage regulator and starter interlock switch as described in this chapter.

6. Disconnect the battery lead at the starter-generator support.

7. Remove the screws holding the support to the power head. Remove support.

8. Remove the magneto flywheel spacer from the crankshaft.

9. Press starter-generator bearing from support with special tool part No. T 8937. See **Figure 63**.

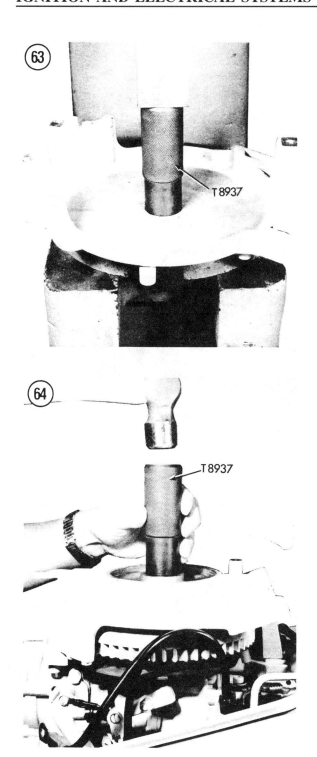

2. Install starter-generator support but do not tighten screws at this time.

3. Wipe outer diameter of starter-generator bearing with Loctite No. 75. Install bearing on support with special tool part No. T 8937. See **Figure 64**. Drive bearing into support until it is flush with the top surface of the bearing boss.

4. Tighten support screws and drive the bearing further onto the crankshaft until the magneto flywheel is seated and the bearing presses tightly against the flywheel spacer.

5. Install the battery lead to the support.

6. Install the starter interlock switch and voltage regulator as described in this chapter.

7. Install the circuit breaker bracket to the support.

8. Install the armature and starter-generator housing as described in this chapter.

9. Install the engine cover and reconnect the negative battery cable.

**Starter-generator
Support and Bearing Removal
(20 hp)**

1. Disconnect the negative battery cable.

2. Remove the engine cover.

3. Remove the starter-generator housing and armature as described in this chapter.

4. Remove the 2 screws and lockwashers holding the starter-generator support to the power head.

5. Remove the screw and lockwasher holding the support to the starter-generator bracket. Remove support from power head.

6. Remove the magneto flywheel spacer from the crankshaft.

7. Press starter-generator bearings from support with special tool part No. T 8937. See **Figure 65**.

**Starter-generator
Support and Bearing Installation
(9.2-15 hp)**

1. Install the magneto flywheel spacer on the crankshaft.

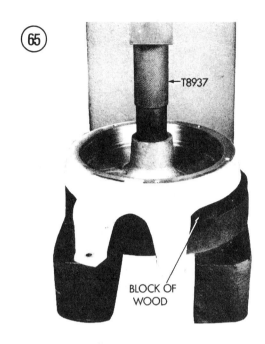

BLOCK OF WOOD

Starter-generator
Support and Bearing Installation
(20 hp)

1. Install the magneto flywheel spacer on the crankshaft.

2. Install starter-generator support. Wipe mounting screw thread with Loctite H and install with lockwashers. Do not tighten screws at this time.

3. Wipe outer diameter of starter-generator bearing with Loctite No. 75. Install bearing on support with special tool part No. T 8937. See **Figure 64** (typical). Drive bearing into support until it is flush with the top surface of the bearing boss, then tighten the support mounting screws.

4. Wipe outer diameter of second bearing with Loctite No. 75. Position bearing on crankshaft above the one installed in Step 3.

5. Using tool part No. T 8937, drive both bearings further on the crankshaft until the bottom bearing presses tightly against the flywheel spacer.

6. Install the armature and starter-generator housing as described in this chapter.

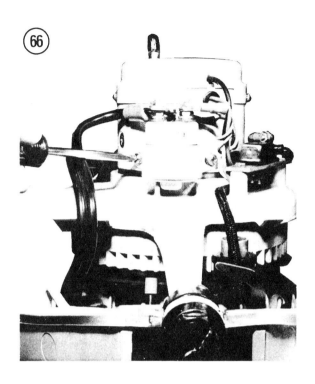

7. Install the engine cover and reconnect the negative battery cable.

Light Switch Replacement

1. Disconnect the negative battery cable.
2. Remove the engine cover.
3. Remove the nut holding the switch to the support plate.
4. Remove the switch and disconnect the 2 leads from the terminals at the rear of the switch.
5. Installation is the reverse of removal. Switch should be positioned with its end terminal facing the support plate (20 hp) or side terminal facing the starboard side of the engine (all others).

Starter Solenoid
(Relay) Replacement

This procedure covers the solenoid (relay) attached to the starter-generator housing. For models equipped with a remote solenoid (relay), see *Starter Motor System, Starter Solenoid (Relay) Replacement* in this chapter.

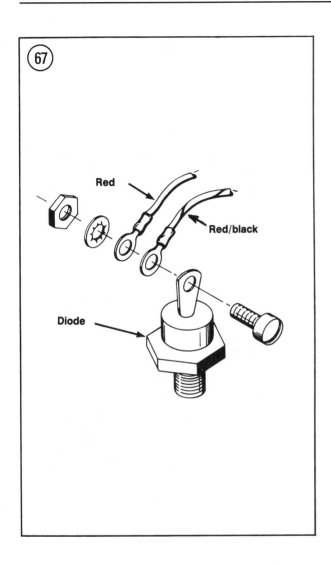

1. Disconnect the negative battery cable.
2. Remove the engine cover.
3. Note the wire colors connected to each terminal for reinstallation reference.
4. Remove the nut from each terminal and disconnect the wire(s) from the terminals.
5. Remove the 2 screws holding the solenoid (relay) to the starter housing. See **Figure 66**. Remove solenoid (relay) with bracket.
6. Installation is the reverse of removal. Be sure to reattach the correct wire(s) to each terminal as noted in Step 3.

Interlock Switch Replacement

The interlock switch is mounted on the magneto control bracket (20 hp) or

starter-generator support (all others). See *Starter Motor System Interlock Switch Replacement* in this chapter.

Terminal Block Replacement (20 hp)

See *Starter Motor System Terminal Block Replacement* in this chapter.

Circuit Breaker Replacement

The circuit breaker is mounted on a terminal board on Autolectric models with the diode (rectifier) and resistor. See *Alternator Charging System Circuit Breaker Replacement* in this chapter.

Diode (Rectifier) Replacement

The diode (rectifier) is mounted on a terminal board with the circuit breaker and resistor.
1. Disconnect the negative battery cable.
2. Remove the engine cover.
3. Disconnect the leads at the diode. See **Figure 67** (typical).
4. Remove the nut and washer holding the diode to the terminal board. Remove the diode.
5. Installation is the reverse of removal. On 20 hp engines, diode terminal should face the rear of the engine.

Resistor Removal/Installation

The resistor is mounted on a terminal board with the circuit breaker and diode (rectifier).
1. Disconnect the negative battery cable.
2. Remove the engine cover.
3. Disconnect the wires at the resistor (**Figure 68**).
4. Remove the screw holding the resistor clamp to the top (20 hp) or rear (all others) terminal board mounting boss. Remove the resistor and clamp.

7

5. Bend clamp slightly and remove from resistor.

6. Installation is the reverse of removal.

Voltage Regulator
Removal/Installation

1. Disconnect the negative battery cable.

2. Remove the engine cover.

3. 20 hp—Disconnect the voltage regulator-to-diode leads.

NOTE
Removal and installation of the regulator on 20 hp models is easier if Step 4 and Step 5 are reversed.

4. Disconnect the lead wires at the voltage regulator.

5. Remove the voltage regulator mounting screws. Remove the voltage regulator.

6. Installation is the reverse of removal. Connect the voltage regulator lead wires as shown in **Figure 69** (20 hp) or **Figure 70** (all others).

IGNITION SYSTEMS

The outboards covered in this manual use one of the following ignition systems:
   a. Magneto breaker point.
   b. Battery breaker point.
   c. Distributor breaker point CD.
   d. Distributor breakerless CD.
   e. Alternator (magneto) CD.
Refer to Chapter Three for troubleshooting and test procedures.

MAGNETO BREAKER
POINT IGNITION

Manual start models use a magneto ignition with a combined primary/secondary ignition coil, condenser and one set of breaker points for each cylinder mounted on a stator plate. The primary section of the coil connects to the stationary breaker point and

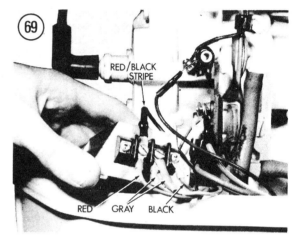

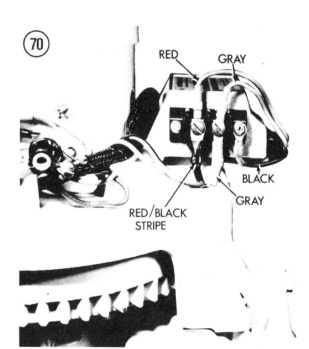

is grounded to the stator plate. The secondary section of the coil connects to the spark plug and is also grounded to the stator plate.

See Chapter Three for a schematic of a 2-cylinder magneto ignition showing the location and relationship of the components. Troubleshooting and test procedures are also given in Chapter Three.

Operation

As the flywheel rotates, magnets around its outer diameter create a current that flows through the closed breaker points. This flow of current through the coil primary winding builds a strong magnetic field in the coil secondary winding. When the cam opens the No. 1 point set, the magnetic field collapses, inducing a high voltage (approximately 18,000 volts) in the coil secondary winding

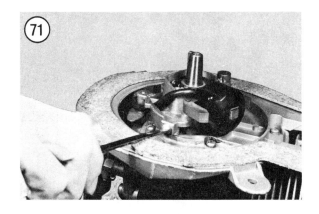

SPARK PLUG LEAD WIRE

that is sent to the No. 1 spark plug. The condenser absorbs any residual current remaining in the primary windings. This eliminates arcing at the points and produces a stronger spark at the plug. The breaker points close and the flywheel continues to rotate, duplicating the sequence in 2-cylinder engines for the No. 2 point set and ignition coil to fire the No. 2 spark plug.

Stator Plate Removal/Installation (1966-1969 3.5 hp)

1. Remove the fuel tank. See Chapter Six.
2. Remove the flywheel. See Chapter Eight.
3. Remove the 2 screws holding the throttle cam and magneto control lever bracket to the stator plate.
4. Remove the clamp which holds the spark plug lead wire to the support plate.
5. Disconnect the spark plug lead from the spark plug.
6. Loosen the friction shoe assembly about 1/4 in. See **Figure 71**.
7. Remove stator plate from support plate.
8. To reinstall, route the spark plug lead around the throttle cam bosses (**Figure 72**) and then position the stator plate on the support plate.
9. Install throttle cam and magneto control lever bracket as follows:
   a. Move stator plate to full retard position (throttle cam bosses on stator toward starboard side).
   b. Install screw, throttle cam and lever bracket to starboard side boss on stator. See **Figure 73**.

NOTE
Note position of lever bracket with throttle cam directly under head of screw—particularly recessed area on bracket casting between boss for magneto control lever and throttle cam surface.

c. Do not tighten screw until port side screw is installed.

d. Move stator to full advance position and install port side screw to hold throttle cam and lever bracket. Tighten the screw securely.

e. Move stator back to full retard position and tighten starboard side screw.

10. Install spark plug lead clamp. Move stator to full advance position and secure clamp to support plate with screw.

11. Adjust stator friction by turning friction shoe assembly (**Figure 71**) in stator. A definite drag must be felt when advancing or retarding stator plate.

12. Adjust the breaker points. See Chapter Four.

13. Connect the spark plug lead.

14. Install the flywheel (Chapter Eight) and fuel tank (Chapter Six).

**Stator Plate Removal/Installation
(1970-1977 3.6 hp; 1976-1978 4 hp;
1974-1976 4.9-5 hp)**

1. Remove the engine cover.
2. Remove the flywheel. See Chapter Eight.
3. Disconnect the throttle link end from the stator ball pivot.
4. Disconnect the spark plug lead at the spark plug.
5. Loosen the friction shoe assembly about 1/4 in. See **Figure 74**.
6. Remove stator plate from support plate.
7. To reinstall, route the spark plug lead as shown in **Figure 75**, then position the stator plate on the support plate.
8. Adjust stator friction by turning friction shoe assembly (**Figure 74**) in stator. A definite drag must be felt when advancing or retarding stator plate.
9. Connect the throttle link end to the stator ball pivot.

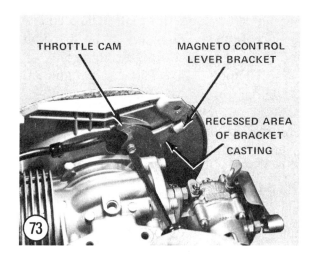

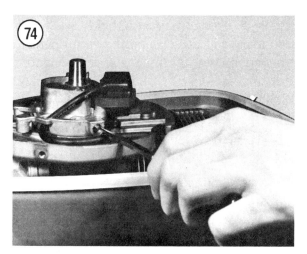

10. Adjust the breaker points. See Chapter Four.

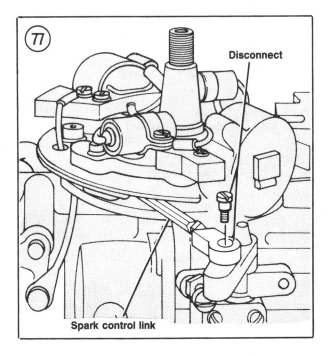

5. Loosen the friction screw at the front of the stator plate about 1/4 turn.

6. Disconnect the ground leads at the stop switch and crankcase cover.

7. Remove the stator plate from the power head.

8. Invert stator plate and remove the throttle cam and lever assembly.

9. Installation is the reverse of removal. Adjust throttle pick-up and stator friction screw to prevent stator from creeping at wide-open throttle.

Stator Plate Removal/Installation (2-cylinder Engines)

1. Remove the engine cover.

2. Remove the flywheel. See Chapter Eight.

3. Disconnect the spark plug leads at the spark plugs. Remove plug leads from exhaust port cover clip, if so equipped.

NOTE
On some engines, the ground lead is attached under one of the stator mounting screws.

4. Disconnect the stator ground lead at the fuel pump or stator, as appropriate.

5. Remove the stator mounting screws. See **Figure 76** for typical location.

6. Disconnect stop switch wires at front of support plate, if so equipped.

7A. Tower shaft models—Disconnect the spark control link swivel from the tower shaft. See **Figure 77**.

7B. Throttle cam models—Lift stator up and to port side to disconnect it from the throttle cam link.

8. Remove the stator plate.

9. If equipped with a throttle cam, invert stator plate. Remove throttle cam screws, bend cam tab enough to free the spark plug wires and remove the cam and ground wire from the stator.

11. Connect the spark plug lead.

12. Install the flywheel (Chapter Eight) and engine cover.

Stator Plate Removal/Installation (1980-on 3.5 hp; 1979-on 4 hp)

1. Remove the engine cover.

2. Remove the fuel tank. See Chapter Six.

3. Remove the flywheel (Chapter Eight).

4. Disconnect the spark plug lead at the spark plug.

7

10. Installation is the reverse of removal. Adjust the breaker points (Chapter Four). Adjust the throttle pick-up and ignition timing as required. See Chapter Five.

Breaker Point and Condenser Replacement

See *Tune-up,* Chapter Four.

Coil Removal/Installation

1. Remove the engine cover.
2. Remove the flywheel. See Chapter Eight.
3. Remove the stator plate as described in this chapter.
4. Disconnect the coil lead wires at the breaker point and ground. See **Figure 78** (typical).
5. Use a pair of pliers and straighten the coil lamination which holds the coil. See **Figure 79** (typical).
6. Pry the lip of the coil wedge spring from the coil laminations with a screwdriver, then pull coil free of laminations. Be sure to note how the spark plug lead is routed.
7. Remove the sparkie cover and spring from the end of the coil spark plug lead.
8. Pull the spark plug lead through the stator plate grommet and remove the coil.
9. To reinstall the coil, position it at the stator plate lamination with spark plug lead routed as noted in Step 6.
10. Fit coil wedge spring in coil with small lip facing downward toward crankshaft.
11. Insert coil halfway into lamination, making sure that the spark plug wire remains properly routed. Insert end of wire through grommet in stator plate.
12. Push down on coil until the wedge spring snaps into place at the rear of the laminations. The large lip on the spring must be positioned over the front surface of the coil.

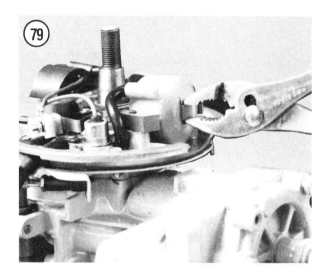

13. Pull any excess plug wire through the stator plate, then bend the bottom lamination up with pliers to retain the coil in place.
14. Connect the coil lead wires to the breaker point set and ground.
15. Reinstall the sparkie spring and cover on the end of the plug wire, making sure that the spring contacts the wire inside the insulation.
16. Reverse Steps 1-3 to complete installation.

BATTERY BREAKER POINT IGNITION

Some 2-cylinder electric start models use a battery ignition with a condenser and one set

of breaker points for each cylinder mounted on a breaker plate. Each cylinder has an externally-mounted ignition coil. The primary section of the coil connects to the stationary breaker point and the battery. The secondary section of the coil connects to the spark plug and is also grounded to the stator plate.

Some early 75 and 105 hp electric start models use a distributor battery ignition. Mounted on the power head, the distributor contains the breaker points and is belt-driven by the crankshaft. A condenser is mounted on the distributor and a single ignition coil serves all cylinders.

Troubleshooting and test procedures are given in Chapter Three.

Operation (2-cylinder)

When the ignition switch is in the RUN position, current flows from the battery through the primary winding of the ignition coil to ground while the breaker points are closed. This flow of current through the coil

primary winding builds a strong magnetic field in the coil secondary winding. When the breaker plate opens the No. 1 point set, the magnetic field collapses, inducing a high voltage (approximately 15,000 volts) in the coil secondary winding that is sent to the No. 1 spark plug. The condenser absorbs residual current remaining in the primary windings. This eliminates arcing at the points and produces a stronger spark at the plug. When the breaker points close, the sequence is duplicated for the No. 2 point set and ignition coil to fire the No. 2 spark plug.

Operation (3- and 4-cylinder)

Battery ignitions using a distributor work essentially the same as those with a breaker plate, except that the distributor determines which spark plug will receive the firing voltage. All cylinders are fired by the operation of a single breaker point set through one ignition coil.

Breaker Plate
Removal/Installation

Refer to **Figure 80** for this procedure.
1. Remove the engine cover.
2. Disconnect the spark plug leads at the spark plugs.
3. Remove the flywheel. See Chapter Eight.
4. Pull the breaker point leads from the spade terminals.
5. Remove the 2 screws holding the spade terminals in place. Remove the terminals.
6. Remove the 3 screws holding the stator to the bearing cage.
7. Lift the stator up and over the crankshaft end. Carefully place stator to one side.
8. Remove the 3 clips, disconnect the spark control link and remove the breaker plate.
9. Installation is the reverse of removal. Position clips so their long tabs curl upward. The hole in each tab should align with the

7

stator mounting screw holes. Adjust ignition timing (Chapter Five).

Distributor Removal/ Installation

1. Disconnect the negative battery cable.
2. Remove the engine cover.
3. Remove the flywheel. See Chapter Eight.
4. Disconnect the spark plug leads at the spark plugs.
5. Disconnect the white/black and blue leads from the distributor housing terminals. See A, **Figure 81**.
6. Disconnect the distributor-to coil lead at the coil. See B, **Figure 81**.
7. Loosen the screw on each distributor cap retaining clip (C, **Figure 81**), then remove distributor cap assembly from housing.
8. Disconnect ground lead. See D, **Figure 81**.
9. Remove the 2 bolts holding the distributor bracket to the power head (E, **Figure 81**).
10. Slip distributor belt off pulley and remove the distributor bracket and housing from the engine.
11. Installation is the reverse of removal. Adjust ignition timing (Chapter Five).

Breaker Point and Condenser Replacement

See *Tune-up,* Chapter Four.

Ignition Coil Removal/Installation (2-cylinder Engine)

1. Disconnect the negative battery cable.
2. Remove the engine cover.
3. Disconnect the spark plug leads at each coil.
4. Disconnect the blue and white leads from the top coil. Disconnect the blue and brown leads from the bottom coil.

NOTE
A crowsfoot socket wrench or a thin open-end wrench must be used to remove the lower coil clamp screws in Step 5.

5. Remove the 2 upper and 2 lower coil clamp mounting screws. See **Figure 82**.
6. Remove the coils.
7. To reinstall, connect a blue lead to the positive (+) terminal of each coil.
8. Connect the white lead to the negative (–) terminal of the top coil and the brown lead to the negative (–) terminal of the bottom coil.
9. Fit the coils in place and install the coil clamp. Be sure to reinstall the rectifier ground lead under the upper left hand bracket screw.
10. Connect the spark plug leads to their respective coils.

Ignition Coil Removal/Installation (3- and 4-cylinder Engine)

1. Disconnect the negative battery cable.
2. Remove the engine cover.
3. Disconnect the distributor-to-coil lead at the coil.

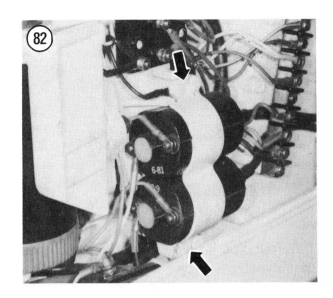

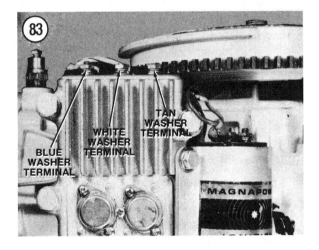

4. Disconnect the blue, white and white/black leads at the coil terminals.

5. Unbolt the coil mounting bracket and remove bracket and coil assembly.

6. Installation is the reverse of removal.

DISTRIBUTOR CD IGNITION

Distributor CD ignitions may use a breaker point or breakerless distributor. The distributor (belt-driven by the crankshaft) is mounted on the power head along with a capacitor-discharge (CD) module and a single ignition coil. Terminal block connections are shown on the exhaust port cover decal.

Troubleshooting and test procedures are given in Chapter Three. Adjust ignition timing after replacing any components.

Operation

Battery current is stored in the CD module. It is also sent to the distributor where it flows through the closed breaker points. When the distributor cam opens the breaker points, it interrupts the current flow and causes the CD module to discharge its stored current into the ignition coil.

Breakerless distributors use a preamplifier instead of breaker points. The preamplifier is a switching device which contains a light-emitting diode and a light-sensitive switch. The current to the distributor energizes the light-emitting diode. A cup on the distributor rotor contains a window for each cylinder. As the distributor shaft rotates the rotor cup, these windows allow the diode to trigger the light-sensitive switch. The switch signals the CD module to discharge its stored current into the ignition coil.

In each system, the coil steps up the current to 25,000-32,000 volts which is then sent to the appropriate spark plug through the distributor.

Distributor Removal/ Installation

See *Battery Breaker Point Ignition* in this chapter.

Breaker Point Replacement

See *Tune-up,* Chapter Four.

Delta CD Module Removal/Installation

1. Disconnect the negative battery cable.

2. Remove the engine cover.

3. Refer to **Figure 83** and disconnect the blue, white and white/black leads from the blue, white and tan washer terminals.

7

4. Disconnect the black coil lead from the side of the CD module.

5. Disconnect the blue and white leads from the coil positive (+) and negative (–) terminals.

6. Remove the 3 bolts holding the CD module to the power head. Remove the CD module and coil assembly.

7. Separate the coil from the CD module by removing the 4 coil clamp screws.

> *NOTE*
> *If washers are not installed correctly in Step 8, the coil will deliver a weak spark.*

8. Installation is the reverse of removal, plus the following:

   a. Install the coil clamp hex head screws with a steel washer and fiber washer under the head of each screw.

   b. Install the plastic washer between the clamp and CD module.

   c. Tighten mounting bolts to specifications (**Table 4**).

Delta Ignition Coil
Removal/Installation

1. Disconnect the negative battery cable.

2. Remove the engine cover.

3. Disconnect the blue and white leads from the coil positive (+) and negative (–) terminals.

4. Disconnect the black coil lead from the side of the CD module.

5. Disconnect the distributor-to-coil lead at the coil.

6. Remove the 2 hex head screws from the coil clamp, then pull the coil from the clamp.

> *NOTE*
> *If washers are not installed correctly in Step 7, the coil will deliver a weak spark.*

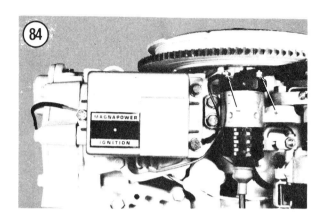

7. Installation is the reverse of removal, plus the following:

   a. Install the coil clamp hex head screws with a steel washer and fiber washer under the head of each screw.

   b. Install the plastic washer between the clamp and CD module.

Motorola CD Module
(Part No. 404301 and 404301-1)
Removal/Installation

1. Disconnect the negative battery cable.

2. Remove the engine cover.

3. Disconnect the CD module blue and white lead wires at the terminal block.

4. Disconnect the gray lead wire at the coil positive (+) terminal. See **Figure 84**.

5. Disconnect the white/black lead wire at the distributor terminal stud.

6. Remove the screws holding the CD module to the bracket. Remove the CD module.

7. Installation is the reverse of removal, plus the following:

   a. Install the black coil ground wire under the bottom right screw.

   b. Route the blue and white lead wires behind the CD module bracket and wrap around the wiring harness to prevent them from contacting the flywheel.

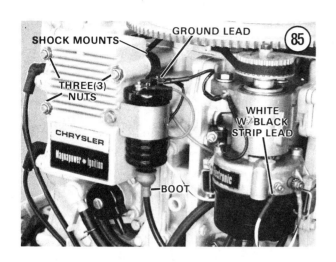

SHOCK MOUNTS — GROUND LEAD ⑧⑤
THREE(3) NUTS — WHITE W/BLACK STRIP LEAD — CHRYSLER Magnapower Ignition — BOOT

Motorola Ignition Coil
(Part No. 404301 and 404301-1)
Removal/Installation

1. Disconnect the negative battery cable.
2. Remove the engine cover.
3. Disconnect the gray lead wire at the coil positive (+) terminal. See **Figure 84**.
4. Disconnect the black lead wire at the coil negative (–) terminal.
5. Disconnect the distributor-to-coil lead at the coil.
6. Remove the screws holding the coil clamp to the CD module bracket. Remove coil, clamp and spacer from power head.
7. Remove coil from clamp.

NOTE
If washers are not installed correctly in Step 8, the coil will deliver a weak spark.

8. Installation is the reverse of removal, plus the following:
   a. Install rubber band on top and bottom portion of coil that fits under the clamp.
   b. Install the coil clamp screws with a steel washer and phenolic washer under the head of each screw.
   c. Install the spacer between the ends of the coil clamp.

Motorola CD Module
(Part No. K404301-2, and A523301-1)
Removal/Installation

Refer to **Figure 85** for this procedure.
1. Disconnect the negative battery cable.
2. Remove the engine cover.
3. Disconnect the white/black lead wire at the distributor housing terminal stud.
4. Disconnect the CD module blue, red and white lead wires at the terminal block.
5. Disconnect the distributor-to-coil ground lead at the distributor.
6. Disconnect the distributor-to-coil lead at the coil.
7. Remove the fasteners holding the CD module to the power head. Remove the CD module.
8. Installation is the reverse of removal, plus the following:
   a. Route the white/black lead behind the tower shaft.
   b. Route the blue, red and white leads over the power head, under the flywheel and down to the terminal block.
   c. Tighten mounting fasteners to specifications (**Table 4**).

Motorola Ignition Coil
(Part No. K404301-2 and A523301-1)
Removal/Installation

Refer to **Figure 86** for this procedure.
1. Remove the CD module as described in this chapter.
2. Disconnect the lead wires at the coil terminals.
3. Remove the screws and lockwashers holding the coil and clamp to the CD module. Separate coil and clamp from CD module.
4. Installation is the reverse of removal, plus the following:
   a. Install rubber band on top and bottom portion of coil that fits under the clamp.

7

b. Position coil with the positive (+) terminal facing away from the power head.

MAGNAPOWER II IGNITION

While Chrysler Marine used the Magnapower II designation rather loosely, this section pertains specifically to the system used on 1977-1979 55 hp, 1974-1976 60 hp, 1977-1978 65 hp and 1975-1977 105, 120 and 135 hp engines.

Manufacture of the CD module (part No. S474301-1) was discontinued in 1981. Other system components are available but should the module fail, the only repair option is to install a retrofit kit which converts the ignition to a Motorola Electronic Breakerless Ignition. This kit was made available for 55-65 hp (part No. K1124) and 105-135 hp (part No. K1054) models and is still available from some Chrysler/U.S. Marine dealers.

Troubleshooting and test procedures are given in Chapter Three. Adjust ignition timing after replacing any components.

CD Module and Timing Ring
Retainer Removal/Installation

1. Disconnect the negative battery cable.
2. Remove the engine cover.
3. Remove the flywheel. See Chapter Eight.
4. Remove the nuts and washers holding the timing ring retainer in place (**Figure 87**). Remove the retainer.
5. Disconnect the lead wires at the module. See **Figure 88**.
6. Remove the 2 studs holding the CD module to the bearing cage (**Figure 89**). Remove the CD module.
7. To reinstall the module, position it on the bearing cage and align the module mounting holes with the bearing cage holes.
8. Install the 2 studs, push down on module to seat it on bearing cage and tighten the studs securely.

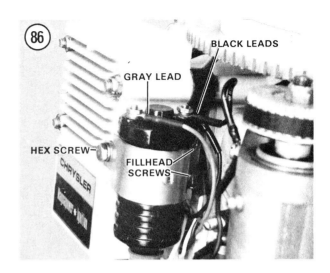

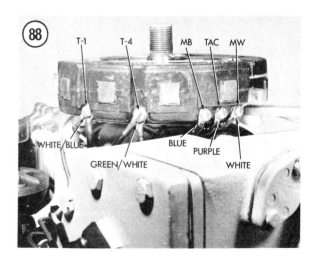

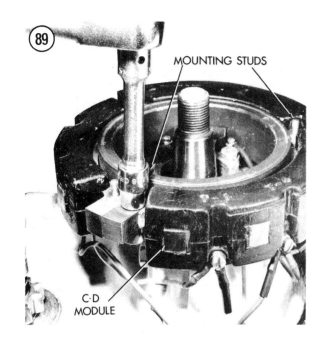

MOUNTING STUDS

C-D MODULE

9. Reconnect the lead wires at the module. See **Figure 88**.

10. Install the washers, timing ring retainer and nuts. Place a 0.003 in. feeler gauge between the retainer and timing ring at the retainer legs. When nuts are tightened down evenly, there should be no drag on the feeler gauge. See **Figure 90**.

11. With tower shaft disconnected from pivot arm, connect a spring scale to the pivot arm outer hole (**Figure 91**). Rotate timing ring manually to check friction. Scale should read 4-6 pounds (55-65 hp) or 3-5 pounds (105-135 hp).

12. If scale reading in Step 11 is not within specifications, tighten or loosen nuts (Step 10) evenly (but no more than 1/8 turn) to obtain the proper friction.

13. Reverse Steps 1-3 to complete installation.

7

Alternator Stator Module
Removal/Installation

1. Disconnect the negative battery cable.
2. Remove the engine cover.
3. Remove the flywheel. See Chapter Eight.
4. Disconnect the lead wires at the alternator stator module. See **Figure 92**.

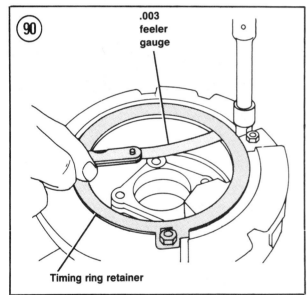

.003 feeler gauge

Timing ring retainer

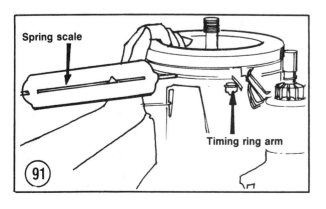

Spring scale

Timing ring arm

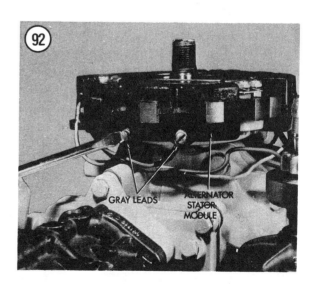

GRAY LEADS

ALTERNATOR STATOR MODULE

5. Remove the nuts and washers holding the timing ring retainer in place (**Figure 87**). Remove the retainer.

6. Remove the 2 studs holding the alternator stator module to the bearing cage (**Figure 93**). Remove the alternator stator module.

7. To reinstall the module, complete Steps 7-13 of *CD Module and Timing Ring Retainer Removal/Installation* in this chapter. Connect the lead wires as shown in **Figure 92**.

Capacitor Module
Removal/Installation

1. Disconnect the negative battery cable.
2. Remove the engine cover.
3. Remove the flywheel. See Chapter Eight.
4. Remove the nuts and washers holding the timing ring retainer in place (**Figure 87**). Remove the retainer.
5. Disconnect the lead wires at the capacitor module. See **Figure 94**.
6. Remove the 2 studs holding the capacitor module to the bearing cage (**Figure 95**). Remove the capacitor module.
7. To reinstall the module, complete Steps 7-13 of *CD Module and Timing Ring Retainer Removal/Installation* in this chapter. Connect the lead wires as shown in **Figure 94**.

Regulator-rectifier Module
Removal/Installation
(60 hp and 105-135 hp)

1. Remove the capacitor module as described in this chapter.
2. Disconnect the 2 regulator-rectifier leads at the alternator stator module. See **Figure 92**.
3. Disconnect the regulator-rectifier lead at the terminal block.

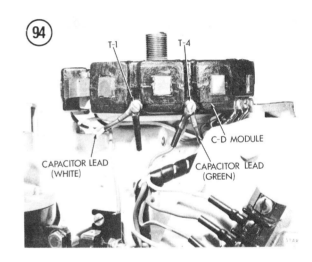

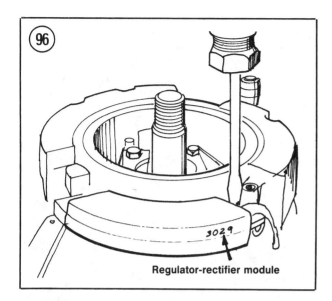

Regulator-rectifier module

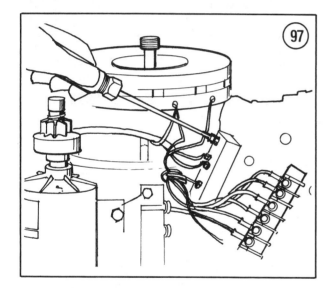

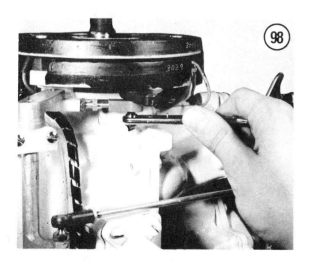

4. Remove the 2 screws holding the regulator-rectifer to the bearing cage (**Figure 96**). Remove the regulator-rectifier.

5. Installation is the reverse of removal.

Regulator-rectifier
Removal/Installation
(55 and 65 hp)

1. Disconnect the negative battery cable.
2. Remove the engine cover.
3. Remove the flywheel. See Chapter Eight.
4. Disconnect the leads at the regulator-rectifier. See **Figure 97**.
5. Remove the 2 screws holding the regulator-rectifier to the power head. Remove the regulator-rectifier.
6. Installation is the reverse of removal.

7

Timing Ring
Removal/Installation

1. Disconnect the negative battery cable.
2. Remove the engine cover.
3. Remove the flywheel. See Chapter Eight.
4. Disconnect the spark control link from the timing ring arm. See **Figure 98** (typical).
5. 55-65 hp—Remove the capacitor module as described in this chapter.
6. 105-135 hp—Remove the trigger module as described in this chapter.
7. 60 and 105-135 hp—Remove the regulator- rectifier module as described in this chapter.
8. Remove the timing ring from the engine (**Figure 99**).
9. Installation is the reverse of removal, plus the following:
   a. Coat bearing cage and timing ring retainer with Rykon No. 2EP grease.
   b. 55-65 hp—Position timing ring in bearing cage with arm to starboard side.
   c. 105-135 hp—Position timing ring in bearing cage with arm to port side.

Trigger Assembly and Module Removal/Installation (55-65 hp)

The trigger assembly consists of 2 individual trigger modules.

1. Remove the timing ring as described in this chapter.

2. Disconnect the lead wires at the trigger modules. See **Figure 100**.

3. Remove the 4 screws and ground lead holding the trigger assembly to the bearing cage.

4. Pry trigger assembly free of bearing cage with a screwdriver (**Figure 101**) and remove from engine.

5. To remove one module from the mounting plate, bend the metal tab down (**Figure 102**) and slip the module from the plate.

6. Installation is the reverse of removal. Be sure to reinstall the ground lead under the head of the rear starboard mounting screw.

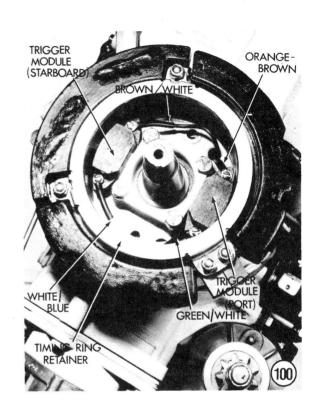

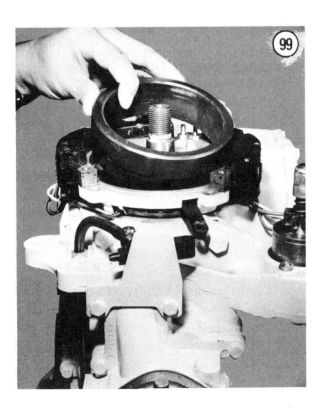

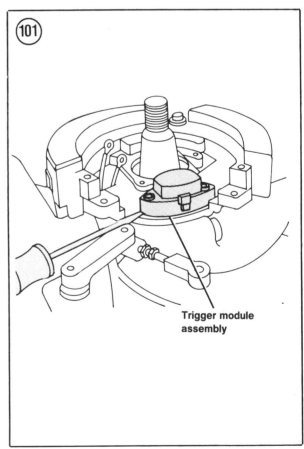

Trigger module assembly

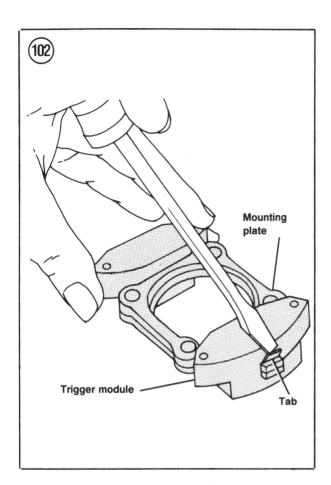

102

Mounting plate

Trigger module

Tab

Trigger Assembly and Module Removal/Installation (105-135 hp)

The trigger assembly consists of 4 individual trigger modules.

1. Disconnect the negative battery cable.
2. Remove the engine cover.
3. Remove the flywheel. See Chapter Eight.
4. Disconnect the lead wires at the trigger modules. See **Figure 103**.
5. Remove the 4 screws holding the trigger assembly to the bearing cage. See **Figure 104**.
6. Remove trigger assembly from bearing cage.
7. To remove one module from the mounting plate, bend the metal tab down (**Figure 105**) and slip the module from the plate.
8. Installation is the reverse of removal.

Ignition Coil Removal/Installation

When removing more than one coil on a 3- or 4-cylinder engine, it is a good idea to draw

7

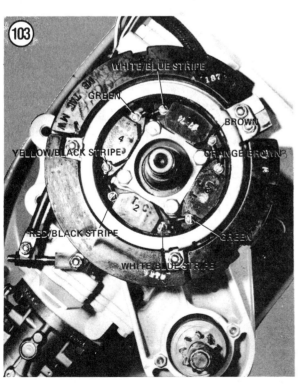

103

WHITE/BLUE STRIPE

GREEN

BROWN

YELLOW/BLACK STRIPE

ORANGE/BROWN

RED/BLACK STRIPE

GREEN

WHITE/BLUE STRIPE

104

a diagram of the wiring to assure proper reconnection of the leads. See **Figure 106** (4-cylinder engine).

1. Disconnect the negative battery cable.

2. Remove the engine cover.

3. Disconnect the spark plug lead at the coil.

4. Remove the screws and washers holding the coil to the power head. Remove the coil.

5. Disconnect the lead wires at the coil terminals.

6. Repeat Steps 3-5 for each coil to be removed.

7. Installation is the reverse of removal. Tighten coil mounting screws to specifications (**Table 4**).

ALTERNATOR CD IGNITION

This system is used on electric start 7.5-35 hp engines. The major components include the flywheel, alternator stator, trigger stator, stator ring or throttle cam, CD module, ignition coils, spark plugs and connecting wiring. See **Figure 107** (typical).

Ignition timing should be adjusted whenever a component is replaced.

Operation

The outer rim of the flywheel contains a series of magnets which create a magnetic field during rotation. This magnetic field cuts through the stator windings and produces an alternating current (AC) of 225 volts. This voltage is sent to the CD module where it is changed into direct current (DC) by an internal rectifier and stored in a capacitor.

Timing magnets in the flywheel hub rotate inside the trigger stator and create a magnetic field at the trigger coils. As the flywheel continues to rotate, this magnetic field collapses and induces a maximum 2-volt pulse in the No. 1 trigger coil. This pulse causes an electronic switch in the CD module to close, discharging the stored voltage into the ignition coil. The coil increases the

TRIGGER MODULE   RETAINING CLIP

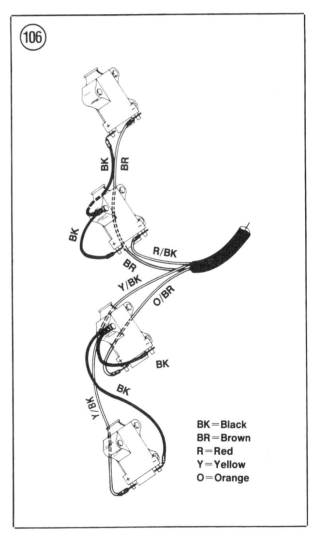

BK = Black
BR = Brown
R = Red
Y = Yellow
O = Orange

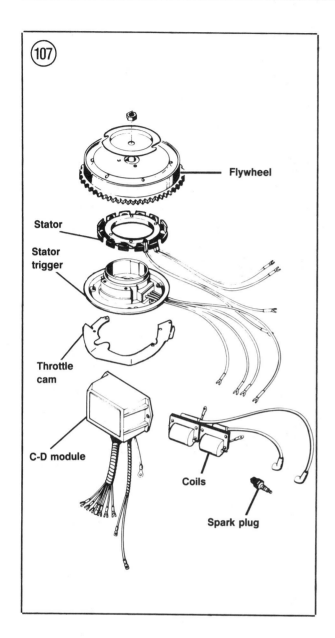

107

Flywheel

Stator

Stator
trigger

Throttle
cam

C-D module

Coils

Spark plug

voltage to 25,000-32,000 volts and sends it to the spark plug. The flywheel continues to rotate, building up another charge at the stator and a magnetic field at the No. 2 trigger coil. This sequence repeats itself as long as the flywheel rotates.

Depressing the stop switch shorts the stator ends (7.5 hp) or CD module (all others) to ground and shuts the engine off.

Alternator Stator
Removal/Installation (7.5-15 hp)

1. Disconnect the negative battery cable.
2. Remove the engine cover.
3. Remove the flywheel. See Chapter Eight.
4. Disconnect the stator leads at their connectors.
5. Remove the screws holding the stator to the trigger stator. See **Figure 108**. Remove the stator.
6. Installation is the reverse of removal.

Alternator Stator
Removal/Installation (20-35 hp)

1. Disconnect the negative battery cable.
2. Remove the engine cover.
3. Remove the flywheel. See Chapter Eight.
4. Remove the screws holding the stator to the trigger stator. See **Figure 109**.

7

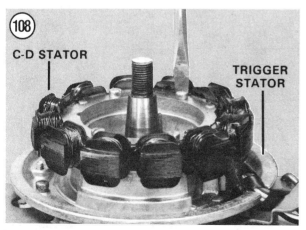

108

C-D STATOR

TRIGGER
STATOR

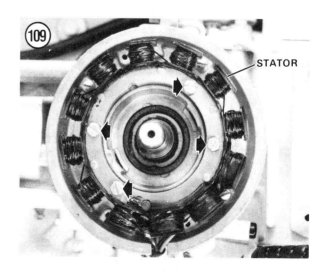

109

STATOR

5. Disconnect the 2 green/yellow leads at the rectifier (**Figure 110**).

6. Unwrap the wiring harness cover to separate the stator and trigger leads.

7. Disconnect the stator lead connectors (**Figure 111**).

8. Remove the stator.

9. Installation is the reverse of removal.

Trigger Stator
Removal/Installation (7.5-15 hp)

1. Remove the alternator stator as described in this chapter.

2A. 7.5 hp—Disconnect the trigger lead wires at the 2 connectors.

2B. 9.6-15 hp—Disconnect the trigger lead wires at the terminal block. Disconnect the ground lead at the power head.

3. Remove the trigger stator mounting screws.

4A. 7.5 hp—Disconnect trigger stator link connector at the ball-joint stud and remove the stator.

4B. 9.6-15 hp—Lift trigger stator off to one side and disconnect the throttle cam link. Remove the stator.

5. Installation is the reverse of removal.

Trigger Stator
Removal/Installation (20-35 hp)

1. Remove the alternator stator as described in this chapter.

2. Remove the trigger stator mounting screws.

3. Remove the ground lead mounting screw. See **Figure 112**.

4. Remove the trigger stator.

5. Installation is the reverse of removal.

CD Module
Removal/Installation (7.5 hp)

1. Disconnect the negative battery cable.

2. Remove the engine cover.

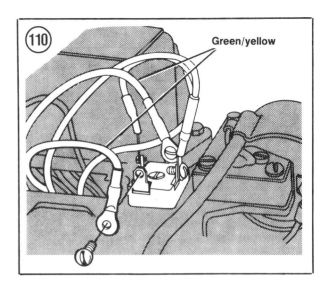

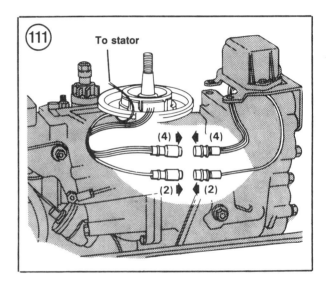

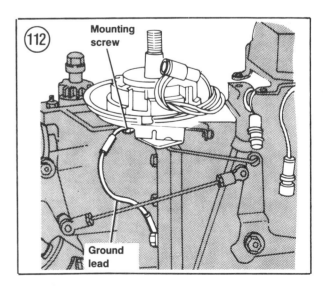

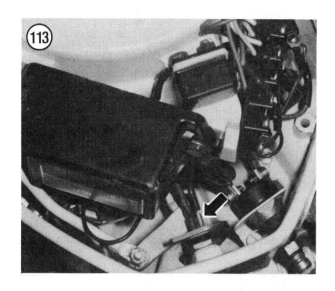

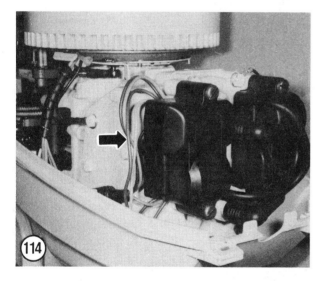

3. Disconnect the module lead wires at the connectors.

4. Disconnect the module ground wire.

5. Disconnect one spade terminal from the stop switch (**Figure 113**).

6. Disconnect the spark plug lead.

7. Remove the 2 module mounting screws (**Figure 114**). Remove the module.

8. Repeat Steps 3-7 to remove the other CD module, if necessary.

9. Installation is the reverse of removal.

CD Module Removal/Installation (9.6-15 hp)

1. Disconnect the negative battery cable.

2. Remove the engine cover.

3. Disconnect the module lead wires at the terminal block.

4. Disconnect one spade terminal from the stop switch (**Figure 113**).

5. Disconnect the brown and white lead wires at the coils.

6. Remove the 2 module mounting screws (**Figure 115**). Note that the CD module ground lead is attached to the starboard mounting screw and a battery cable lead is attached to the port screw.

7. Remove the module.

8. Installation is the reverse of removal.

CD Module Removal/Installation (20-35 hp)

1. Disconnect the negative battery cable.

2. Remove the engine cover.

3. Remove the 2 CD module mounting screws. Remove the module from the electrical component mounting bracket.

4. Disconnect both CD lead connectors.

5. Disconnect the brown lead at the upper coil and the white lead at the lower coil. See **Figure 116**.

6. Disconnect the terminal block blue lead, circuit breaker red lead and starter relay black lead. See **Figure 117**.

7

7. Remove the CD module.

8. Installation is the reverse of removal.

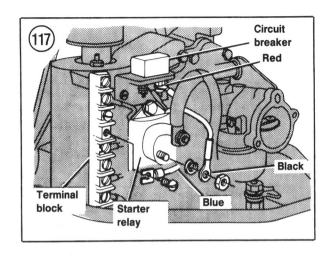

PRESTOLITE CD IGNITION

Chrysler Marine introduced this ignition on some 1980 55 hp engines. Its use was extended to the 1982-1984 125 hp engine and some 1983-1984 75 hp and 85 hp engines.

The ignition components are similar to those of the alternator CD ignition, consisting of a flywheel, stator, trigger housing, CD module(s), an ignition coil for each cylinder, spark plugs and connecting wiring. See **Figure 118**.

Ignition timing should be adjusted whenever a component is replaced.

Operation

See *Alternator CD Ignition Operation* in this chapter.

Stator Removal/Installation

1. Disconnect the negative battery cable.

2. Remove the engine cover.

3. Remove the flywheel. See Chapter Eight.

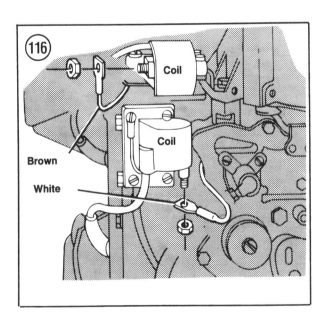

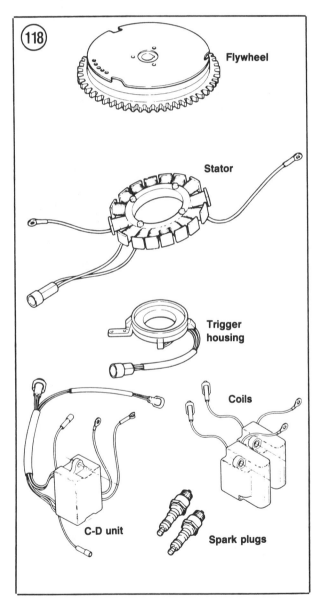

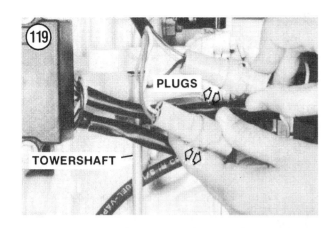

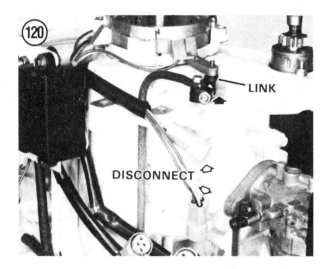

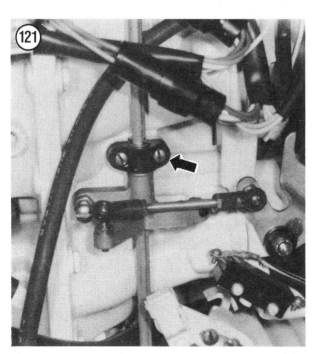

4. Disconnect the 2 yellow/green stator leads at the rectifier.

NOTE
3-cylinder engines have 2 connector plugs side-by-side to disconnect in Step 5.

5. Disconnect the stator lead at the connector plug.

6. Remove the 4 screws holding the stator to the trigger housing. Lift stator from trigger housing.

7. Installation is the reverse of removal.

**Trigger Housing
Removal/Installation**

1. Remove the stator as described in this chapter.

NOTE
*3-cylinder engines have 2 connector plugs side-by-side to disconnect in Step 2. See **Figure 119**.*

2. Disconnect the trigger housing lead at the connector plug. See **Figure 120**.

3A. 55 and 125 hp—Disconnect the spark control link at the tower shaft. See **Figure 120** (typical).

3B. 75 and 85 hp—Remove the tower shaft support (**Figure 121**). Remove spark control link fastener (**Figure 122**). Lift tower shaft

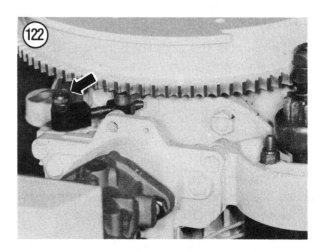

about 1/2 in. and disconnect spark control link.

4. Remove trigger housing from bearing cage.

5. Installation is the reverse of removal.

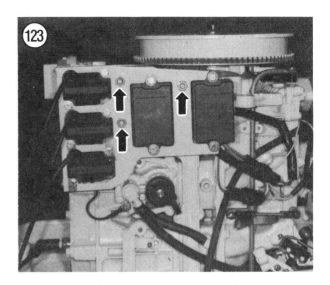

Ignition Coil
Removal/Installation
(55 and 125 hp)

1. Disconnect the negative battery cable.

2. Remove the engine cover.

3. Disconnect the coil lead at the spark plug.

4. Disconnect the electrical connector between the coil and the CD module.

5. Remove the 2 screws and lockwashers or nuts holding the coil to the cylinder head.

6. Repeat Steps 3-5 for each coil to be removed.

7. Installation is the reverse of removal, plus the following:

    a. Reinstall the ground lead under the top mounting screw on each coil.

    b. Connect the orange wire connector plug from the CD module to the No. 1 (55 hp) or No. 1 and No. 3 (125 hp) coil(s).

    c. Connect the red wire connector plug from the CD module to the No. 2 (55 hp) or No. 2 and No. 4 (125 hp) coil(s).

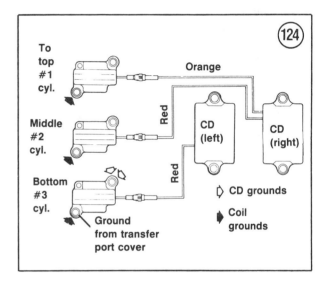

Ignition Coil Removal/
Installation (75 and 85 hp)

1. Disconnect the negative battery cable.

2. Remove the engine cover.

3. Disconnect the coil leads at the spark plugs.

4. Remove the 3 fasteners holding the electrical components bracket to the power head. See **Figure 123**.

5. Disconnect the electrical plug between each coil and the CD module. See **Figure 124**.

6. Remove the fasteners holding each coil to be replaced. Remove the coil(s).

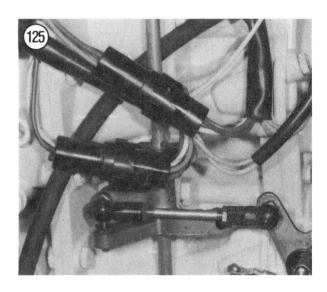

7. Installation is the reverse of removal. Be sure to clean and reconnect the ground leads properly.

CD Module Removal/Installation (55 and 125 hp)

The 2-cylinder engine uses 1 CD module; the 4-cylinder engine has 2 modules.
1. Disconnect the negative battery cable.
2. Remove the engine cover.
3. Disconnect the 2 large connector plugs (**Figure 125**).

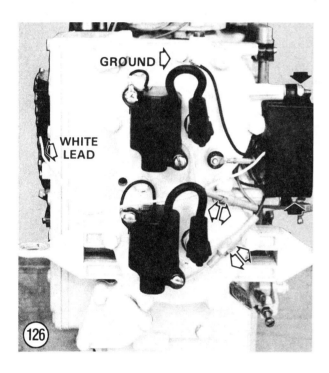

4. Disconnect the 2 small connector plugs. See **Figure 126** (55 hp).
5. Disconnect the white CD module lead at the terminal block (**Figure 126**).

NOTE
If the engine has been previously overhauled, it is possible that the ground lead in Step 6 was relocated. If it cannot be easily located, proceed to Step 7. When the module is removed, the ground lead positioning will become evident.

6. Remove the top center cylinder head bolt which holds the black CD ground lead. Disconnect the ground lead. See **Figure 126** for the 55 hp engine; the 125 hp is similar.
7. Remove the 2 nuts, washers, bushings and bolts holding the CD module.
8. 125 hp—Repeat Step 7 to remove the other CD module, if necessary.
9. Installation is the reverse of removal. Install the black ground lead under the cylinder head bolt and tighten bolt to 225 in.-lb.

CD Module Removal/Installation (75 and 85 hp)

1. Disconnect the negative battery cable.
2. Remove the engine cover.
3. Disconnect the 2 large connector plugs (A and B, **Figure 127**).
4. Remove the 3 fasteners holding the electrical components bracket to the power head. See **Figure 123**.
5. Disconnect the electrical plug between each coil and the CD module. See **Figure 128**.
6. Disconnect the white CD module leads at the terminal board.
7. Remove the right mounting screw from the bottom (No. 3) coil and disconnect the CD module ground leads. See **Figure 124**.
8. Remove the 2 nuts, washers, bushings and bolts holding the CD module. Repeat this

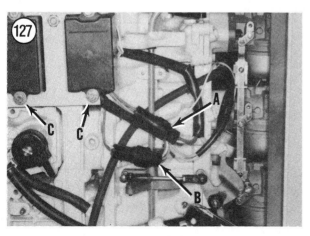

7

step to remove the other CD module, if necessary.

9. Installation is the reverse of removal.

Ignition Switch
Removal/Installation

1. Disconnect the negative battery cable.

2. Unscrew the hex nut holding the switch to the control box or instrument panel.

3. Remove the switch and disconnect the lead wires. See **Figure 129** (Motorola) or **Figure 130** (Prestolite).

4. Installation is the reverse of removal.

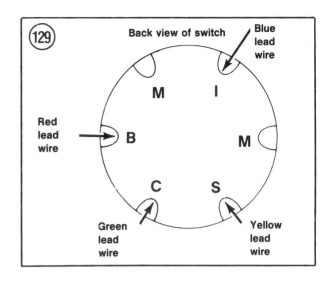

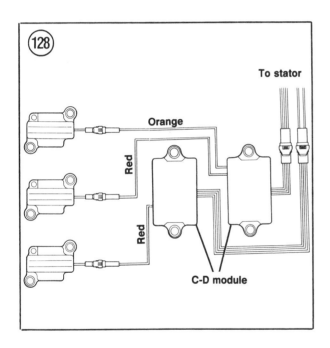

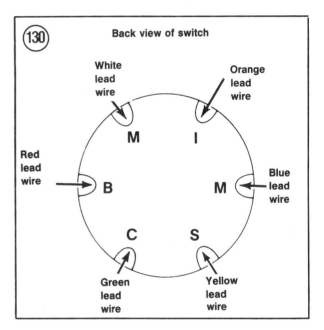

Table 1 BATTERY CAPACITY (HOURS)

Accessory draw	Provides continuous power for:	Approximate recharge time
80 AMP-HOUR BATTERY		
5 amps	13.5 hours	16 hours
15 amps	3.5 hours	13 hours
25 amps	1.8 hours	12 hours
105 AMP-HOUR BATTERY		
5 amps	15.8 hours	16 hours
15 amps	4.2 hours	13 hours
25 amps	2.4 hours	12 hours

Table 2 SELF-DISCHARGE RATE

Temperature	Approximate allowable self-discharge per day for first 10 days (specific gravity)
100° F (37.8° C)	0.0025 points
80° F (26.7° C)	0.0010 points
50° F (10.0° C)	0.0003 points

Table 3 BATTERY CHARGE PERCENTAGE

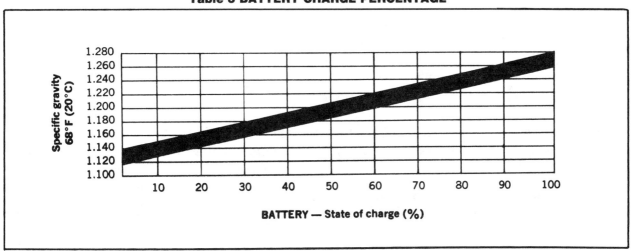

Table 4 TIGHTENING TORQUES

Fastener	in.-lb.
Coil mounting screws	70
CD module mounting screws	
Delta	160
Motorola	90
Starter through-bolts	95-100
Standard torque values	
(screw or nut size)	
6-32	9
8-32	20
10-24	30
10-32	35
12-24	45
1/4-20	70
5/16-18	160
3/8-16	270

7

Chapter Eight

Power Head

This chapter covers the basic repair of Chrysler outboard power heads, except for the air-cooled 3.5 hp (1966-1969) and 3.6 hp.

The procedures involved are similar from model to model, with minor differences. Some procedures require the use of special tools, which can be purchased from a dealer. Certain tools may also be fabricated by a machinist, often at substantial savings. Power head stands are available from specialty shops such as Bob Kerr's Marine Tool Co. (P.O. Box 1135, Winter Garden, FL 32787).

Work on the power head requires considerable mechanical ability. You should carefully consider your own capabilities before attempting any operation involving major disassembly of the engine.

Much of the labor charge for dealer repairs involves the removal and disassembly of other parts to reach the defective component. Even if you decide not to tackle the entire power head overhaul after studying the text and illustrations in this chapter, it can be cheaper to perform the preliminary operations yourself and then take the power

head to your dealer. Since many marine dealers have lengthy waiting lists for service (especially during the spring and summer season), this practice can reduce the time your unit is in the shop. If you have done much of the preliminary work, your repairs can be scheduled and performed much quicker.

Repairs go much faster and easier if your motor is clean before you begin work. There are special cleaners for washing the motor and related parts. Just spray or brush on the cleaning solution, let it stand, then rinse it away with a garden hose. Clean all oily or greasy parts with fresh solvent as you remove them.

WARNING

Never use gasoline as a cleaning agent. It presents an extreme fire hazard. Be sure to work in a well-ventilated area when using cleaning solvents. Keep a fire extinguisher rated for gasoline and oil fires nearby in case of emergency.

Once you have decided to do the job yourself, read this chapter thoroughly until

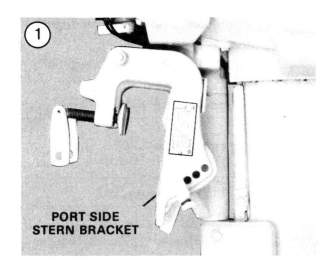

PORT SIDE STERN BRACKET

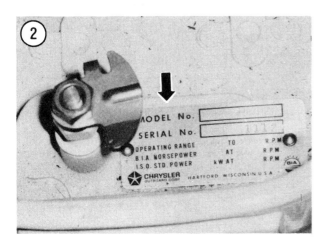

procedures are somewhat generalized to accommodate all models. Where individual differences occur, they are specifically pointed out. The power heads shown in the accompanying pictures are current designs. While it is possible that the components shown in the pictures may not be identical with those being serviced, the step-by-step procedures may be used with all models covered in this manual.

Tables 1-3 are at the end of the chapter.

you have a good idea of what is involved in completing the overhaul satisfactorily. Make arrangements to buy or rent any special tools necessary and obtain replacement parts before you start. It is frustrating and time-consuming to start an overhaul and then be unable to complete it because the necessary tools or parts are not at hand.

Before beginning the job, re-read Chapter Two of this manual. You will do a better job with this information fresh in your mind.

Remember that new engine break-in procedures should be followed after an engine has been overhauled. Refer to your owner's manual for specific instructions.

Since this chapter covers a large range of models over a lengthy time period, the

ENGINE SERIAL NUMBER

Chrysler outboards are identified by engine serial number and model number. These numbers are stamped on a plate riveted to the port side stern bracket (**Figure 1**) or to the starboard side of the support plate (**Figure 2**). This information identifies the outboard and indicates if there are unique parts or if internal changes have been made during the model run. The serial and model numbers should be used when ordering any replacement parts for your outboard.

From 1970-on, Chrysler also stamps a block identification number on production and service replacement engine blocks. This is useful in identifying the engine if stolen or if the identification plate containing the serial and model numbers has been removed. The number usually starts with a letter followed by 6-7 digits about 1/8 in. high. An example would be B422422. The number is located as follows:

a. 4 hp—Starboard transfer port blister.
b. 6-7.5 hp—Top of exhaust chest, port side.
c. 35-45 hp—Top of No. 1 transfer port cavity, starboard side.
d. All others—Rear of cylinder next to cylinder head.

8

FASTENERS AND TORQUE SPECIFICATIONS

Always replace a worn or damaged fastener with one of the same size, type and torque requirement.

Power head tightening torques are given in **Table 1**. Where a specification is not provided for a given bolt, use the standard bolt and nut torque according to fastener size.

To prevent cylinder head warpage, the head bolts should be tightened to 50 in.-lb. (3.5-7.5 hp) or 75 in.-lb. (9.2-140 hp), then in steps of 25 in.-lb. (3.5-7.5) or 50 in.-lb. (9.2-140 hp) until the specified torque is reached.

Other power head fasteners should be tightened in 2 steps. Tighten to 50 percent of the torque value in the first step, then to 100 percent in the second step.

Retighten the cylinder head bolts after the engine has been run for 15 minutes and allowed to cool. It is a good idea to retorque them again after 10 hours of operation.

To retighten the power head mounting fasteners properly, back them out one turn and then tighten to specifications.

When spark plugs are reinstalled after an overhaul, tighten to the specified torque. Warm the engine to normal operating temperature, let it cool down and retorque the plugs.

FLYWHEEL

Removal/Installation
(1966-1969 3.5 hp; 1970-1977 3.6 hp; 1976-1978 4 hp; 1974-1976 4.9-5 hp)

1. Remove the fuel tank. See Chapter Six.
2. Remove the bolt holding the starter rewind cup and flywheel to the crankshaft.
3. Remove rewind cup and thread knock-off bolt (part No. T 2919) into crankshaft as far as possible, then back it off 2 full turns.

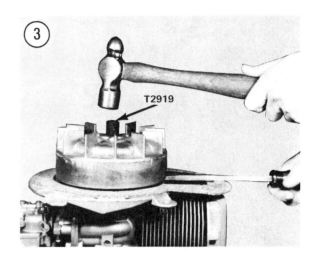

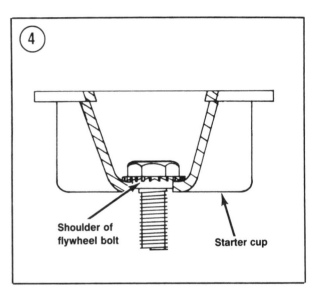

4. Insert a large screwdriver between the flywheel and top of power head at the rear of the engine. See **Figure 3** (3.5 hp shown).

CAUTION
Striking the knock-off nut with excessive force in Step 5 can cause crankshaft and/or bearing damage.

5. Pry upwards on the flywheel while tapping the knock-off nut with a 16 ounce hammer.
6. When flywheel comes loose, remove the pry tool. Unscrew the knock-off nut and remove the flywheel from the crankshaft.
7. Inspect flywheel carefully as described in this chapter.

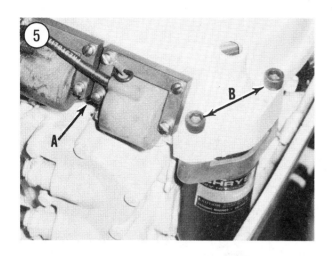

8. Install flywheel key, flywheel, starter cup and bolt. Locate shoulder on flywheel bolt into starter cup hole by lifting cup up against bolt head (seating shoulder fully into cup) while tightening bolt. See **Figure 4**. Tighten nut to specifications (**Table 1**).

9. Install the fuel tank. See Chapter Six.

Removal/Installation (9.2-15 hp Electric Start)

1. Disconnect the negative battery cable.
2. Remove the engine cover.

NOTE
Steps 3-5 do not apply to Autolectric models.

3. Refer to Chapter Seven and remove the following components from the starter support, if required. See **A, Figure 5** for coil location and **Figure 6** for the other components (typical):

   a. Ignition coils.
   b. Rectifier.
   c. Terminal block.
   d. Circuit breaker.
   e. CD module.

4. Remove the 2 screws holding the starter motor to the starter support. See **B, Figure 5**. Place starter inside support plate.

5. Remove 2 screws holding the rear of the starter support to the powerhead, then remove 2 screws holding the starter support to the support bracket. Remove the starter support.

NOTE
Step 6 applies only to Autolectric models.

6. Remove the starter-generator cover, starter-generator and support assembly. See Chapter Seven.

7. Remove the flywheel nut with an appropriate size socket.

8. Thread knock-off nut (part No. T 2909) on crankshaft as far as possible, then back it off 2 full turns.

CAUTION
Be sure that pry tool does not contact stator or other components in Step 9 that might be damaged by prying in Step 10.

9. Slide flywheel removal wedge (part No. T 2989) or a large screwdriver between the flywheel and power head. Wedge or screwdriver position should align with flywheel keyway and contact only the first 1/8 in. of the flywheel.

8

CAUTION
Do not strike knock-off nut with excessive force in Step 10 or crankshaft and/or bearing damage may result.

10. Pry upwards on the flywheel while tapping the knock-off nut with a 16 ounce hammer. See **Figure 7**.

11. When flywheel comes loose, remove the pry tool. Unscrew the knock-off nut and remove the flywheel from the crankshaft.

12. Inspect flywheel carefully as described in this chapter.

13. Inspect crankshaft and flywheel tapers. They must be perfectly dry and free of oil. Wipe tapered surfaces with solvent and blow dry with compressed air.

14. Install flywheel key with outer edge of key parallel to the crankshaft centerline. See **Figure 8**.

15. Install the flywheel on the crankshaft.

16. Install flywheel nut and tighten to specifications (**Table 1**).

17. Reverse Step 1, Step 2 and Step 6 (Autolectric) or Steps 1-5 (all others) to complete installation.

Removal/Installation
(All Other 3.5-55 hp Models
Except 1979-on 45-55 hp)

1. Remove the engine cover.

2. Disconnect the negative battery cable or the spark plug leads to prevent accidental starting of the engine.

3. Autolectric—Remove the starter-generator cover, starter-generator and support assembly. See Chapter 7.

4. Remove the flywheel nut (**Figure 9**) with an appropriate size socket.

5. Remove the emergency starter collar from the flywheel, if so equipped.

6. Thread appropriate knock-off nut (see **Table 2**) on crankshaft as far as possible, then back it off 2 full turns.

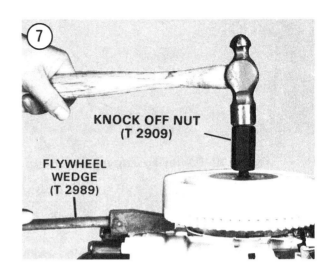

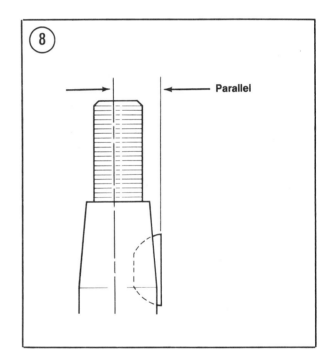

CAUTION
Be sure that pry tool does not contact stator or other components in Step 7 that might be damaged by prying in Step 8.

7. Slide flywheel removal wedge (part No. T 2989) or a large screwdriver between the flywheel and power head. Wedge or screwdriver position should align with flywheel keyway and contact only the first 1/8 in. of the flywheel.

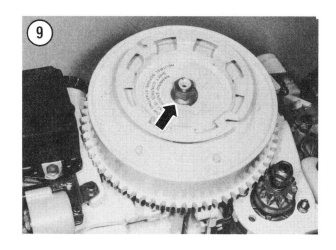

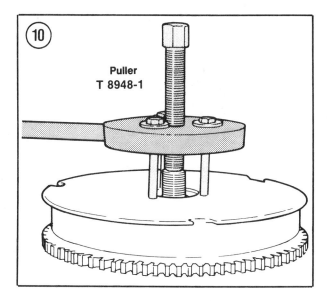

Puller
T 8948-1

CAUTION
Do not strike knock-off nut with excessive force in Step 8 or crankshaft and/or bearing damage may result.

8. Pry upwards on the flywheel while tapping the knock-off nut with a 16 ounce hammer.

9. When flywheel comes loose, remove the pry tool. Unscrew the knock-off nut and remove the flywheel from the crankshaft.

10. Inspect flywheel carefully as described in this chapter.

11. Inspect crankshaft and flywheel tapers. They must be perfectly dry and free of oil. Wipe tapered surfaces with solvent and blow dry with compressed air.

12. Install flywheel key with outer edge of key parallel to the crankshaft centerline. See **Figure 8**.

13. Install the flywheel on the crankshaft.

14. Install emergency starter collar (if so equipped). Collar legs must engage flywheel holes to prevent rotation.

15. Install flywheel nut and tighten to specifications (**Table 1**).

16. Autolectric—Install the starter-generator support, starter-generator and cover. See Chapter Seven.

17. Reconnect spark plug leads or negative battery cable. Install the engine cover.

**Removal/Installation
(All 60-140 hp Models and
1979-on 45-55 hp)**

1. Remove the engine cover.

2. Disconnect the negative battery cable or the spark plug leads to prevent accidental starting of the engine.

3. Remove the flywheel nut with an appropriate size socket.

4. Install puller (part No. T 8948-1) on flywheel with its flat side facing up. See **Figure 10**.

5. Hold puller body with puller handle and tighten center screw. If flywheel does not pop from the crankshaft taper, pry up on the rim of the flywheel with a large screwdriver while tapping the puller center screw with a brass hammer.

6. Remove puller from flywheel. Remove flywheel from crankshaft.

7. Inspect the flywheel as described in this chapter.

8. Install flywheel key with its outer edge parallel to the crankshaft centerline. See **Figure 8**.

9. Make sure distributor belt is properly installed on distributor ignition models.

10. Install the flywheel and flywheel nut. Tighten nut to specifications (**Table 1**).

8

Inspection

1. Check the flywheel carefully for cracks or breaks.

> *WARNING*
> *A cracked or chipped flywheel must be replaced. A damaged flywheel may fly apart at high rpm, throwing metal fragments over a large area. Do not attempt to repair a damaged flywheel.*

2. Check tapered bore of flywheel and crankshaft taper for signs of fretting or working. If this defect is apparent or if the engine has had 25 or more hours of operation, the flywheel bore must be lapped as described in this chapter.

Lapping

1. Apply a light coat of a water base valve lapping compound (240 grit or finer) to the tapered portion of the crankshaft.

> *NOTE*
> *Do not spin flywheel completely around crankshaft in this procedure.*

2. Place flywheel on crankshaft and gently rotate it back and forth several times within a 90 degree range. See **Figure 11** (position 1).
3. Rotate the flywheel 90° and repeat Step 2. See **Figure 11** (position 2).
4. Remove the flywheel and inspect the bore for surface contact with crankshaft taper. The flywheel bore taper must have a minimum of 90 percent (3.5-50 hp) or 80 percent (55-140 hp) surface contact:

   a. To check percentage of contact, use a lead pencil (No. 2 or 2B grade lead) and draw 3 vertical lines on crankshaft taper 120° apart. See **Figure 12**.

   b. Mount flywheel on crankshaft and rotate it 360° or 1 full turn while exerting light downward pressure.

   c. Remove flywheel and check percentage of pencil lines that have been rubbed

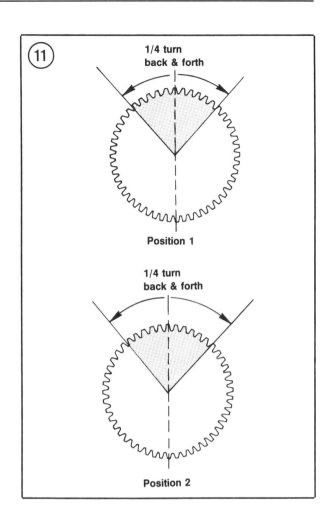

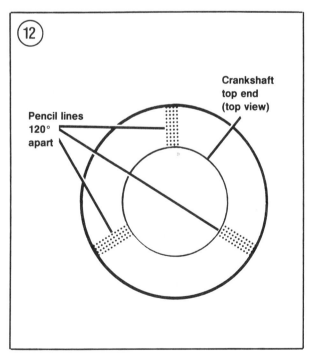

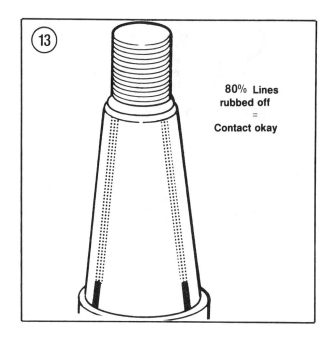

80% Lines
rubbed off
=
Contact okay

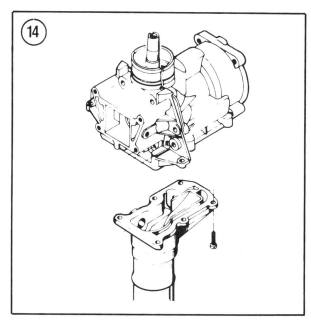

off. If 90 percent (3.5-50 hp) or 80 percent (55-140 hp) or more of the pencil lines have been removed and no rocking is evident (**Figure 13**), proceed to Step 5.

d. Repeat the lapping procedure if less than the minimum lines have been removed. If 3 lappings cannot correct flywheel taper contact to crankshaft,

replace the flywheel. If a new flywheel does not lap in properly, replace the crankshaft.

5. Clean all compound from the flywheel bore, crankshaft taper and keyway.

6. Reinstall flywheel as described in this chapter.

POWER HEAD REMOVAL/INSTALLATION

When removing any power head, it is a good idea to make a sketch or take an instant picture of the location, routing and positioning of electrical wiring, brackets and J-clamps for reassembly reference. Take notes as you remove wires, washers and engine grounds so they may be reinstalled in their correct position. Unless specified otherwise, install lockwashers on the engine side of the electrical lead to assure a good ground.

1980-on 3.5 hp, 1976-on 4 hp

1. Remove the engine covers.
2. Remove the fuel tank. See Chapter Six.
3. Remove the carburetor and fuel shut-off valve. See Chapter Six.
4. Disconnect the shorting switch.
5. Remove the screws holding the support plate. Remove the support plate.
6. Remove the flywheel as described in this chapter.
7. Remove the stator plate. See Chapter Seven.
8. Remove the screws holding the power head to the motor leg (**Figure 14**). Remove the power head.
9. Remove and discard the crankshaft spline seal.
10. Installation is the reverse of removal, plus the following:
    a. Use a new motor leg gasket.
    b. Align motor leg drive shaft splines with power head crankshaft splines.

8

c. Coat motor leg screw threads with RTV sealant and tighten to specifications (**Table 1**).

d. Complete engine synchronization and linkage adjustments. See Chapter Five.

4.4-15 hp

1. Remove the engine cover.

2. Disconnect the spark plug leads to prevent accidental starting of the engine.

3. Remove the carburetor and fuel pump. See Chapter Six.

4. Disconnect the stop switch leads.

5A. Manual models—Remove the manual starter assembly. See Chapter Ten.

5B. Autolectric—Remove the starter-generator cover, starter-generator and support assembly. See Chapter Seven.

6. Remove the flywheel as described in this chapter.

7A. Magneto ignition—Loosen the magneto shaft retainer setscrew. Position steering handle up. Pull magneto control shaft with gear from steering arm boss. Slip control rod link from control shaft as shown in **Figure 15**.

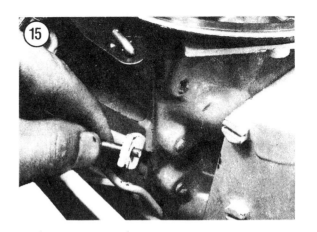

NOTE
On early 4.4, 5, 6.5 and 7 hp models, this link is attached to the magneto control lever with a swivel and stop nut.

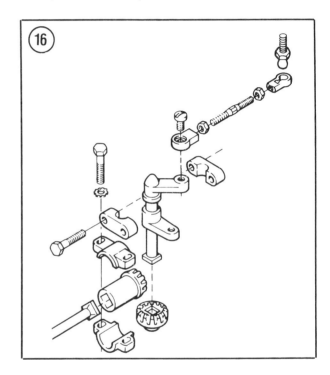

Remove stop nut to disconnect link from magneto control lever.

7B. CD ignition—Disconnect the spark control link at the tower shaft. Remove tower shaft assembly (**Figure 16**).

8. Remove the stator plate (magneto ignition) or the alternator and trigger stators (CD ignition). See Chapter Seven.

9. Remove the CD module (CD ignition).

10. Remove the shift interlock bearing screw and disconnect the interlock rod from the crankcase cover lever. See **Figure 17**.

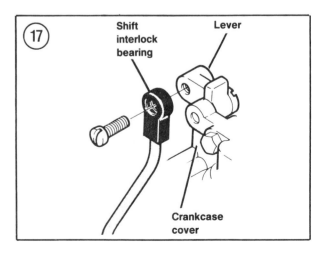

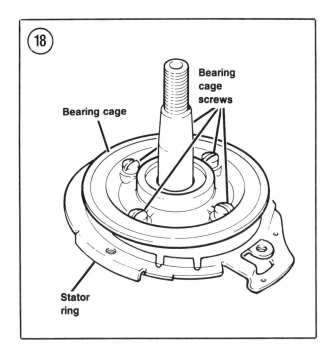

18

Bearing cage

Bearing cage screws

Stator ring

Disconnect interlock rod from gear shift shaft lever.

11. Remove the 4 screws holding the bearing cage to the power head. Remove the bearing cage, stator ring and bearing cage gasket. See **Figure 18**. Discard the gasket.

NOTE
At this point, there should be no linkage, ground leads or other electrical wiring connecting the power head to the support plate. Recheck to make sure that nothing will hamper power head removal.

12. Remove the motor leg screws holding the power head to the support plate. See **Figure 19** (typical). Remove the power head from the support plate and place on a clean workbench.

13. Remove and discard the cylinder mounting gasket. See **Figure 19** (typical).

14. Remove and discard the crankshaft spline seal (**Figure 20**).

15. Clean cylinder mounting and support plate gasket surfaces of all gasket residue.

16. Installation is the reverse of removal, plus the following:

    a. Use new cylinder mounting and bearing cage gaskets.

8

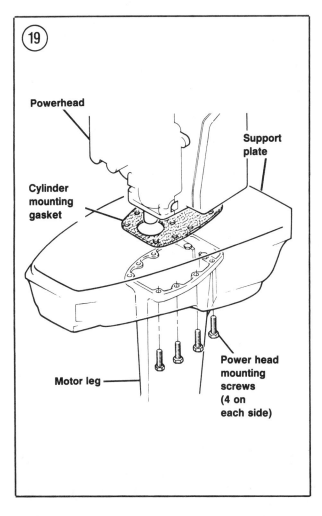

19

Powerhead

Cylinder mounting gasket

Support plate

Motor leg

Power head mounting screws (4 on each side)

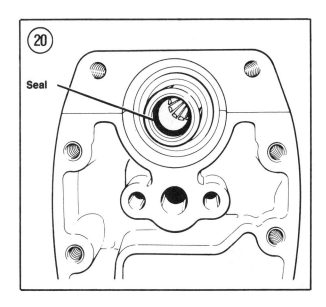

20

Seal

b. Coat drive shaft splines with a liberal amount of anti-seize lubricant.

c. Install a new bearing cage seal with lip facing the power head.

d. Apply Rykon No. 2EP to stator ring between ring and bearing cage.

e. Coat bearing cage and motor leg screw threads with RTV sealant.

f. Tighten all fasteners to specifications (**Table 1**).

g. Complete engine synchronization and linkage adjustments. See Chapter Five.

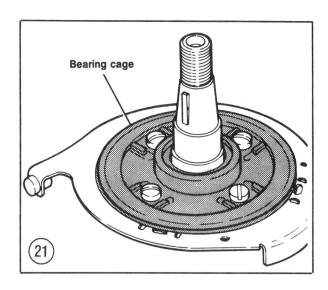

1966-1976 20 hp

1. Disconnect the negative battery cable.
2. Remove the engine cover.
3. Disconnect the spark plugs.
4A. Manual model—Remove the manual starter assembly. See Chapter Ten.
4B. Autolectric—Remove the starter-generator cover, starter-generator and support assembly. See Chapter Seven.
5. Remove the flywheel as described in this chapter.
6. Remove the carburetor. See Chapter Six.

> *NOTE*
> *Steps 7-9 do not apply to Autolectric models.*

7. Disconnect magneto control connector at the control lever.
8. Remove the stator plate. See Chapter Seven.
9. Remove the 4 screws holding the bearing cage to the power head. Remove bearing cage and stator ring. Remove and discard the gasket or O-ring.
10. Remove and discard the cotter pins at each end of the gear shift rod.
11. Remove pivot screw holding interlock linkage to gear shift arm. Remove shift rod.
12. Remove the 4 screws holding the bearing cage.

13. Remove the motor leg covers.

> *NOTE*
> *At this point, there should be no linkage, ground leads or other electrical wiring connecting the power head to the support plate. Recheck to make sure that nothing will hamper power head removal.*

14. Remove the 6 screws holding the motor leg to the power head.
15. Lift the power head from the motor leg and remove from the support plate. Place the power head on a clean workbench.
16. Clean all gasket residue from the cylinder mounting and support plate surfaces.
17. Installation is the reverse of removal, plus the following:

a. Use new cylinder mounting and bearing cage gaskets.

b. Coat drive shaft splines with a liberal amount of anti-seize lubricant.

c. Rotate flywheel or propeller as required to align crankshaft and drive shaft splines.

d. Install a new bearing cage seal with lip facing the power head.

e. Insert seal protector (part No. T 2908) in bearing cage to protect seal when reinstalling bearing cage to power head.

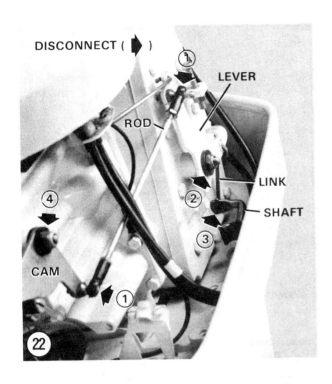

DISCONNECT (▶)

LEVER

ROD

LINK

SHAFT

CAM

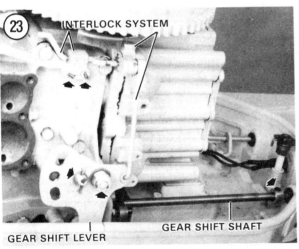

INTERLOCK SYSTEM

GEAR SHIFT SHAFT

GEAR SHIFT LEVER

f. Apply Rykon No. 2EP between stator ring and bearing cage.

g. Coat bearing cage and motor leg screw threads with Permatex No. 2 or equivalent sealant.

h. Tighten all fasteners to specifications (**Table 1**).

i. Complete engine synchronization and linkage adjustments. See Chapter Five.

1979-on 20 hp; All 25-30 hp; 1976-on 35 hp

1. Remove the engine cover.
2. Electric start—Disconnect the negative battery cable.
3. Disconnect spark plug leads. Remove the spark plugs.
4. Remove the manual starter housing and bracket. See Chapter Ten.
5. Electric start—Disconnect electrical leads at the following components and remove the components. See Chapter Seven:
   a. Choke solenoid lead.
   b. Support plate ground lead.
   c. Starter solenoid (relay).
   d. Interlock switch.
   e. Rectifier.
   f. Electrical components bracket.
   g. Wiring harness plugs.
6. Remove the flywheel as described in this chapter.
7. Remove the stator plate (manual) or alternator stator and trigger assembly (electric).
8. Remove the 4 screws holding the bearing cage to the power head (**Figure 21**). Lift bearing cage and stator ring off power head and disconnect stator ring from magneto control link. Remove and discard bearing cage gasket.
9. Disconnect the throttle cam rod. Remove the magneto control shaft lever. Disconnect the control shaft and remove the throttle cam on manual start models. See **Figure 22**.
10. Remove the carburetor, fuel pump and reed valve assembly. See Chapter Six.
11. Refer to **Figure 23** and remove the shift components (electric start models have no gear shift shaft).

NOTE
At this point, there should be no hoses, wires or linkage connecting the power head to the exhaust housing. Recheck this to make sure nothing will hamper power head removal.

8

12. Remove the 6 rear motor leg cover screws. Remove the cover.

13. Remove the 2 nuts holding the upper side shock mounts between the kingpin and spacer plates.

14. Remove the side and front mounts.

15. Engage the reverse lock mechanism, then pull the motor leg back far enough to remove the front motor leg cover.

16. Remove the 6 screws holding the motor leg to the power head.

17. Remove the power head and place it on a clean workbench.

18. Clean all gasket residue from the power head and spacer plate surfaces.

19. Installation is the reverse of removal, plus the following:

   a. Use new power head and bearing cage gaskets.

   b. Coat drive shaft splines with a liberal amount of anti-seize lubricant.

   c. Rotate flywheel or propeller as required to align crankshaft and drive shaft splines.

   d. Install a new bearing cage seal with lip facing the power head. Use seal installer part No. T 8971 and drive seal in until its top is 0.70 ±0.005 in. from bottom of bearing cage chamfer. See **Figure 24**.

   e. Lubricate seal lip with grease and install on seal protector (part No. T 8967). Garter spring must face small end of protector. Install seal as shown in **Figure 25**.

   f. Coat bearing cage screw threads with EC-750 sealant.

   g. Tighten all fasteners to specifications **(Table 1)**.

   h. Complete engine synchronization and linkage adjustments. See Chapter Five.

1966-1975 35 hp, All 45 and 50 hp; 1967-1976 55 hp

1. Remove the engine cover.

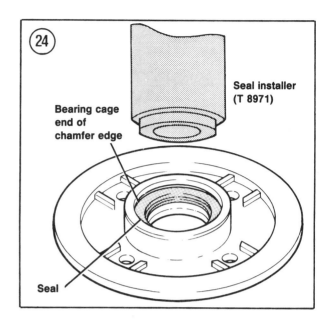

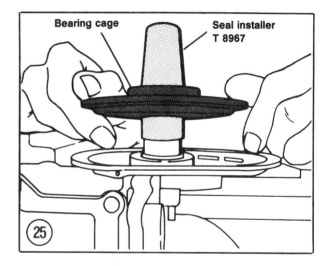

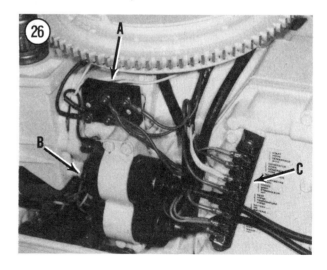

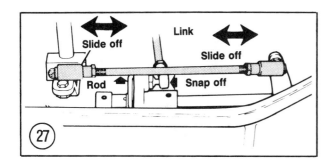

(27)

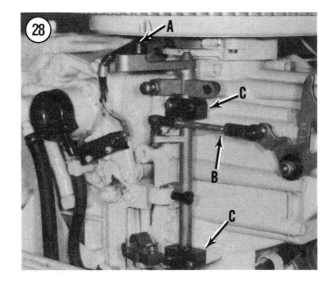

(28)

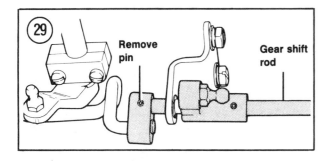

(29)

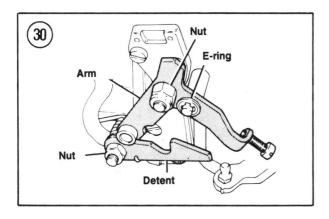

(30)

2. Electric start—Disconnect the negative battery cable.

3. Disconnect spark plug leads. Remove the spark plugs.

4. Manual start—Remove the manual starter assembly. See Chapter Ten.

5. Remove the flywheel as described in this chapter.

6. Electric start—Disconnect electrical leads at the following components and remove the components. See Chapter Seven:

    a. Interlock switch.
    b. Rectifier (A, **Figure 26**).
    c. Ignition coils (B, **Figure 26**).
    d. Terminal block (C, **Figure 26**).
    e. Circuit breaker.
    f. Starter.
    g. Choke solenoid.
    h. Starter solenoid (relay).
    i. Battery ground lead (at power head).
    j. Wiring harness.

7A. Manual start—Disconnect magneto control rod connectors and unsnap rod from interlock link (**Figure 27**).

7B. Electric start—Remove screw holding outer end of link to top of tower shaft (A, **Figure 28**) and unscrew spark control link from breaker plate. Disconnect throttle cam link (B, **Figure 28**) and remove tower shaft brackets (C, **Figure 28**). Remove the tower shaft.

8. Remove the breaker plate (manual) or alternator stator and stator plate (electric).

9. Manual start—Remove roll pin holding gear shift arm to gear shift rod. See **Figure 29**. Remove rod and both arms.

10. Remove stop nut and washer holding gear shift arm to cylinder drain cover. Remove E-ring holding arm to gear shift rod. See **Figure 30**. On manual start models, unscrew hex nut and remove the shift detent.

11. Remove the choke solenoid (electric), carburetor, reed valve assembly and fuel pump. See Chapter Six.

8

12. Disconnect recirculation line at top and bottom of power head.

> *NOTE*
> *At this point, there should be no hoses, wires or linkage connecting the power head to the motor leg. Recheck this to make sure nothing will hamper power head removal.*

13. Remove the 6 screws holding the motor leg covers.

14. Remove the 7 motor leg-to-power head screws. Remove the motor leg exhaust cover, then remove the remaining motor leg-to-power head screw located under the cover.

15. Remove the air baffle and 2 screws holding the upper shock mount cover to the upper shock mount.

16. Remove power head from motor leg and place it on a clean workbench.

17. Remove the 4 screws holding the crankshaft bearing cage to the power head. Remove the bearing cage and stator ring from the power head. Remove and discard the gasket.

18. Clean all gasket residue from the power head and spacer plate surfaces.

19. Installation is the reverse of removal, plus the following:

   a. Use new power head and bearing cage gaskets.

   b. Coat drive shaft splines with a liberal amount of anti-seize lubricant.

   c. Rotate flywheel or propeller as required to align crankshaft and drive shaft splines.

   d. Install a new bearing cage seal with installer part No. T 3512. See **Figure 31**.

   e. Lubricate seal lip with grease and install on seal protector (part No. T 3512). Garter spring must face the power head. Install seal as shown in **Figure 32**.

   f. Tighten all fasteners to specifications (**Table 1**).

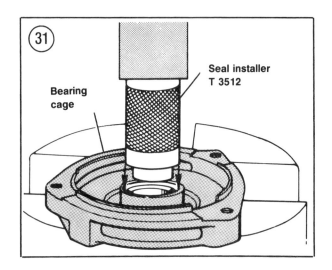

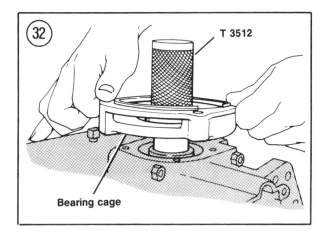

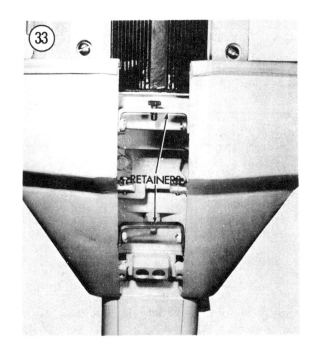

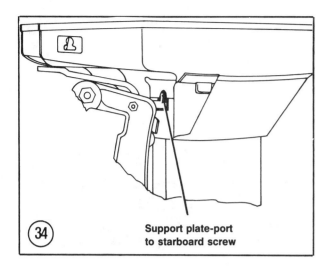

Support plate-port
to starboard screw

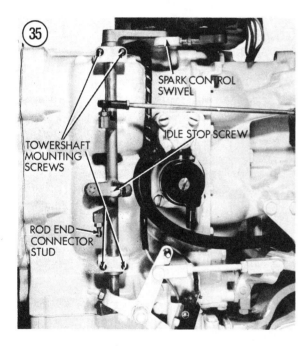

SPARK CONTROL SWIVEL

IDLE STOP SCREW

TOWERSHAFT MOUNTING SCREWS

ROD END CONNECTOR STUD

b. Remove the screws holding the support plate cover. Remove the cover.

c. Loosen the screw holding the support plate cover rear to the top support plate retainer.

d. Remove 2 screws holding the support plate cover rear to the port and starboard support plate. Remove the cover.

e. Remove the idle relief cover grommet from the support plate cover rear.

f. Remove the 4 screws holding the support plate retainers to the port and starboard support plate. See **Figure 33**.

g. Remove the support plate port-to-starboard screw shown in **Figure 34**. Remove the support plate.

4. Remove the flywheel as described in this chapter.

5. Remove the following ignition components. See Chapter Seven:

    a. CD module.

    b. Alternator stator module.

    c. Capacitor module.

    d. Regulator-rectifier module.

    e. Timing ring.

    f. Trigger module.

6. Remove the 2 screws holding the bearing cage to the power head. Remove the bearing cage and gasket. Discard the gasket.

7. Remove the carburetor adapter with carburetors. Remove the fuel pump. See Chapter Six.

8. Remove the starter motor, circuit breaker and starter relay. See Chapter Seven.

9. Remove the tower shaft adapter mounting screws. Remove the tower shaft. See **Figure 35**.

10. Disconnect the gear shift link couplers. Remove the link. See **Figure 36**.

11. Remove the gear shift arm. Remove the interlock switch bracket.

12. Remove the remote control bracket.

13. Disconnect the upper gear shift rod at the gear shift pivot.

8

g. Complete engine synchronization and linkage adjustments. See Chapter Five.

1977-on 55 hp; All 60 and 65 hp

1. Disconnect the negative battery cable.

2. Remove the engine cover.

3. Remove the port and starboard support plate as follows:

    a. Remove the fasteners and support plate mounts holding the plate to the front and rear stabilizers.

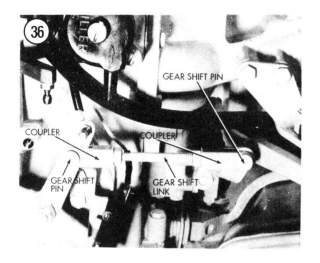

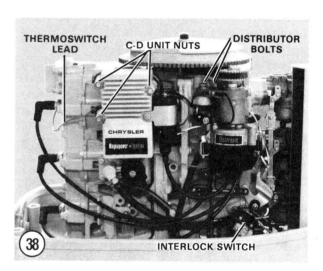

14. Remove the 4 screws holding the front stabilizer to the crankcase. Remove the stabilizer.

NOTE
At this point, there should be no hoses, wires or linkages connecting the power head to the motor leg. Recheck this to make sure nothing will hamper power head removal.

15. Remove the 8 stop nuts holding the power head to the motor leg. Remove the stop nut holding the power head to the spacer plate.

16. Remove power head from the spacer plate and place it on a clean workbench.

17. Clean all gasket residue from the power head and spacer plate surfaces.

18. Installation is the reverse of removal, plus the following:

   a. Use new power head and bearing cage gaskets.

   b. Coat drive shaft splines with a liberal amount of anti-seize lubricant.

   c. Rotate flywheel or propeller as required to align crankshaft and drive shaft splines.

   d. Install a new bearing cage seal with installer part No. T 8985. Seal lip must face flywheel.

   e. Apply a small bead of RTV sealing around the bearing cage face. Lubricate seal lip with Rykon No. 2EP.

   f. Tighten all fasteners to specifications (**Table 1**).

   g. Complete engine synchronization and linkage adjustments. See Chapter Five.

70-140 hp

1. Disconnect the negative battery cable.

2. Remove the engine cover.

3. Disconnect the spark plug leads. Remove the spark plugs.

4. Remove the flywheel as described in this chapter.

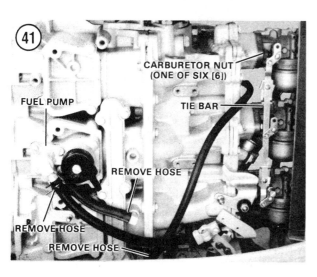

FUEL PUMP

CARBURETOR NUT
(ONE OF SIX [6])

TIE BAR

REMOVE HOSE

REMOVE HOSE

REMOVE HOSE

5. Remove the stator and terminal block without disconnecting the electrical leads. See Chapter Seven.

6. Remove the starter, starter relay and voltage regulator (**Figure 37**). See Chapter Seven.

7. Disconnect the ground cable at the power head.

8. Disconnect the green choke solenoid lead.

9. Remove interlock switch with leads attached (Chapter Seven) and pull it under the carburetors to the port side.

10. Disconnect the orange lead at the cylinder head thermoswitch.

11A. Motorola ignition—Remove the CD unit/coil assembly and distributor without disconnecting the electrical leads (**Figure 38**). See Chapter Seven.

11B. Prestolite ignition—Remove the CD module and coil assembly (**Figure 39**). See Chapter Seven.

12. Carefully lift the CD unit/coil and distributor (Motorola) or CD module and coil assembly (Prestolite) over the top of the power head as an assembly. Remove the electrical components and wiring harness from the engine as an assembly.

13. Remove the carburetor air intake (**Figure 40**).

14. Remove the carburetors, fuel pump(s) and reed valve assembly (**Figure 41**). See Chapter Six.

15. Remove the 2 brackets holding the tower shaft to the power head. Remove the tower shaft.

NOTE
At this point, there should be no hoses, wires or linkages connecting the power head to the motor leg. Recheck this to make sure nothing will hamper power head removal.

16. Remove the 3 screws on each side holding the front and rear motor leg covers together. Remove the 3 screws on each side

8

holding the rear motor leg cover to the support plate. Remove the rear motor leg cover. See **Figure 42**.

17. Remove the 7 screws holding the support plate to the front motor leg cover. Remove the support plate. See **Figure 43**.

18. Remove the 6 screws holding the spacer plate to the motor leg. Remove the 6 nuts holding the power head to the motor leg.

> *WARNING*
> *If a hoist is not available for use in Step 19, have an assistant help with power head removal to avoid possible serious personal injury.*

19. Attach a lifting hook (part No. T 8933) to the crankshaft. Connect a hoist to the hook and remove the power head from the motor leg. Place the power head on a clean workbench, disconnect the hoist and remove the lifting hook.

20. Clean all gasket or RTV sealant residue from the power head and motor leg sealing surfaces.

21. Installation is the reverse of removal, plus the following:

    a. Apply RTV sealant to motor leg sealing surface.

    b. Coat drive shaft splines with a liberal amount of anti-seize lubricant.

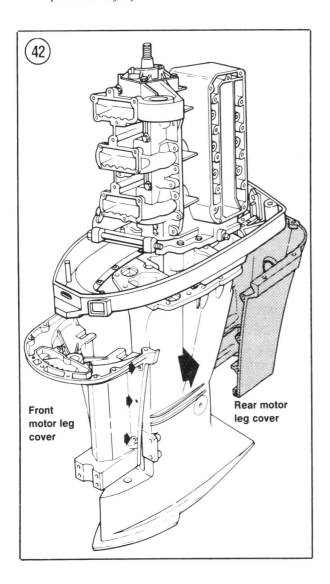

Front motor leg cover

Rear motor leg cover

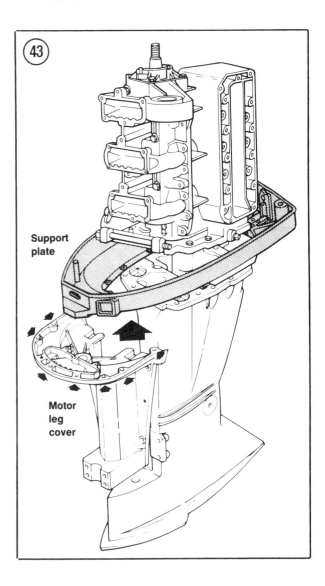

Support plate

Motor leg cover

c. Rotate flywheel or propeller as required to align crankshaft and drive shaft splines.

d. Tighten all fasteners to specifications (**Table 1**).

e. Complete engine synchronization and linkage adjustments. See Chapter Five.

POWER HEAD DISASSEMBLY

1980-on 3.5 hp; 1976-on 4 hp

Refer to **Figure 44** for this procedure.

1. Remove the 4 cylinder head screws. Remove the cylinder head and gasket. Discard the gasket.

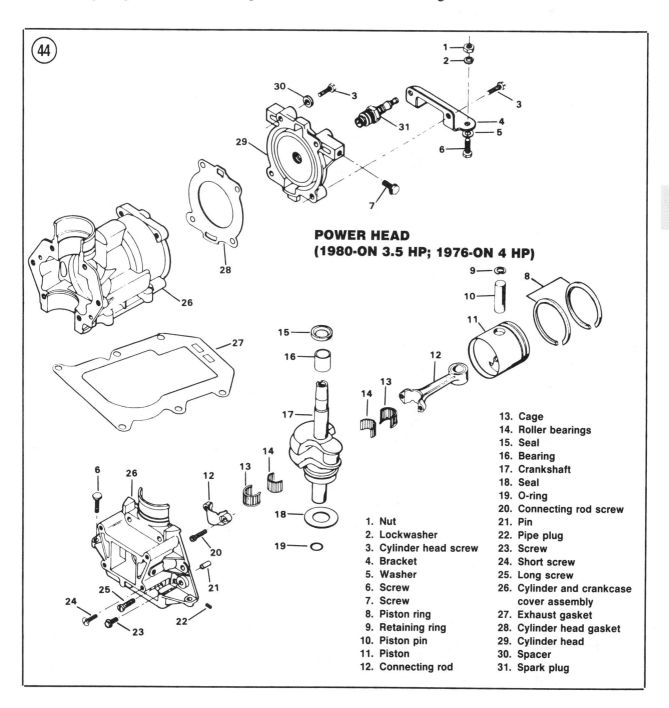

**POWER HEAD
(1980-ON 3.5 HP; 1976-ON 4 HP)**

1. Nut
2. Lockwasher
3. Cylinder head screw
4. Bracket
5. Washer
6. Screw
7. Screw
8. Piston ring
9. Retaining ring
10. Piston pin
11. Piston
12. Connecting rod
13. Cage
14. Roller bearings
15. Seal
16. Bearing
17. Crankshaft
18. Seal
19. O-ring
20. Connecting rod screw
21. Pin
22. Pipe plug
23. Screw
24. Short screw
25. Long screw
26. Cylinder and crankcase cover assembly
27. Exhaust gasket
28. Cylinder head gasket
29. Cylinder head
30. Spacer
31. Spark plug

8

2. With crankcase and cylinder block on a solid surface, drive out the 2 locating pins with a pin punch. See **Figure 45**.

3. Remove the 6 screws holding the crankcase cover to the cylinder.

4. Carefully separate crankcase cover from cylinder by prying at the pry points with a putty knife or similar instrument.

5. Remove the connecting rod cap screws with a ratchet and suitable hex head driver. See **Figure 46**.

6. Remove the connecting rod cap. Remove the needle bearings and bearing cages from the rod cap and crankshaft. Place bearings and cages in a clean container.

7. Remove the crankshaft assembly from the cylinder block. Remove the upper crankshaft bearing and lower crankshaft seal.

8. Reinstall connecting rod cap on connecting rod. Push the piston toward the cylinder head end of the crankcase until the piston rings can be seen.

9. Remove and discard the piston rings with ring expander tool part No. T 8926 or equivalent. See **Figure 47**.

10. Remove the piston and connecting rod through the carburetor end of the crankcase.

11. If the piston is to be removed from the connecting rod, remove the piston pin retaining rings with snap ring pliers.

12. Place piston in pillow block (part No. T 2990) and remove the piston pin with tool part No. T 8919. See **Figure 48**.

4.4-15 hp

Refer to **Figure 49** or **Figure 50** for this procedure.

1. Remove the screws holding the exhaust tube to the power head, if so equipped.

2. Remove the cylinder head bolts and washers. Remove the cylinder head and gasket. Discard the gasket.

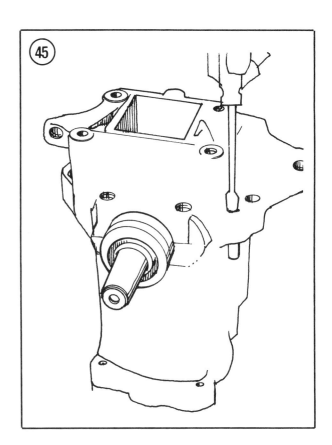

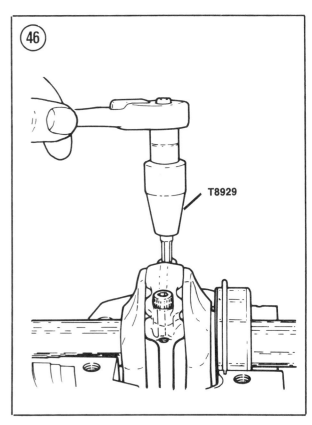

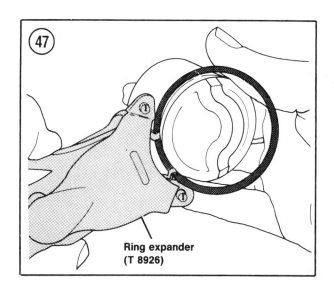

Ring expander
(T 8926)

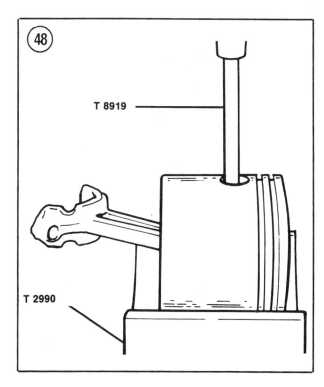

T 8919

T 2990

3. Remove the cylinder drain cover screws. Remove the cover and gasket. Discard the gasket.

4. Remove the transfer port cover screws, if so equipped. Remove the cover and gasket. Discard the gasket.

5. With crankcase and cylinder block on a solid surface, drive out the 2 locating pins with a suitable pin punch.

6. Remove the crankcase cover-to-cylinder block screws.

NOTE
*If the top half of the main bearing liner comes off when the cover is removed in Step 7, retrieve any roller bearings from the cylinder with a swab covered with grease. Do **not** use a magnet.*

7. Carefully pry the cover and block apart using a suitable pry tool at the pry points provided. Remove the cover from the block.

8. Remove the main bearing liner and roller bearings. Place in a clean container.

9. Lift the crankshaft slightly at the lower main journal. Slide seal and lower main bearing off the crankshaft.

10. Mark the connecting rod and cap. Remove each connecting rod cap and needle bearings. Place in a clean container.

11. Remove the crankshaft from the cylinder block.

12. Remove the remaining connecting rod and main bearing roller bearings and place in separate containers.

13. Reinstall the rod cap to its respective connecting rod. Remove each piston and connecting rod assembly from its cylinder. Mark the cylinder number on the top of the piston with a felt-tipped pen.

14. Remove and discard the piston rings with ring expander tool part No. T 8926 or equivalent. See **Figure 47**.

15. If the piston is to be removed from the connecting rod, remove the piston pin retaining rings with snap ring pliers part No. T 1749 or equivalent.

16. Place piston in pillow block (part No. T 2990) and remove the piston pin with tool part No. T 8919. See **Figure 48**.

8

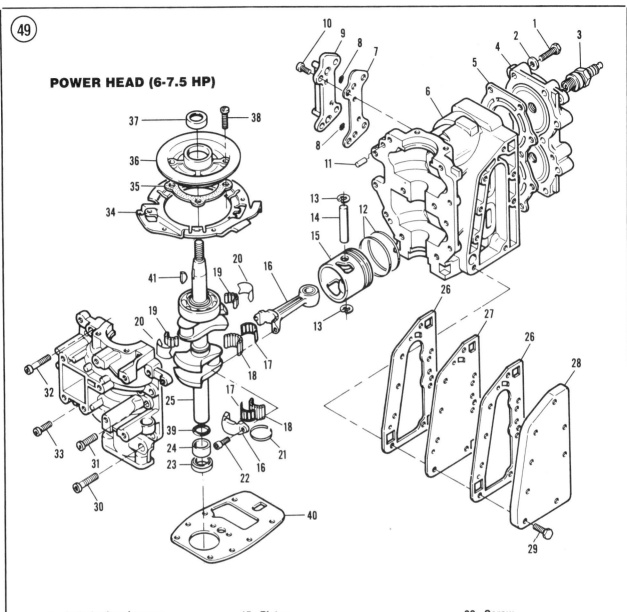

POWER HEAD (6-7.5 HP)

1. Cylinder head screw	15. Piston	29. Screw
2. Plain washer	16. Connecting rod	30. Screw
3. Spark plug	17. Bearing cage	31. Screw
4. Cylinder head	18. Roller bearings	32. Screw
5. Head gasket	19. Center main roller bearings	33. Screw
6. Cylinder and crankcase cover assembly	20. Center main bearing line	34. Stator ring
7. Drain cover gasket	21. Crankcase seal	35. Bearing cage gasket
8. Drain screen	22. Connecting rod screw	36. Bearing cage
9. Drain cover	23. Lower crankshaft seal	37. Upper crankshaft seal
10. Screw	24. Lower main bearing	38. Screw
11. Locating pin	25. Crankshaft and upper main bearing assembly	39. Crankshaft spline seal
12. Piston ring	26. Exhaust plate gasket	40. Cylinder mounting gasket
13. Piston pin retainer	27. Exhaust plate	41. Magneto key
14. Piston pin	28. Exhaust cover	

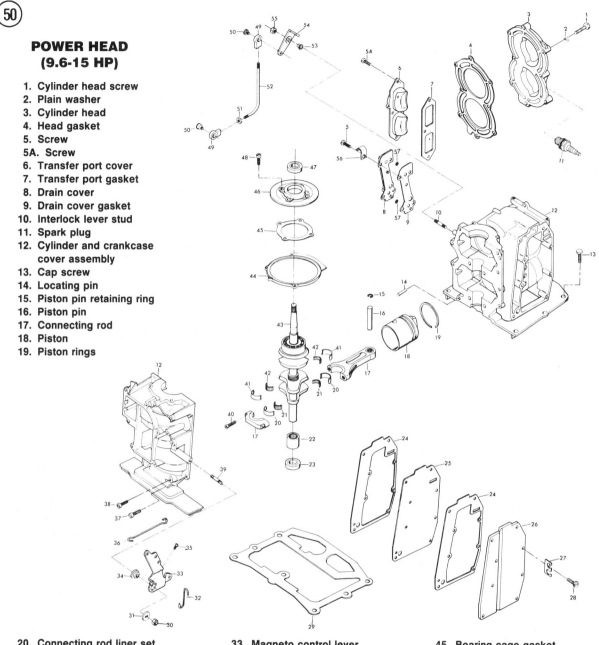

50

POWER HEAD
(9.6-15 HP)

1. Cylinder head screw
2. Plain washer
3. Cylinder head
4. Head gasket
5. Screw
5A. Screw
6. Transfer port cover
7. Transfer port gasket
8. Drain cover
9. Drain cover gasket
10. Interlock lever stud
11. Spark plug
12. Cylinder and crankcase
    cover assembly
13. Cap screw
14. Locating pin
15. Piston pin retaining ring
16. Piston pin
17. Connecting rod
18. Piston
19. Piston rings

8

20. Connecting rod liner set
21. Roller bearings
22. Lower main crankshaft bearing
23. Crankshaft lower seal
24. Exhaust cover gasket
25. Exhaust plate
26. Exhaust cover
27. Lead wire clip
28. Screw
29. Cylinder mounting gasket
30. Stop nut
31. Washer
32. Magneto control rod link

33. Magneto control lever
34. Control lever bearing
35. Screw
36. Throttle cam link
37. Screw
38. Screw
39. Magneto lever stud
40. Connecting rod screw
41. Main bearing liner set
42. Roller bearings
43. Crankshaft and upper main
    bearing assembly
44. Magneto stator ring

45. Bearing cage gasket
46. Bearing cage
47. Crankshaft upper seal
48. Screw
49. Interlock lever bearing
50. Interlock rod bearing swivel
51. Hex nut
52. Interlock rod
53. Interlock lever bearing
54. Interlock lever
55. Stop nut
56. Hose clamp
57. Screen

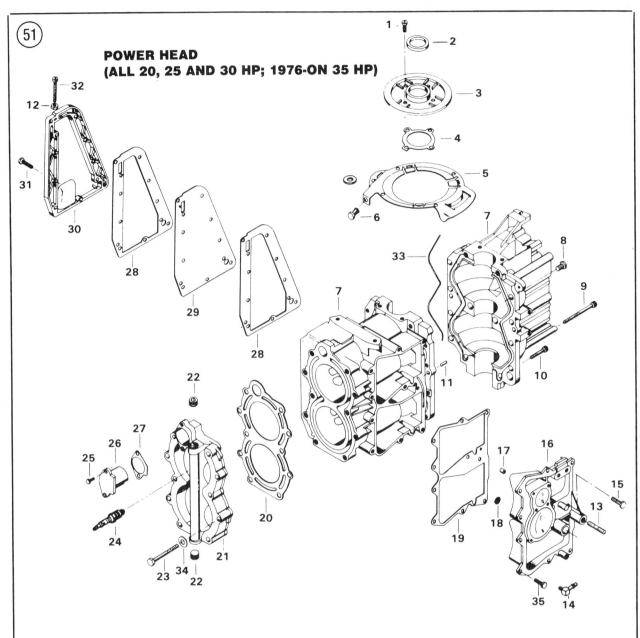

**POWER HEAD
(ALL 20, 25 AND 30 HP; 1976-ON 35 HP)**

1. Screw
2. Crankshaft upper main seal
3. Crankshaft bearing
4. Bearing cage gasket
5. Magneto stator ring
6. Plastic rivet
7. Cylinder and crankcase
   cover assembly
8. Screw
9. Socket head screw (long)
10. Socket head screw (short)
11. Locating pin
12. Hex nut
13. Gear shift lever stud
14. Drain hose elbow
15. Screw
16. Transfer port cover
17. Check valve
18. Cylinder drain screen
19. Transfer port cover gasket
20. Cylinder head gasket
21. Cylinder head
22. Plug
23. Cylinder head screw
24. Spark plug
25. Cap screw
26. Thermostat cover
27. Gasket
28. Exhaust port plate gasket
29. Exhaust port plate
30. Exhaust port cover
31. Screw
32. Screw
33. Spaghetti seal
34. Plain washer
35. Cap screw

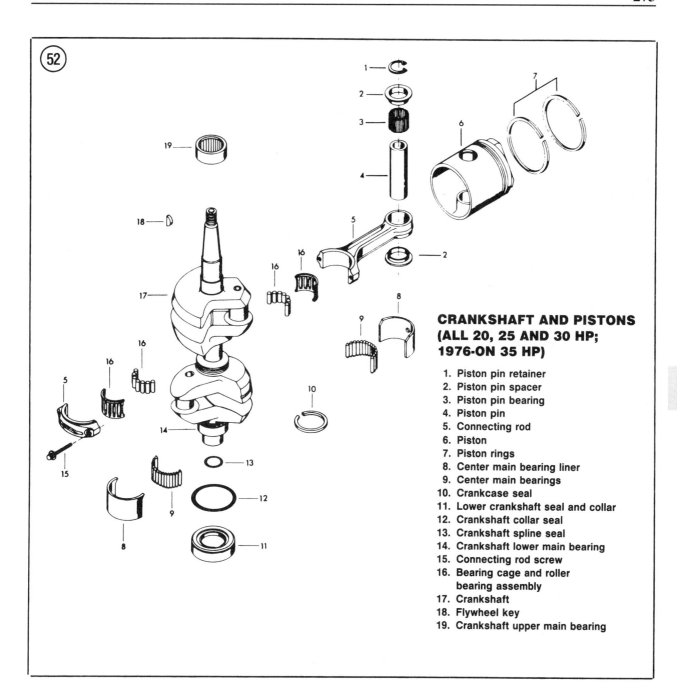

(52)

**CRANKSHAFT AND PISTONS
(ALL 20, 25 AND 30 HP;
1976-ON 35 HP)**

1. Piston pin retainer
2. Piston pin spacer
3. Piston pin bearing
4. Piston pin
5. Connecting rod
6. Piston
7. Piston rings
8. Center main bearing liner
9. Center main bearings
10. Crankcase seal
11. Lower crankshaft seal and collar
12. Crankshaft collar seal
13. Crankshaft spline seal
14. Crankshaft lower main bearing
15. Connecting rod screw
16. Bearing cage and roller
    bearing assembly
17. Crankshaft
18. Flywheel key
19. Crankshaft upper main bearing

8

**All 20, 25 and 30 hp;
1976-on 35 hp**

Refer to **Figure 51** and **Figure 52** for this procedure.

1. Remove the exhaust port cover screws. Remove the cover, plate and 2 gaskets. Discard the gaskets.

2. Remove the transfer port cover screws. Remove the cover and gasket. Discard the gasket.

3. Remove the cylinder head thermostat cover screws. Remove the cover and gasket. Discard the gasket.

4. Remove the cylinder head bolts. Remove the cylinder head and gasket. Discard the gasket.

5. Remove the crankcase cover screws. Note that 2 of the screws are inside the reed plate opening. See **Figure 53**.

6. With crankcase and cylinder block on a solid surface, drive out the 2 locating pins with a suitable pin punch.

7. Carefully pry the cover and block apart using a suitable pry tool at the pry points provided. See **Figure 54**. Remove the cover from the block.

8. Remove the main bearing race half and roller bearings. Place in a clean container.

9. Mark the connecting rods and caps. Remove each connecting rod cap, roller bearings and bearing cage. Place bearings and cage from each rod in separate containers.

10. Remove the crankshaft from the cylinder block.

11. Remove the remaining connecting rod and main bearing roller bearings and place in separate containers.

12. Reinstall each rod cap to its respective connecting rod. Remove each piston and connecting rod assembly from its cylinder. Mark the cylinder number on the top of the piston with a felt-tipped pen.

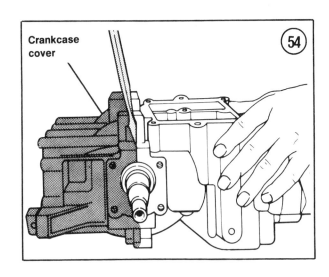

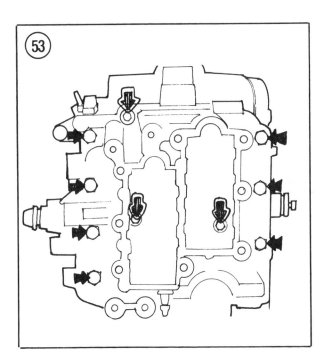

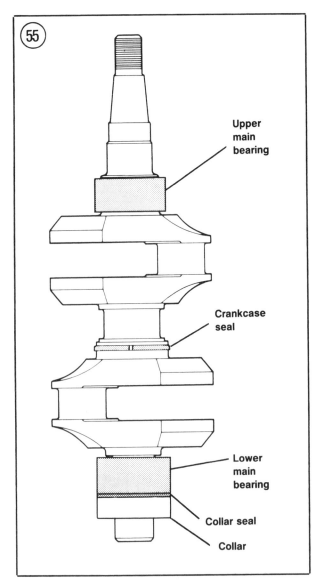

13. Remove the crankshaft upper main bearing, lower main bearing collar seal and crankcase seal ring. See **Figure 55**.

14. Remove and discard the piston rings with ring expander tool part No. T 8926 or equivalent. See **Figure 47**.

15. If the piston is to be removed from the connecting rod, remove the piston pin retaining rings with snap ring pliers.

16. Place piston in pillow block (part No. T 2990) and remove the piston pin with a suitable driver. See **Figure 56**.

1966-1975 35 hp; All 45 and 50 hp; 1967-1976 55 hp

Refer to **Figure 57** and **Figure 58** for this procedure.

1. Use pliers to slide each thermostat water tube hose clamp away from the fitting enough to disconnect the tube. Pull the tube off the exhaust port cover and thermostat cover fittings. See **Figure 59**.

2. Remove the screws holding the thermostat cover to the cylinder head. Remove the cover, gasket, thermostat and grommet. See **Figure 60**.

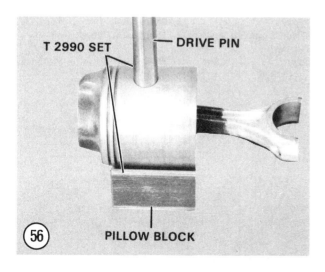

T 2990 SET — DRIVE PIN

(56) PILLOW BLOCK

3. Remove the cylinder head bolts. Remove the cylinder head and gasket. Discard the gasket.

4. Remove the exhaust port cover bolts and lockwashers. Remove the cover and gasket. Discard the gasket.

5. Electric start—remove the 2 screws holding the coil bracket. Remove the bracket.

6. Remove the fuel pump cover and gasket. Discard the gasket.

7. Remove the 2 transfer port covers and gaskets. Discard the gaskets.

8. Remove the cylinder drain cover screws. Remove the cylinder drain cover, gasket, reed plate and gasket assembly. See **Figure 61**. Discard the gasket.

9. Remove the 5 screws holding the shock mount cover (**Figure 62**). Remove the cover.

10. Remove the crankcase cover screws installed along the crankcase parting line.

11. With crankcase and cylinder block on a solid surface, drive out the 2 locating pins with a suitable pin punch.

12. Remove the crankcase cover bolts. Note that one bolt is inside the reed plate opening. See **Figure 63**.

13. Carefully pry the cover and block apart using a suitable pry tool at the pry points provided. Remove the cover from the block.

14. Remove the spaghetti seal from each side of the crankcase. See **Figure 64**.

15. Mark the connecting rods and caps. Remove each connecting rod cap, roller bearings and bearing cage. Place bearings and cage from each rod in separate containers.

16. Remove the crankshaft from the cylinder block.

17. Remove the remaining connecting rod roller bearings and place in separate containers.

18. Reinstall each rod cap to its respective connecting rod. Remove each piston and connecting rod assembly from its cylinder. Mark the cylinder number on the top of the piston with a felt-tipped pen.

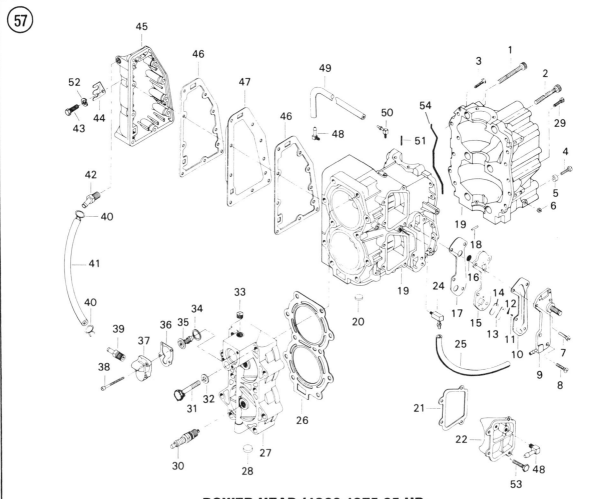

**POWER HEAD (1966-1975 35 HP;
ALL 45 AND 50 HP; 1967-1976 55 HP)**

1. Socket head cap screw (long)
2. Socket head cap screw (short)
3. Slot head screw
4. Screw
5. Bumper
6. Stop nut
7. Pan head screw
8. Screw
9. Drain cover
10. Drain cover gasket
11. Screw
12. Plain washer
13. Drain reed stop
14. Drain reed
15. Drain reed plate
16. Drain reed plate screen
17. Drain reed plate gasket
18. Locating pin
19. Cylinder and crankcase
    cover assembly
20. Welch plug
21. Transfer port cover gasket
22. Transfer port cover
23. Fuel line elbow
24. Elbow
25. Hose
26. Cylinder head gasket
27. Cylinder head
28. Welch plug
29. Screw (short)
30. Spark plug
31. Screw (long)
32. Plain washer
33. Pipe plug
34. Thermostat grommet
35. Thermostat
36. Thermostat cover gasket
37. Thermostat cover
38. Screw
39. Thermostat fitting
40. Clamp
41. Thermostat water tube
42. Thermostat fitting
43. Screw
44. Lead wire clip
45. Exhaust port cover
46. Exhaust port plate gasket
47. Exhaust port plate
48. Fitting
49. Elbow
50. Elbow with metering cup
51. Roll pin
52. Spring lockwasher
53. Screw
54. Spaghetti seal

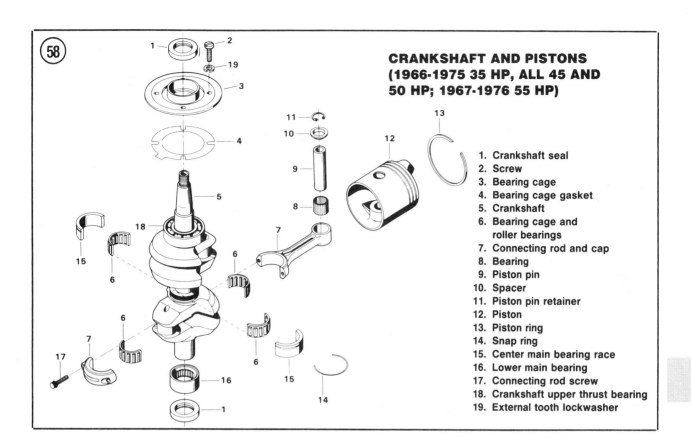

58

**CRANKSHAFT AND PISTONS
(1966-1975 35 HP, ALL 45 AND
50 HP; 1967-1976 55 HP)**

1. Crankshaft seal
2. Screw
3. Bearing cage
4. Bearing cage gasket
5. Crankshaft
6. Bearing cage and
   roller bearings
7. Connecting rod and cap
8. Bearing
9. Piston pin
10. Spacer
11. Piston pin retainer
12. Piston
13. Piston ring
14. Snap ring
15. Center main bearing race
16. Lower main bearing
17. Connecting rod screw
18. Crankshaft upper thrust bearing
19. External tooth lockwasher

8

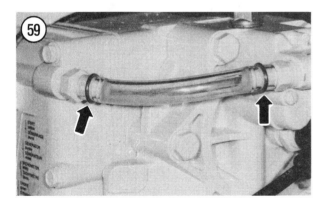

59

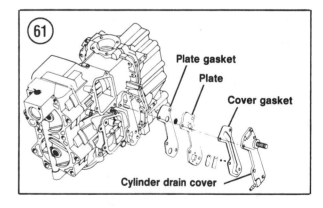

61

Plate gasket

Plate

Cover gasket

Cylinder drain cover

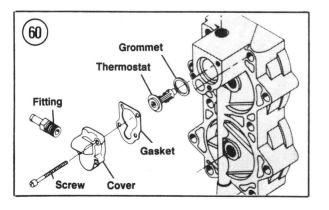

60

Grommet

Thermostat

Fitting

Gasket

Screw   Cover

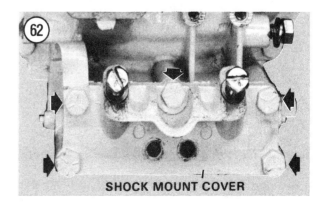

62

SHOCK MOUNT COVER

19. Slide the seal and lower main bearing from the crankshaft.

20. Carefully pry the snap ring from the center main bearing race. Expand ring just enough to remove it. Remove the bearing race halves and the roller bearings. Place in a clean container.

21. If crankshaft upper main bearing (**Figure 65**) requires replacement, remove with an arbor press and suitable mandrel.

22. Remove and discard the piston rings with ring expander tool part No. T 8926 or equivalent. See **Figure 66**.

23. If the piston is to be removed from the connecting rod, remove the piston pin retaining rings with snap ring pliers.

24. Place piston in pillow block (part No. T 2990) and remove the piston pin with a suitable driver. See **Figure 56**.

1977-on 55 hp; All 60 and 65 hp

Refer to **Figure 67** and **Figure 68** for this procedure.

1. Remove the bypass valve cover at the upper left of the cylinder head. Remove the cover and gasket. Discard the gasket.

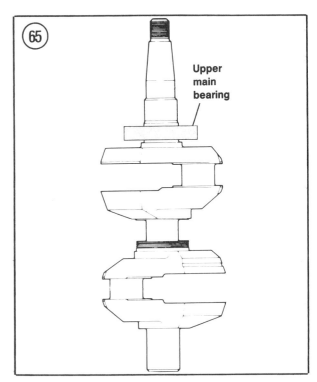

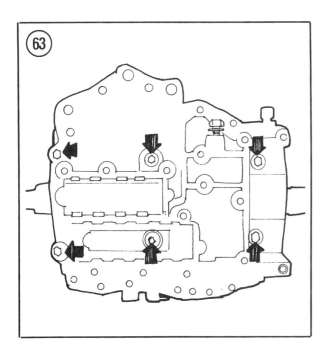

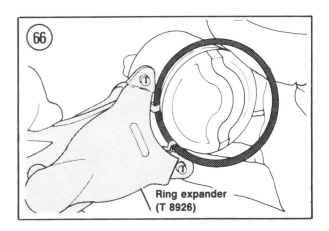

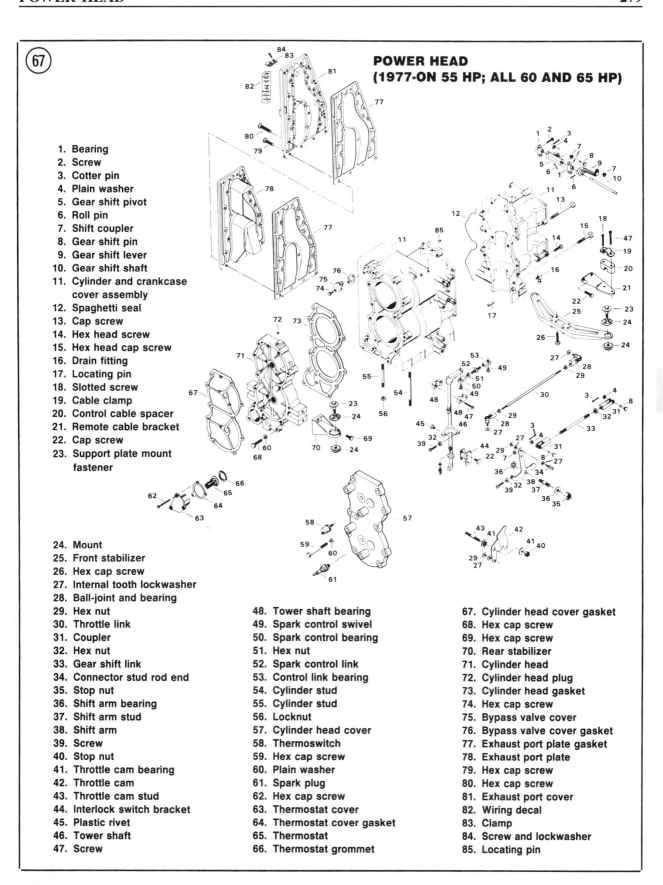

**POWER HEAD
(1977-ON 55 HP; ALL 60 AND 65 HP)**

1. Bearing
2. Screw
3. Cotter pin
4. Plain washer
5. Gear shift pivot
6. Roll pin
7. Shift coupler
8. Gear shift pin
9. Gear shift lever
10. Gear shift shaft
11. Cylinder and crankcase cover assembly
12. Spaghetti seal
13. Cap screw
14. Hex head screw
15. Hex head cap screw
16. Drain fitting
17. Locating pin
18. Slotted screw
19. Cable clamp
20. Control cable spacer
21. Remote cable bracket
22. Cap screw
23. Support plate mount fastener
24. Mount
25. Front stabilizer
26. Hex cap screw
27. Internal tooth lockwasher
28. Ball-joint and bearing
29. Hex nut
30. Throttle link
31. Coupler
32. Hex nut
33. Gear shift link
34. Connector stud rod end
35. Stop nut
36. Shift arm bearing
37. Shift arm stud
38. Shift arm
39. Screw
40. Stop nut
41. Throttle cam bearing
42. Throttle cam
43. Throttle cam stud
44. Interlock switch bracket
45. Plastic rivet
46. Tower shaft
47. Screw

48. Tower shaft bearing
49. Spark control swivel
50. Spark control bearing
51. Hex nut
52. Spark control link
53. Control link bearing
54. Cylinder stud
55. Cylinder stud
56. Locknut
57. Cylinder head cover
58. Thermoswitch
59. Hex cap screw
60. Plain washer
61. Spark plug
62. Hex cap screw
63. Thermostat cover
64. Thermostat cover gasket
65. Thermostat
66. Thermostat grommet

67. Cylinder head cover gasket
68. Hex cap screw
69. Hex cap screw
70. Rear stabilizer
71. Cylinder head
72. Cylinder head plug
73. Cylinder head gasket
74. Hex cap screw
75. Bypass valve cover
76. Bypass valve cover gasket
77. Exhaust port plate gasket
78. Exhaust port plate
79. Hex cap screw
80. Hex cap screw
81. Exhaust port cover
82. Wiring decal
83. Clamp
84. Screw and lockwasher
85. Locating pin

8

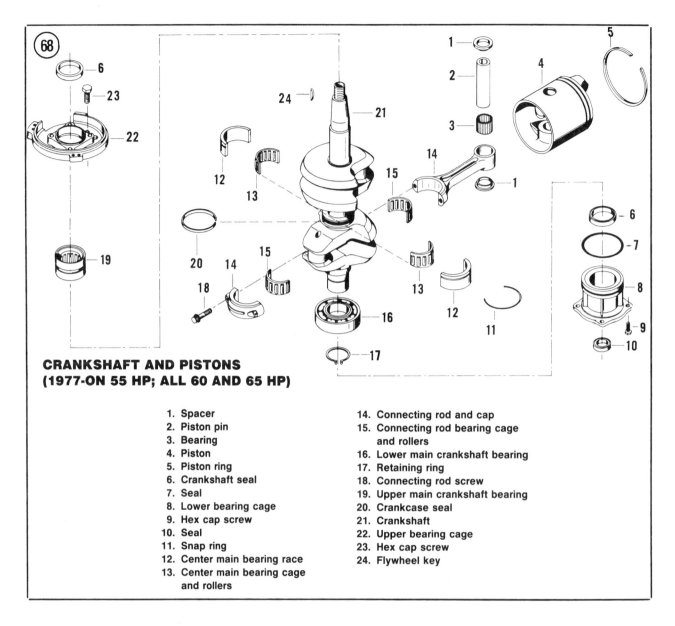

CRANKSHAFT AND PISTONS
(1977-ON 55 HP; ALL 60 AND 65 HP)

1. Spacer
2. Piston pin
3. Bearing
4. Piston
5. Piston ring
6. Crankshaft seal
7. Seal
8. Lower bearing cage
9. Hex cap screw
10. Seal
11. Snap ring
12. Center main bearing race
13. Center main bearing cage and rollers
14. Connecting rod and cap
15. Connecting rod bearing cage and rollers
16. Lower main crankshaft bearing
17. Retaining ring
18. Connecting rod screw
19. Upper main crankshaft bearing
20. Crankcase seal
21. Crankshaft
22. Upper bearing cage
23. Hex cap screw
24. Flywheel key

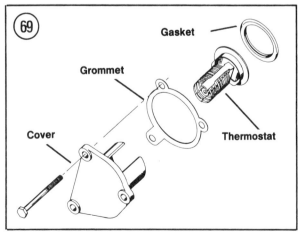

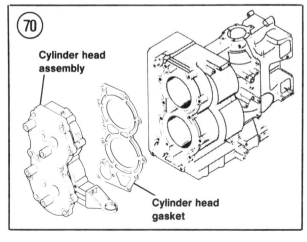

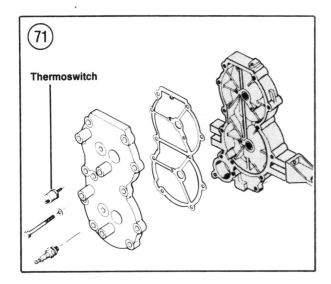

71

Thermoswitch

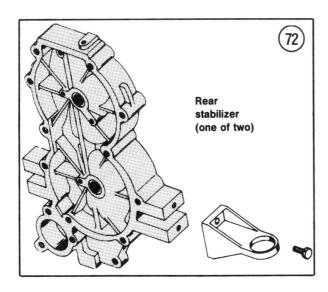

72

Rear stabilizer (one of two)

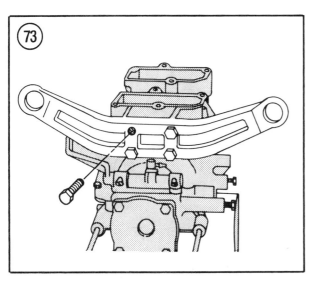

73

2. Remove the thermostat cover, gasket, thermostat and grommet. See **Figure 69**. Discard the gasket.

3. Remove the thermoswitch from the cylinder head cover.

4A. 1977-1980—Remove the cylinder head and cover assembly from the power head (**Figure 70**). Remove the 3 screws holding the cover to the cylinder head. Separate the cover and head. Discard the cover and head gaskets.

4B. 1981-on—Remove the 2 screws holding the cover to the cylinder head (**Figure 71**). Remove the cover and discard the gasket. Remove the cylinder head bolts. Remove the cylinder head and gasket. Discard the gasket.

5. Remove the 2 screws holding each rear stabilizer to the cylinder head. Remove the stabilizer (**Figure 72**).

6. Remove the exhaust port cover screws. Remove the cover and gasket. Discard the gasket.

7. If the forward stabilizer was not removed during power head removal, remove the 4 bolts shown in **Figure 73**. Remove the stabilizer.

8. Remove the lower bearing cage.

9. Remove the gearshift lever and pivot (**Figure 74**).

8

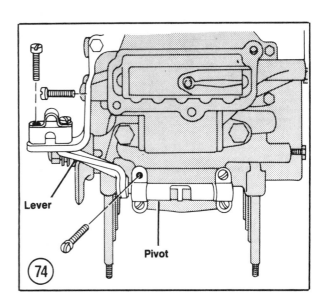

Lever

Pivot

74

10. Remove the remote cable assembly.

11. Remove the crankcase cover bolts and screws. Note that one of the bolts is inside the intake manifold opening. See arrows, **Figure 75**.

12. With crankcase and cylinder block on a solid surface, drive out the 2 locating pins (1 and 2, **Figure 75**) with a suitable pin punch.

13. Carefully pry the cover and block apart using a suitable pry tool at the pry points provided. Remove the cover from the block.

14. Remove the spaghetti seal from each side of the crankcase. See **Figure 76**.

15. Mark the connecting rods and caps. Remove each connecting rod cap, roller bearings and bearing cage. Place bearings and cage from each rod in separate containers.

16. Remove the crankshaft from the cylinder block.

17. Remove the remaining connecting rod and main bearing roller bearings and place in separate containers.

18. Reinstall each rod cap to its respective connecting rod. Remove each piston and connecting rod assembly from its cylinder. Mark the cylinder number on the top of the piston with a felt-tipped pen.

19. Slide the upper main bearing off the crankshaft. Remove the crankcase seal. See **Figure 77**.

20. Carefully pry the snap ring from the center main bearing race (**Figure 77**). Expand ring just enough to remove it. Remove the bearing race halves and the roller bearings. Place in a clean container.

21. If crankshaft lower main bearing requires replacement, remove the snap ring, then press bearing off with a suitable mandrel.

22. Remove and discard the piston rings with ring expander tool part No. T 8926 or equivalent. See **Figure 66**.

NOTE
Do not attempt further disassembly of the piston. The piston, connecting rod, bearings and spacers are serviced as a complete assembly.

70-140 hp

A large number of bolts and screws of different lengths are used to secure the various covers and components. It is a good idea to use a cupcake tin or similar compartmented container to hold the various fasteners removed from each cover or component. This will make reassembly easier and faster.

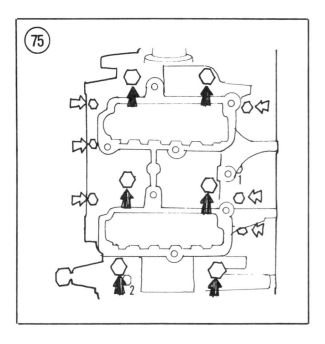

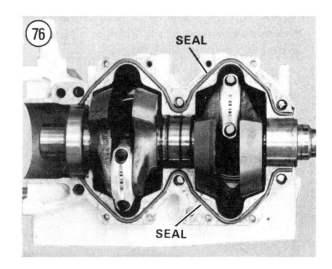

Refer to **Figure 78** (3-cylinder) or **Figure 79** (4-cylinder) for this procedure.

1. Remove the CD module studs and shock mounts.
2. Remove the cylinder drain cover, reed plate and gaskets. Discard the gaskets.
3. Remove the transfer port covers and gaskets. Discard the gaskets.
4. Remove the exhaust port cover, plate and 2 gaskets. Discard the gaskets.
5. Remove the bypass cover, gasket, spring and valve.
6. Remove the thermoswitch from the cylinder head. Remove the thermostat cover, grommet, thermostat and gasket. See **Figure 80**. Discard the gasket.
7. Remove the bolts holding the cylinder head cover, cylinder head and head gasket to the power head. Remove the assembly and discard the gasket.
8. Remove the 4 screws holding the head cover to the head. Separate the cover from the head.
9. Remove the 4 screws holding the exhaust tube and seal to the power head spacer plate. Remove the tube and seal.
10. Remove the 4 nuts and washers holding the spacer plate and upper exhaust gasket to the power head. Remove the spacer plate and discard the gasket.

11. Remove the timing pointer from the crankcase cover.
12. Remove the 4 crankshaft bearing cage screws. Pry the bearing cage from the power head.
13. Remove the bearing seal from the bearing cage. Remove the bearing cage seal from the power head groove.
14. Remove the main bearing bolts and crankcase cover screws installed along the crankcase split line.
15. With crankcase and cylinder block on a solid surface, drive out the 2 locating pins with a suitable pin punch.

NOTE
Pry points are located on the same corners as the locating pins on 3-cylinder blocks. On 4-cylinder blocks, they are on the side of the block opposite the locating pins.

16. Carefully pry the cover and block apart using a suitable pry tool. The pry points are located on opposite ends of the block. Remove the cover from the block.
17. Mark the connecting rods and caps. Remove each connecting rod cap, roller bearings and bearing cage. Place bearings and cage from each rod in separate containers.
18. Remove the crankshaft from the cylinder block.
19. Remove the remaining connecting rod and main bearing roller bearings and place in separate containers.
20. Reinstall each rod cap to its respective connecting rod.
21. Insert a screwdriver between the lower main bearing and the lower crankshaft seal. Pry seal from bearing cage.
22. Slide lower main bearing and seal from crankshaft.
23. Carefully pry the snap ring from each center main bearing race. Expand ring just

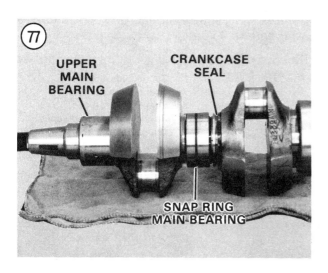

⑰

UPPER MAIN BEARING

CRANKCASE SEAL

SNAP RING MAIN BEARING

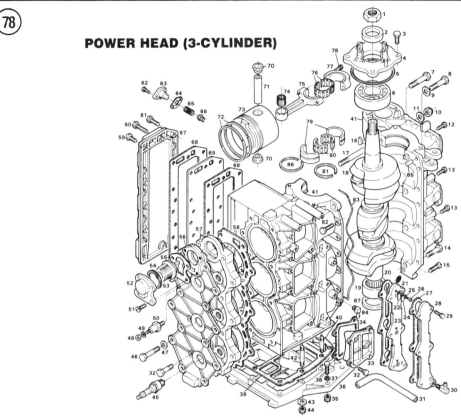

POWER HEAD (3-CYLINDER)

1. Flywheel nut
2. Seal
3. Hex cap screw
4. Crankshaft bearing cage
5. Crankshaft bearing cage seal
6. Upper crankshaft thrust bearing
7. Hex bolt
8. Hex bolt
9. Plain washer
10. Hex nut
11. Plain washer
12. Hex cap screw
13. Hex cap screw
14. Hex bolt
15. Screw
16. Flywheel key
17. Stud
18. Crankshaft
19. Crankshaft lower main bearing
20. Drain reed plate gasket
21. Screen
22. Reed
23. Drain reed plate
24. Reed stop
25. Plain washer
26. Pan head screw
27. Drain cover gasket
28. Drain cover
29. Hex head screw

30. Fitting
31. Hose
32. Hex head screw
33. Transfer port cover
34. Transfer port cover gasket
35. Stop nut
36. Spacer plate
37. Spring lockwasher
38. Hex bolt
39. Upper cylinder exhaust gasket
40. Forward stud
41. Cylinder and crankcase
    cover assembly
42. Rear stud
43. Plain washer
44. Stop nut
45. Spark plug
46. Hex bolt
47. Plain washer
48. Hex nut
49. Spring lockwasher
50. Thermostat
51. Screw
52. Thermostat cover
53. Grommet
54. Thermostat
55. Thermostat cover gasket
56. Cylinder head cover
57. Cylinder head

58. Cylinder head gasket
59. Hex bolt
60. Hex head screw
61. Hex head screw
62. Hex head screw
63. Bypass valve cover
64. Bypass valve cover gasket
65. Bypass valve spring
66. Bypass valve
67. Exhaust port cover
68. Exhaust port plate gasket
69. Exhaust port plate
70. Piston pin spacer
71. Piston pin
72. Piston rings
73. Piston
74. Piston pin bearing
75. Connecting rod
76. Bearing cage and rollers
77. Connecting rod cap
78. Connecting rod cap screw
79. Crankshaft main bearing race
80. Bearing cage and rollers
81. Crankcase seal
82. Hex head screw
83. Spaghetti seal
84. Elbow
85. Locating pin
86. Snap ring
87. Crankshaft lower seal

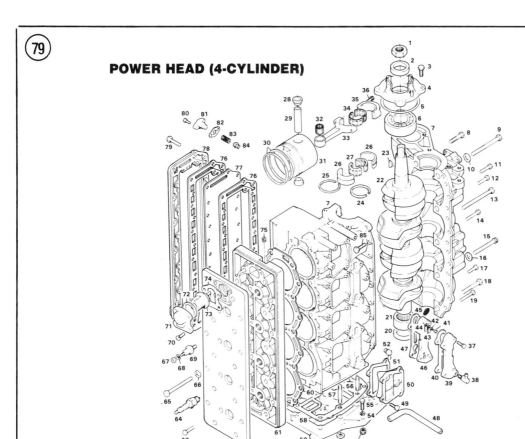

⑦⑨

POWER HEAD (4-CYLINDER)

1. Flywheel nut	29. Piston pin	58. Cylinder exhaust upper gasket
2. Seal	30. Piston ring	59. Spacer plate
3. Hex cap screw	31. Piston	60. Cylinder head gasket
4. Crankshaft bearing cage	32. Piston pin bearings	61. Cylinder head
5. Seal	33. Connecting rod	62. Cylinder head cover
6. Crankshaft upper bearing	34. Bearing cage with rollers	63. Hex cap screw
7. Cylinder and crankcase	35. Connecting rod cap	64. Spark plug
cover assembly	36. Connecting rod screw	65. Hex cap screw
8. Hex head screw (long)	37. Hex head screw	66. Plain washer
9. Hex head screw (short)	38. Lower cylinder drain elbow	67. Hex nut
10. Plain washer	39. Cylinder drain cover	68. Lockwasher
11. Hex head screw	40. Cylinder drain cover gasket	69. Thermoswitch
12. Hex head screw	41. Screw	70. Slotted head cap screw
13. Hex cap screw	42. Plain washer	71. Thermostat cover
14. Head head screw	43. Reed stop	72. Thermostat grommet
15. Socket head cap screw	44. Cylinder drain reed	73. Thermostat
16. Plain washer	45. Screen	74. Gasket
17. Hex head screw	46. Cylinder drain reed plate	75. Plug
18. Hex cap screw	47. Gasket	76. Exhaust port plate gasket
19. Screw	48. Hose	77. Exhaust port plate
20. Crankshaft lower seal	49. Hex cap screw	78. Exhaust port cover
21. Crankshaft lower main bearing	50. Transfer port cover	79. Hex bolt
22. Crankshaft	51. Transfer port cover gasket	80. Hex cap screw
23. Flywheel key	52. Elbow	81. Bypass valve cover
24. Crankcase seal	53. Stop nut	82. Bypass valve cover gasket
25. Snap ring	54. Lockwasher	83. Bypass valve spring
26. Crankshaft main bearing race	55. Hex cap screw	84. Bypass valve
27. Bearing cage with rollers	56. Stud	85. Hex head screw
28. Piston pin spacer	57. Stud	

8

enough to remove it. Remove the bearing race halves and the caged roller bearings. Place in a separate clean containers.

24. Remove the crankcase seal from the crankshaft grooves next to the main bearing journals. Place each seal in the container with its main bearing.

25. Remove the spaghetti seal from each side of the crankcase.

26. Remove each piston and connecting rod assembly from its cylinder. Mark the cylinder number on the top of the piston with a felt-tipped pen.

27. Remove and discard the piston rings with ring expander tool part No. T 8926 or equivalent. See **Figure 66**.

28. If the piston is to be removed from the connecting rod, use special tool part No. T 2990 and refer to **Figure 81** for the following steps:

   a. Insert the shim bar end marked "295" between the large diameter of the spacer and roller bearings.

   b. Position piston in pillow block and align piston pin with hole in the block.

   c. Position the stepped end of the drive pin in the piston pin inner diameter.

   d. Press the piston pin out.

   e. Separate the connecting rod, piston pin spacers and roller bearings from the piston.

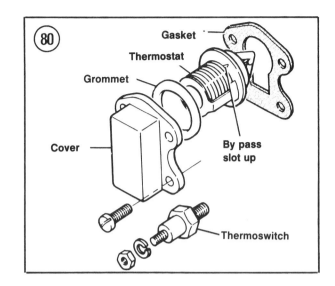

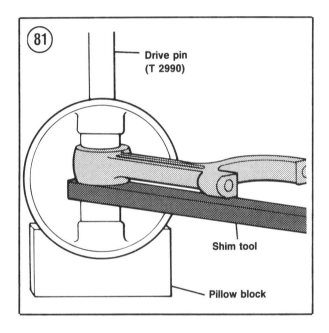

POWER HEAD CLEANING/INSPECTION

Cylinder Block and Crankcase

Chrysler outboard cylinder blocks and crankcase covers are matched and line-bored assemblies. For this reason, you should not attempt to assemble an engine with parts salvaged from other blocks. If inspection indicates that either the block or cover requires replacement, replace both as an assembly.

Carefully remove all gasket and sealant residue from the cylinder block and crankcase cover mating surfaces with lacquer thinner. Clean the aluminum surfaces carefully to avoid nicking them. A dull putty knife can be used, but a piece of Lucite with one edge ground to a 45 degree angle is more efficient and will also reduce the possibility of damage to the surfaces. When sealing the crankcase cover and cylinder block, both mating surfaces must be free of all sealant residue, dirt and oil or leaks will develop.

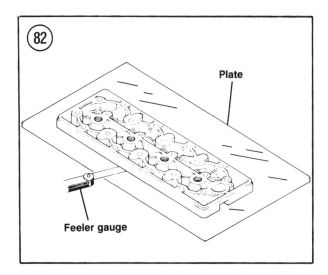

Plate

Feeler gauge

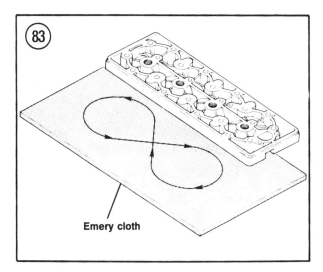

Emery cloth

emery cloth and recheck surface flatness on the pane of glass.

If warpage exists, the high spots will be dull while low areas will remain unchanged in appearance. It may be necessary to repeat this procedure 2-3 times until the entire mating surface has been polished to a dull luster. Do not remove more than a total of 0.010 in. from the cylinder block and head. Finish the resurfacing with No. 180 emery cloth.

1. Clean the cylinder block and crankcase cover thoroughly with solvent and a brush.

2. Carefully remove all gasket and sealant residue from the cylinder block and crankcase cover mating surfaces.

3. Check the cylinder heads and exhaust ports for excessive carbon deposits or varnish. Remove with a scraper or other blunt instrument.

4. Check the block, cylinder head and cover for cracks, fractures, stripped bolt or spark plug holes or other defects.

5. Check the gasket mating surfaces for nicks, grooves, cracks or excessive distortion. Any of these defects will cause compression leakage. Replace as required.

6. Check all oil and water passages in the block and cover for obstructions. Make sure any plugs installed are properly tightened.

7. Inspect the cylinder drain plate reeds for wear or damage. Check the drain screens to make sure they are open. Replace reeds or screens as required.

NOTE
With older engines, it is a good idea to have the cylinder walls lightly honed with a medium stone even if they are in good condition. This will break up any glaze that might reduce compression.

8. Check each cylinder bore for signs of aluminum transfer from the pistons to the cylinder walls. If scoring is present but not excessive, have the cylinders honed by a dealer or qualified machine shop.

Once the gasket surfaces are cleaned, place the mating surface of each component on a large pane of glass. Apply uniform downward pressure on the component and check for warpage at several points by trying to insert a feeler gauge between the glass and component mating surface. See **Figure 82**. Replace the component if more than a slight degree of warpage exists.

In cases where there is a slight amount of warpage, it can often be eliminated by placing a large sheet of No. 120 emery cloth over the glass. Apply a slight amount of pressure and move the component in a figure-8 pattern. See **Figure 83**. Remove the component and

8

9. Measure each cylinder bore at 3 points with an inside micrometer or bore gauge. See **Figure 84**. Measure at the following points:

   a. 1/4 in. down from the top of the bore.

   b. 1/4 in. up from the exhaust port.

   c. 3/16 in. down from lower edge of intake port.

Turn the cylinder 90° and repeat the measurements. If the difference between the largest and smallest measurements exceeds 0.002 in. or if the bore is out-of-round, replace the cylinder block and cover assembly. If bore wear is less than 0.002 in., have the cylinders honed or rebored by a dealer or qualified machine shop.

Crankshaft and Connecting Rod Bearings

Bearings can be reused if they are in good condition. To be on the safe side, however, it is a good idea to discard all bearings and install new ones whenever the engine is disassembled. New bearings are inexpensive compared to the cost of another overhaul caused by the use of marginal bearings.

> *NOTE*
> *Connecting rod bearing faces are silver-plated on 1986 Force 35-125 hp power heads. If bearing replacement is necessary, be sure to use the same type as removed.*

1. Remove old sealer from outer edge of ball bearings with a scraper, then clean bearing surface with Locquic.

2. Place ball bearings in a wire basket and submerge in a suitable container of fresh solvent. The bottom of the basket should not touch the bottom of the container.

3. Agitate basket containing bearings to loosen all grease, sludge and other contamination.

4. Dry ball bearings with dry filtered compressed air. Be careful not to spin the bearings.

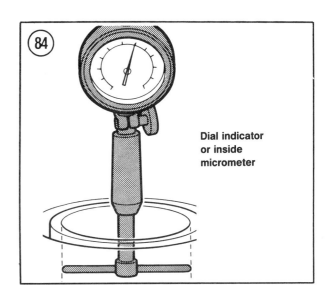

Dial indicator or inside micrometer

5. Lubricate the dry bearings with a light coat of Chrysler Outboard Motor oil and inspect for rust, wear, scuffed surfaces, heat discoloration or other defects. Replace as required.

6. If needle bearings are to be reused, repeat Steps 2-5, cleaning one set at a time to prevent any possible mixup. Check bearings for flat spots. If one needle bearing is defective, replace all in the set with new bearings and liners.

Piston

1. Check the piston(s) for signs of scoring, cracking, cracked or worn piston pin bosses or metal damage. Replace piston and pin as an assembly if any of these defects are noted.

2. Check piston ring grooves for distortion, loose ring locating pins or excessive wear. If the flexing action of the rings has not kept the lower surface of the ring grooves free of carbon, clean with a bristle brush and solvent.

> *NOTE*
> *Do not use an automotive ring groove cleaning tool in Step 3 as it can damage the piston ring locating pin.*

3. Clean the piston skirt, ring grooves and dome with the recessed end of a broken ring to remove any carbon deposits.

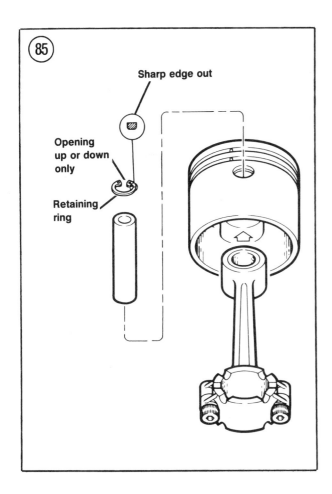

85

Sharp edge out

Opening up or down only

Retaining ring

4. Immerse pistons in a carbon removal solution to remove any carbon deposits not removed in Step 3. If the solution does not remove all of the carbon, carefully use a fine wire brush; avoid burring or rounding of the machined edges. Clean the piston skirt with crocus cloth.

5. Measure piston at right angle to piston pin with a micrometer. Check for wear at skirt and top land. Replace piston and pin as an assembly if piston is out-of-round by more than 0.0025 in.

Crankshaft

1. Clean the crankshaft thoroughly with solvent and a brush. Blow dry with dry filtered compressed air, if available, and lubricate with a light coat of Chrysler Outboard Motor oil.

2. Check the crankshaft journals and crankpins for scratches, heat discoloration or other defects.

3. Check drive shaft splines, flywheel taper threads and keyway for wear or damage. Replace crankshaft as required.

4. If lower crankshaft ball bearing has not been removed, grasp inner race and try to work it back and forth. Replace bearing if excessive play is noted.

5. Lubricate ball bearing with Chrysler Outboard Motor oil and rotate outer race. Replace bearing if it sounds or feels rough or if it does not rotate smoothly.

PISTON AND CONNECTING ROD

Assembly (3.5-15 hp)

If the pistons were removed from the connecting rods, they must be correctly oriented when reassembling. Position piston with locating pins in ring grooves facing up. Assemble connecting rod as shown in **Figure 85** with match marks facing up. On 1966-1979 models, the match marks are a pair of identification "bumps" on one side of the rod near the cap fracture line. On 1980 and later models, the "bumps" are replaced by a bevel or "V" cut on one side of the rod near the fracture line. Piston pin retaining rings must be positioned with the opening facing up or down.

1. Coat piston pin holes, piston pin and connecting rod small end bore with Chrysler Outboard Motor oil.

2. Place piston on pillow block (part No. T 2990) with ring groove locating pins facing upward.

3. Warm piston with a heat gun for one minute to expand the piston pin holes.

4. Position connecting rod as shown in **Figure 85** and install piston pin through piston and connecting rod with a suitable

8

driver (included with pillow block). See **Figure 86**.

5. Install piston pin retaining rings with snap ring pliers. Sharp edge of ring must face outward and ring opening must face directly up or down. See **Figure 85**. Make sure rings seat in piston grooves.

6. Repeat procedure for the other piston on 2-cylinder engines.

Assembly (20-50 hp)

If the pistons were removed from the connecting rods, they must be correctly oriented when reassembling. On 1966-1979 models, the match marks are a pair of identification "bumps" on one side of the rod near the cap fracture line. On 1980 and later models, the "bumps" are replaced by a bevel or "V" cut on one side of the rod near the fracture line. The "bumps" or bevel cut on the connecting rod must face up. Assemble piston to rod with intake baffle side of dome facing starboard. See **Figure 87**.

1. Warm piston with a heat gun for one minute to expand piston pin hole. Handle piston with a glove or wrap it in shop cloths.

2. Fit the alignment pin from piston tool set part No. T 2990 through one side of the piston and install a spacer on the pin with its small diameter facing the inside of the piston. See **Figure 88**.

3. Coat connecting rod small end bore with needle bearing grease and install the roller bearings.

4. Install connecting rod on alignment pin. Make sure piston and rod are properly oriented (**Figure 87**) and insert the other spacer with its small diameter facing the inside of the piston. Push alignment pin through bore on other side of the piston.

5. Position piston assembly on pillow block. Carefully remove alignment pin and insert piston pin in its place. Fit the stepped end of the drive pin into the piston pin and drive piston pin into piston until it is centered.

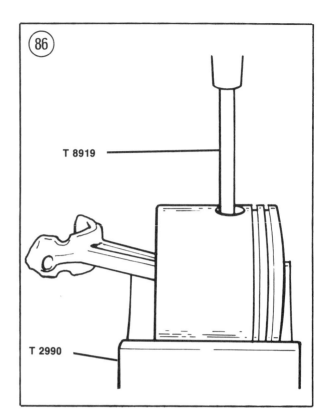

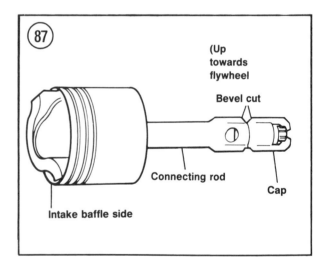

6. Remove drive pin and lubricate piston pin and bearings with Chrysler Outboard Motor oil.

7. Install piston pin retaining rings with snap ring pliers. Sharp edge of ring must face outward and ring opening must face directly up or down. See **Figure 85**. Make sure rings seat in piston grooves.

8. Repeat procedure for the other piston.

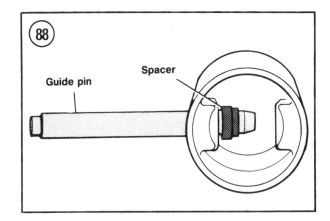

Guide pin   Spacer

Piston pin

Spacer   Alignment pin

2. Coat connecting rod small end bore with needle bearing grease and install the roller bearings.

3. Install connecting rod on alignment pin. Make sure piston and rod are properly oriented (**Figure 87**) and insert the other spacer with its small diameter facing the inside of the piston. Push alignment pin through bore on other side of the piston.

4. Continue pushing the alignment or guide pin through the piston pin bore, then install shim bar with end marked "310" between the connecting rod and the large end of the first spacer installed. The slots in the shim tool have a step to provide for piston pin roller clearance.

5. Place pillow block on a suitable press. Position piston and rod assembly on pillow block.

6. Insert the piston pin in the piston bore. Fit stepped end of drive pin into piston pin. Hold alignment tool and press piston pin in place.

7. Remove drive pin and lubricate piston pin and bearings with Chrysler Outboard Motor oil.

8. Repeat procedure for each remaining piston.

Assembly (70-140 hp)

If the pistons were removed from the connecting rods, they must be correctly oriented when reassembling. On 1966-1979 models, the match marks are a pair of identification "bumps" on one side of the rod near the cap fracture line. On 1980 and later models, the "bumps" are replaced by a bevel or "V" cut on one side of the rod near the fracture line. The "bumps" or bevel cut on the connecting rod must face up. Assemble piston to rod with intake baffle side of dome facing starboard. See **Figure 87**.

1. Fit the alignment pin from piston tool set part No. T 2990 through one side of the piston and install a spacer on the pin with its small diameter facing the inside of the piston. See **Figure 89**.

Piston Ring Installation (All Engines)

1. Check end gap of new rings before installing on piston. Place ring in cylinder bore just above the intake and exhaust ports, then square it up by inserting the bottom of an old piston. Measure the gap with a feeler gauge (**Figure 90**) and compare to specifications (**Table 3**).

2. If ring gap is excessive in Step 1, repeat the step with the ring in the other cylinder. If gap is also excessive in that cylinder, discard and replace with another new ring.

3. If ring gap is insufficient in Step 1, the ends of the ring can be filed slightly. Clean ring thoroughly and recheck gap as in Step 1.

8

NOTE
Piston rings must be installed in Step 4
and Step 5 with the bevel on the inner
edge of the ring facing the piston dome.
See **Figure 91**.

4. Once the ring gaps are correctly established, install the lower ring on the piston with a ring expander. Spread the ring just enough to fit it over the piston head and into position. See **Figure 66**.

5. Repeat Step 4 to install the upper ring.

6. Position each ring so that the piston groove locating pin fits in the ring gap. Proper ring positioning is necessary to minimize compression loss and prevent the ring ends from catching on the cylinder ports.

Piston and Connecting Rod Installation (All Engines)

1. Coat the piston, rings and cylinder bore with Chrysler Outboard Motor oil.

2. Check the piston dome number made during disassembly and match piston with its correct cylinder.

3A. 3.5-15 hp—Place piston assembly in bore with open end of piston pin facing the top of the cylinder. See **Figure 92**.

3B. 70-140 hp—Orient the intake side of piston to intake port side of cylinder block. See **Figure 93**.

4. Insert piston into cylinder bore. Make sure the rings are properly positioned in their grooves and that the locating pins are positioned in the ring gaps.

5. Install a suitable ring compressor over the piston dome and rings. With compressor resting on cylinder head, tighten it until the rings are compressed sufficiently to enter the bore.

6. Hold connecting rod end with one hand to prevent it from scraping or scratching the cylinder bore and slowly push piston into cylinder. See **Figure 93**.

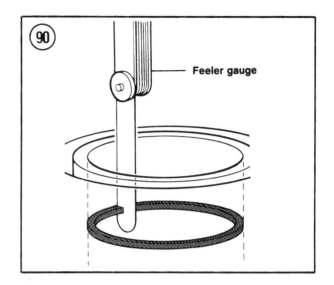

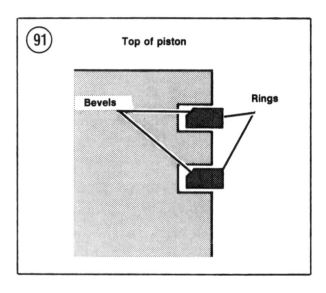

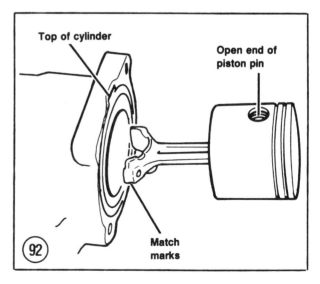

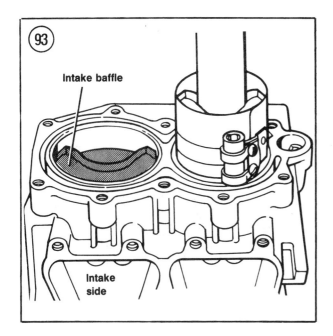

Intake baffle

Intake side

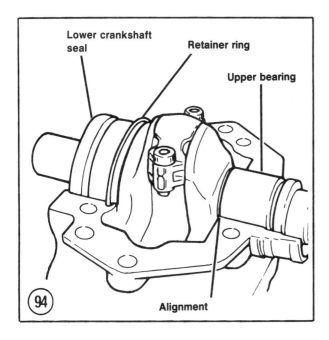

Lower crankshaft seal

Retainer ring

Upper bearing

Alignment

7. Remove the ring compressor tool and repeat the procedure to install remaining pistons.

8. Reach through exhaust port and lightly depress each ring with a pencil point or small screwdriver blade. The ring should snap back when pressure is released. If it does not, the ring was broken during piston installation and will have to be replaced.

CONNECTING ROD AND CRANKSHAFT

Assembly (3.5-4 hp)

Refer to **Figure 94** for this procedure.

1. Lubricate the crankshaft with Chrysler Outboard Motor oil and install in the cylinder block with the retainer ring engaging the cylinder block groove.

2. Install the upper crankshaft bearing (lettered side up) with its bottom even with the inside of the cylinder wall.

3. Slide lower seal on crankshaft until it bottoms. Inner sealing lip with spring must face bearing.

4. Repeat Step 3 to install upper seal.

5. Coat connecting rod bearing cages with Rykon No. 2EP grease. Place one cage in connecting rod and install roller bearings.

6. Position connecting rod under crankshaft and pull it up to the crankpin.

7. Lightly coat exposed part of crankpin with Rykon No. 2EP. Install the other bearing cage and bearings.

8. Coat rod cap screw threads with Loctite D. Install cap and tighten cap screws finger-tight.

CAUTION
The procedure detailed in Step 9 is very important to proper engine operation as it affects bearing action. If not done properly, major engine damage can result. It can also be a time-consuming and frustrating process. Work slowly and with patience. If alignment cannot be achieved, replace the connecting rod.

8

9. Run a dental pick or pencil point along the cap match marks to check cap offset. See **Figure 95**. Rod and cap must be aligned so that the dental pick or pencil point will pass smoothly across the fracture line. If alignment is not correct, gently tap cap with a plastic hammer. When rod and cap are

properly aligned, tighten cap screws to specifications (**Table 1**).

10. Rotate the crankshaft to check for binding. If the crankshaft does not float freely over the full length of the crankpin, loosen the rod cap and repeat Step 9.

Assembly (4.4-55 hp)

1. Remove connecting rod caps. Coat connecting rod bearing surfaces with Rykon No. 2EP grease. Install a bearing retainer half and needle bearings in the rod.

2A. Center main bearing with snap ring—Install bearing cages and bearings, fit races in position and install snap ring so that it covers the 2 fracture lines where the races dovetail.

2B. Center main bearing without snap ring—Install the center main bearing race in the block with its hole engaging the bearing bore peg. Coat race with Rykon No. 2EP grease and install the roller bearings.

3. If crankshaft uses a lower main bearing, slide or press bearing on as required. Install snap ring, if used.

4. If crankshaft uses an upper main bearing, slide or press bearing on as required.

5. If crankshaft uses a seal ring, install it on the crankshaft groove below the bearing.

6. If crankshaft uses a center main bearing with snap ring, position the bearing races so their locating holes will align with the locating pins in the cylinder block.

7. Lubricate crankshaft assembly with Chrysler Outboard Motor oil and lower into the cylinder block. On crankshafts with a slide-on lower main bearing, position bearing as shown in **Figure 96**.

8. Lightly coat the crankshaft journal with Rykon No. 2EP grease. Install the remaining roller bearings and race.

9. Draw connecting rods up around crankshaft crankpin journals. Coat crankpins

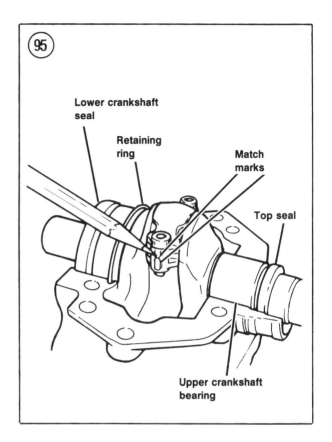

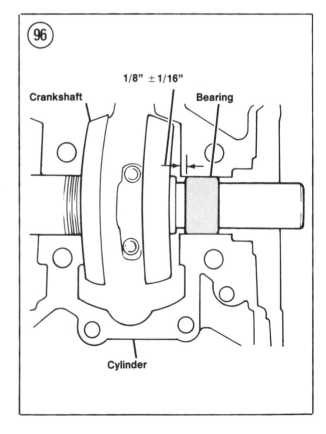

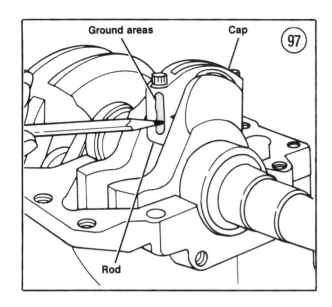

Figure 97 — Ground areas, Cap, Rod

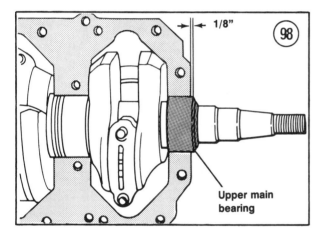

Figure 98 — 1/8", Upper main bearing

11. Run a dental pick or pencil point along the ground areas (**Figure 97**) to check cap offset. Rod and cap must be aligned so that the dental pick or pencil point will pass smoothly across the fracture line. If rod and cap are not properly aligned, gently tap cap with a plastic hammer. When alignment is correct, tighten cap screws to specifications (**Table 1**).

12. Rotate the crankshaft to check for binding. If the crankshaft does not float freely over the full length of the crankpins, loosen the rod caps and repeat Step 11.

13. Position crankshaft seal (if used) with gap facing directly up.

14. If a slide-on upper main bearing is used, position bearing to extend 1/8 in. above top of bore. See **Figure 98**.

15. If upper or lower main bearing has a locating pin, make sure that it is positioned in the pin groove in the block.

8

Assembly (70-140 hp)

1. Install a new seal in the lower main bearing with tool part No. T 8925 until flush with bearing. Seal lip and garter spring must face outward.

2. Coat the cavity between the seal lips with Rykon No. 2EP grease and install lower main bearing and seal assembly on bottom of crankshaft.

3. Press upper main bearing on crankshaft with lettered side of bearing facing up. Coat outer surface of bearing with Loctite D.

4. Coat crankcase seal grooves in crankshaft with Rykon No. 2EP grease and install a seal in each groove.

5. Assemble each main bearing to crankshaft (**Figure 99**). Holes in bearing race should be between snap ring on race and crankcase seal ring.

6. Install snap ring on each main bearing so it covers the 2 fracture lines where the races dovetail. See **Figure 100**.

with Rykon No. 2EP grease and install remaining cage halves and needle bearings on each crankpin.

10. Coat rod cap screw threads with Loctite D. Install cap and tighten cap screws finger-tight.

CAUTION
The procedure detailed in Step 11 is very important to proper engine operation as it affects bearing action. If not done properly, major engine damage can result. It can also be a time-consuming and frustrating process. Work slowly and with patience. If alignment cannot be achieved, replace the connecting rod.

7. Remove connecting rod caps. Coat connecting rod bearing surfaces with Rykon No. 2EP grease. Install a bearing cage and needle bearings in the rod.

8. Lubricate crankshaft assembly with Chrysler Outboard Motor oil and install in cylinder block, aligning the large hole of each main bearing with the bearing bore locating pins.

9. Position the sealing rings with their gaps facing up.

10. Draw connecting rods up around crankshaft crankpin journals. Coat crankpins with Rykon No. 2EP grease and install remaining cage halves and needle bearings on each crankpin.

11. Coat rod cap screw threads with Loctite D. Install cap and tighten cap screws finger-tight.

CAUTION
The procedure detailed in Step 12 is very important to proper engine operation as it affects bearing action. If not done properly, major engine damage can result. It can also be a time-consuming and frustrating process. Work slowly and with patience. If alignment cannot be achieved, replace the connecting rod.

12. Run a dental pick or pencil point along the ground areas (**Figure 97**) to check cap offset. Rod and cap must be aligned so that the dental pick or pencil point will pass smoothly across the fracture line. If rod and cap are not properly aligned, gently tap cap with a plastic hammer. When proper alignment is achieved, tighten cap screws to specifications (**Table 1**).

13. Rotate the crankshaft to check for binding. If the crankshaft does not float freely over the full length of the crankpins, loosen the rod caps and repeat Step 12.

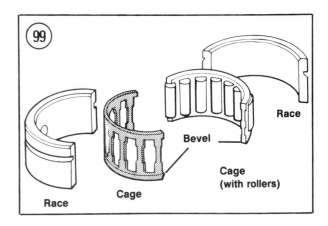

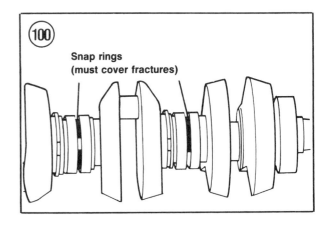

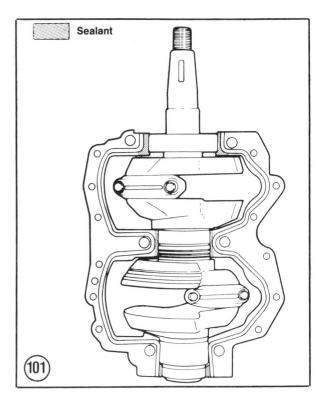

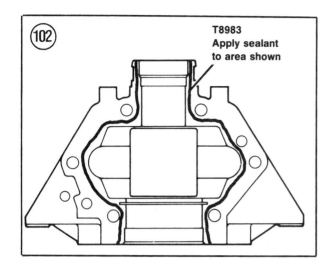

T8983
Apply sealant
to area shown

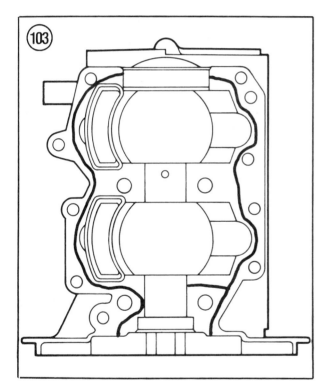

CYLINDER BLOCK AND CRANKCASE ASSEMBLY

The cylinder block face on 35-140 hp models is grooved on each side of the crankshaft opening for the use of a spaghetti or rubber seal. The new seal should be fully seated in the grooves and then cut 1/2 in. longer at each end to assure a good butt seal against both crankcase bearings. Run a thin

bead of EC-750 industrial sealant in the groove before installing the seal. Force the seal into the groove and let it set 15-20 minutes. Trim the ends of the seal with a sharp knife, leaving about 1/32 in. of the seal end to butt against the bearings. Apply additional EC-750 sealant on both sides of the seal groove in the upper, center and lower main bearing areas. See **Figure 101**. Use care not to apply an excessive amount, as it can squeeze over when the parts are mated and may block oil or water passages.

RTV sealant is used on engines with no parting line seal groove. Some engines under 35 hp do have a parting line seal groove, but a spaghetti seal is not used. RTV sealant is recommended for use on such engines. The sealant should be evenly applied as shown in **Figure 102** (1-cylinder) or **Figure 103** (2-cylinder). Make sure sealant reaches lower main seal case and upper main bearing race. On engines with a seal groove, fill the groove completely with enough sealant to extend about 1/16 in. higher than the block surface. As with larger engines, do not apply an excessive amount that will overflow into the crankcase or onto bearings when the parts are mated.

Once EC-750 or RTV sealant has been applied, final torque should be applied to the crankcase bolts within 3 minutes to assure that the sealant does not have an opportunity to take a set.

1980-on 3.5 hp, 1976-on 4 hp

Refer to **Figure 44** for this procedure.
1. Apply RTV sealant to the cylinder block and install the crankcase cover.
2. With crankcase and cylinder block on a solid surface, install the 2 locating pins with a suitable pin punch.
3. Wipe crankcase cover screw threads with RTV sealant. Install screws and tighten to specifications (**Table 1**).

8

4. Rotate the crankshaft several turns to check for binding. If crankshaft does not turn easily, disassemble and correct the interference.

5. Install the cylinder head with a new gasket. Coat head bolt threads with RTV sealant. Install the 2 top screws with spacers and the 2 lower screws with the bracket. Tighten bolts to specifications (**Table 1**).

6. Install the power head as described in this chapter.

4.4-15 hp

Refer to **Figure 49** or **Figure 50** for this procedure.

1. Apply RTV sealant to the cylinder block and install the crankcase cover.

2. With crankcase and cylinder block on a solid surface, install the 2 locating pins with a suitable pin punch.

3. Wipe crankcase cover screw threads with RTV sealant. Install screws and tighten to specifications.

4. Rotate the crankshaft several turns to check for binding. If crankshaft does not turn easily, disassemble and correct the interference.

5. Use a bearing scraper to remove the old lower crankshaft seal stake marks from the block and crankcase cover.

6. Fill the cavity between the lower crankshaft seal lips with Rykon No. 2EP grease. Install seal on crankshaft with spring side facing out.

7. Drive seal fully into bore with an appropriate installer.

8. Use a center punch to stake the outer edge of the seal in place at 2 points 180° apart. Stake marks should be about 1/16-1/8 in. from edge of seal and deep enough to cover the outer seal edge with at least 0.005 in. of metal. See **Figure 104**.

9. Install the cylinder head with a new gasket. Install the head bolts with washers

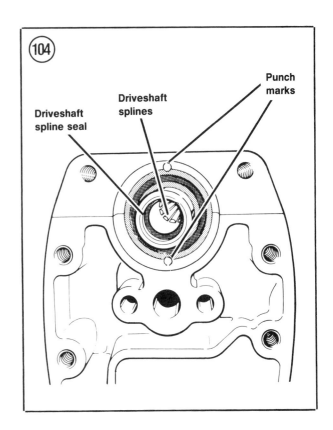

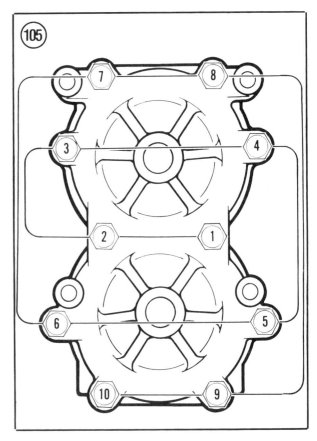

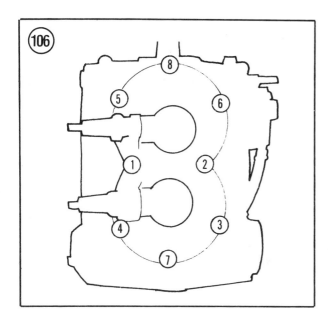

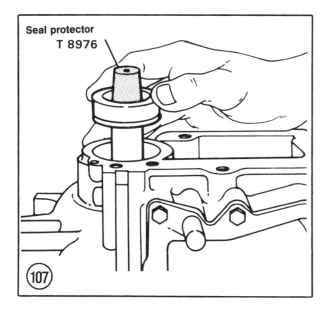

Seal protector
T 8976

assembly, then install screws and lead wire clip (if used). Position assembly on cylinder block and tighten screws to specifications (**Table 1**). Tighten inner screws first, then outer screws.

11. Install the cylinder drain cover with a new gasket. Make sure the screens cover the cover check valves. Install hose clamp with top right hand cover screw, if used. Tighten all screws to specifications (**Table 1**).

12. Install the transfer port covers (if so equipped) with new gaskets. Tighten screws to specifications (**Table 1**).

13. Install exhaust tube to power head, if so equipped.

14. Install the power head as described in this chapter.

All 20, 25 and 30 hp; 1976-on 35 hp

Refer to **Figure 51** for this procedure.

1. Apply RTV sealant to the cylinder block and install the crankcase cover.

2. With crankcase and cylinder block on a solid surface, install the 2 locating pins with a suitable pin punch.

3. Install the crankcase cover screws and tighten to specifications working from the inside out in a circle.

4. Rotate the crankshaft several turns to check for binding. If crankshaft does not turn easily, disassemble and correct the interference.

5. Install lower crankcase seal collar (O-ring side up) and seal protector (part No. T 8976) on end of crankshaft (**Figure 107**). Coat the cavity between the seal lips with Rykon No. 2EP grease. Push seal collar onto crankshaft and remove seal protector.

6. Make sure O-ring is seated in seal collar groove, then drive collar into power head with installer part No. T 8977 until it is about 0.30 in. above the power head surface. See **Figure 108**.

and tighten to specifications (**Table 1**) following the sequence shown in **Figure 105** or **Figure 106** according to head design.

NOTE
Be sure to install trigger stator ground lead under the head of the top front screw in Step 10 on models so equipped.

10. Sandwich the exhaust port plate between new gaskets. Fit cover to plate/gasket

8

7. Install thermostat and grommet in cylinder head. Install cover with new gasket.

8. Install cylinder head with a new gasket. Tighten head bolts to specifications (**Table 1**) following the sequence shown in **Figure 109**.

9. Install transfer port cover with a new gasket. Tighten screws to specifications (**Table 1**) in a spiral pattern starting at the center.

10. Sandwich the exhaust port plate between new gaskets. Fit cover to plate/gasket assembly, then install screws. Position assembly on cylinder block and tighten screws to specifications (**Table 1**) in a spiral pattern starting at the center.

11. Install the power head as described in this chapter.

1966-1975 35 hp; All 45 and 50 hp; 1967-1976 55 hp

Refer to **Figure 57** for this procedure.

1. Install spaghetti seals in the cylinder block grooves. Apply RTV sealant as shown in **Figure 101** and install the crankcase cover.

2. With crankcase and cylinder block on a solid surface, install the 2 locating pins with a suitable pin punch.

3. Install the crankcase cover bolts and screws. Tighten to specifications (**Table 1**).

4. Rotate the crankshaft several turns to check for binding. If crankshaft does not turn easily, disassemble and correct the interference.

5. Install thermostat assembly to cylinder head as shown in **Figure 110**.

6. Install cylinder head with a new gasket. Coat the first 3/4 in. of each head bolt with anti-seize lubricant. Tighten head bolts to specifications (**Table 1**) following the sequence shown in **Figure 105**.

7. Connect the water tube between the thermostat cover and block fitting.

8. Install the cylinder drain assembly as shown in **Figure 111**.

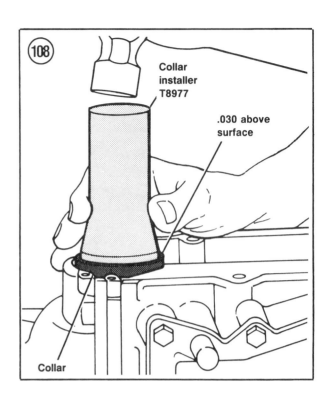

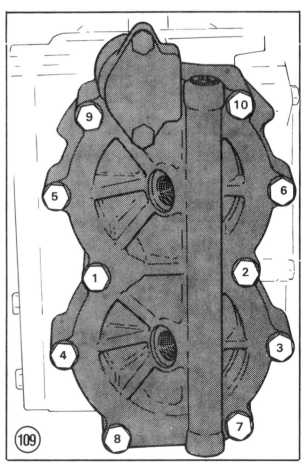

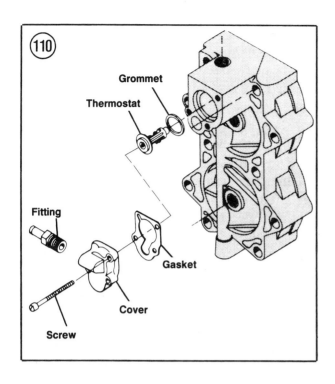

110

Grommet

Thermostat

Fitting

Gasket

Cover

Screw

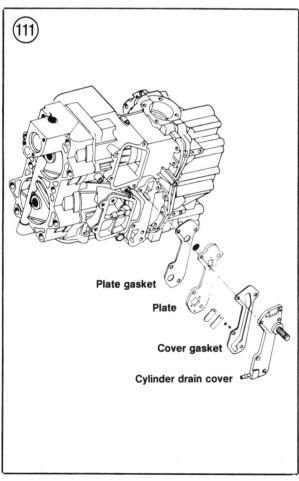

111

Plate gasket

Plate

Cover gasket

Cylinder drain cover

9. Install the transfer port cover with a new gasket. Tighten screws to specifications (**Table 1**).

10. Install the fuel pump cover with a new gasket.

11. Sandwich the exhaust port plate between new gaskets. Fit cover to plate/gasket assembly, then install screws. Position assembly on cylinder block and tighten screws to specifications (**Table 1**) in a spiral pattern starting at the center.

12. Electric start—Install the coil mounting bracket.

13. Install the power head as described in this chapter.

1977-on 55 hp; All 60 and 65 hp

Refer to **Figure 67** for this procedure.

1. Install spaghetti seals in cylinder block grooves. Apply RTV sealant as shown in **Figure 101** and install the crankcase cover.

2. With crankcase and cylinder block on a solid surface, install the 2 locating pins with a suitable pin punch.

3. Install the crankcase cover bolts and screws. Tighten to specifications (**Table 1**).

4. Rotate the crankshaft several turns to check for binding. If crankshaft does not turn easily, disassemble and correct the interference.

5. Install the remote cable assembly, gear shift lever and pivot. See **Figure 74**.

6. Install lower bearing cage. Side marked "Front" should face the front of the power head.

7. Install the forward stabilizer (**Figure 73**).

8. Sandwich the exhaust port plate between new gaskets. Fit cover to plate/gasket assembly, then install screws. Position assembly on cylinder block and tighten screws to specifications (**Table 1**) in a spiral pattern starting at the center.

8

9. Install the rear stabilizer to the cylinder head (**Figure 72**).

10. Install cylinder head cover to cylinder head with a new gasket.

11. Install the thermoswitch in the cylinder head cover.

12. Install cylinder head assembly to power head with a new gasket.

13. Install the thermostat assembly as shown in **Figure 69**.

14. Tighten head bolts to specifications (**Table 1**) following the sequence shown in **Figure 105**.

15. Install the bypass valve cover to the power head with a new gasket.

70-140 hp

Refer to **Figure 78** (3-cylinder) or **Figure 79** (4-cylinder) for this procedure.

1. Install spaghetti seals in the cylinder block grooves. Run a thin bead of RTV sealant along the area between the spaghetti seals and the crankshaft opening. Apply RTV sealant on both sides of the seal grooves at the upper and lower main bearings. Install the crankcase cover.

2. With crankcase and cylinder block on a solid surface, install the 2 locating pins with a suitable pin punch.

NOTE
The 3-cylinder engine has 8 cover bolts; the 4-cylinder engine uses 7 bolts and 1 nut.

3. Coat the crankcase cover bolt threads with RTV sealant. Install the bolts and tighten to specifications (**Table 1**) working from the center out.

4. Install the parting line screws and tighten to specifications (**Table 1**).

5. Rotate the crankshaft several turns to check for binding. If crankshaft does not turn easily, disassemble and correct the interference.

6. Make sure the power head and bearing cage gasket surfaces are clean.

7. Fit a new bearing seal on the bearing cage bore with its major sealing lip facing into the bore. Press seal into bore with a suitable installer.

8. Install a new O-ring on the bearing cage.

9. Run a thin bead of RTV sealant around the mating surface of the bearing cage. Apply sealant at a point midway between the O-ring seal and screw holes.

10. Install seal protector part No. T 8927 on crankshaft. Lubricate the bearing cage seal lips with Rykon No. 2 EP grease, then slide the bearing cage over the seal protector and onto the power head.

11. Wipe bearing cage screw threads with RTV sealant. Install and tighten screws to specifications (**Table 1**), then remove the seal protector.

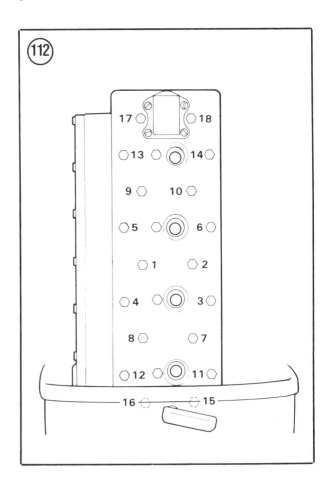

12. Position a new upper cylinder exhaust gasket on the power head and install the spacer plate. Tighten spacer plate nuts to specifications (**Table 1**).

13. Coat the exhaust tube mating area on the spacer plate with RTV sealant. Wipe exhaust tube screw threads with Loctite D and install exhaust tube. Tighten screws to specifications (**Table 1**), then install a new exhaust tube seal.

14. Install transfer port covers with new gaskets and tighten screws to specifications (**Table 1**).

15. Sandwich the cylinder drain reed plate between new gaskets. Fit drain cover to reed plate/gasket assembly and install screws. Position cylinder drain plate assembly on power head and tighten screws to specifications (**Table 1**).

16. If cylinder drain plate fitting was removed, wipe threads with Loctite H and reinstall with nipple pointing directly up.

17. Install the bypass spring and valve on the bypass cover. Install cover and valve assembly with a new gasket.

18. Sandwich the exhaust port plate between new gaskets. Fit cover to plate/gasket assembly, then install screws. Coat screw threads with anti-seize compound. Position assembly on cylinder block and tighten screws to specifications (**Table 1**) in a spiral pattern starting at the center.

19. Coat cylinder head cover side of cylinder head with RTV sealant. Install cover to cylinder head.

NOTE
Figure 112 is the 4-cylinder torque pattern. To use it for 3-cylinder engines, follow the sequence from 1 to 14.

20. Install cylinder head assembly with a new gasket. Tighten the head bolts to specifications (**Table 1**) following the sequence in **Figure 112**.

21. Install thermostat assembly as shown in **Figure 80**. Thermostat bypass slot should face up.

22. Install thermoswitch (**Figure 80**).

23. Install the power head as described in this chapter.

8

Table 1 POWER HEAD TIGHTENING TORQUES[1]

Fastener	in.-lb.	ft.-lb.
Connecting rod screw		
3.5-4 hp	80	
4.4-8 hp		
1968-1976	90	
1977-1979	110	
1980-on	80	
9.2-15 hp		
1966-1976	110	
1977-1979	120	
1980-on	80	
20 hp		
1966-1976	110	
1977-on	180-190	
25-30 hp		
1973-1979	165-175	
1980-on	180-190	

(continued)

Table 1 POWER HEAD TIGHTENING TORQUES[1] (continued)

Fastener	in.-lb.	ft.-lb.
Connecting rod screw		
55 hp	270-280	
70-135 hp		
1966-1979	170	
1980-on	180-190	
All others	180-190	
Crankcase cover bolts		
70-140 hp	270	
Cylinder drain cover bolts		
70-140 hp	70	
Cylinder head bolts		
3.5-4 hp	130	
4.9-5 hp	80	
4.4, 4.5, 6, 6.6 and 8 hp		
1968-1976	125	
1977-1979	110	
1980-on	130	
9.2-15 hp		
1968-1976	120	
1977-1979	125	
1980-on	130	
20 hp		
1966-1976	115-135	
1979-on	190	
25 and 35 hp		
1973-1978	225	
1979-on	190	
30 hp	190	
45-50 hp	270	
55 hp		
1967-1979	270	
1980-on	225	
70-140 hp	225	
CD shock mount		
Stud	20-40	
Nut	90	
Distributor shaft nut	60	
Exhaust cover screw		
9.9-15 hp	90	
20-55 hp	65-75	
Fuel pump cover-to-body screws		
70-140 hp	30-35	
Fuel tank screws	90	
Flywheel nut		
3.5-4 hp		17
4.4-8 hp		40
9.9-15 hp		45-50[2]
20 hp		
1967-1976		45[2]
1979-on		55-65
25, 30 and 35 hp		
1973-1978		45
1979-on	(continued)	55-65

Table 1 POWER HEAD TIGHTENING TORQUES[1] (continued)

Fastener	in.-lb.	ft.-lb.
Flywheel nut		
45-50 hp		80
55 hp		
1967-1979		80
1980-on		90
70-140 hp		90
Gear shift arm pin	30	
Power head-to-motor leg		
9.9-50 hp	160	
All others	225	
Power head-to-spacer plate	270	
Spacer plate-to-motor leg	270	
Steering handle and magneto control shaft setscrew		
9.9-15 hp	55	
All others	40-50	
Steering support tube nuts		50
Stern bracket bolt		
9.9-15 hp	125	
All others	160	
Stern bracket clamp screws		35
Thermoswitch	70-85	
Transfer port cover screws		
9.9-15 hp		90
All others		80
Standard torque values (screw or nut size)		
6-32	9	
8-32	20	
10-24	30	
10-32	35	
12-24	45	
1/4-20	70	
5/16-18	160	
3/8-16	270	

1. Use standard torque values if specific fastener is not listed.
2. Torque Autolectric armature bolt to 25 ft.-lb.

Table 2 FLYWHEEL REMOVAL TOOL

Engine	Part No.
3.5 hp (1966-1969) and 3.6 hp	T 2919
3.5 hp (1980-on) and 4 hp	T 8998
4.4-8 hp (except 7.5 hp)	T 18091
7.5-55 hp (except 8 hp)	
CD ignition	T 2909
Magneto ignition	T 2910
60-140 hp	T 8948-1

8

Table 3 PISTON RING GAP SPECIFICATIONS

Engine	Gap (in.)
3.5-7.5 hp	0.006-0.011
9.9 hp	0.006-0.016
12-15 hp	0.004-0.014
20-25 hp	0.007-0.017
30 hp	
Top ring	0.006-0.016
Bottom ring	0.004-0.014
35-50 hp (except 1983-on 35 hp)	0.006-0.016
1983-on 35 hp	
Top ring	0.006-0.016
Bottom ring	0.004-0.014
55-65 hp	
1967-1980	0.004-0.014
1981 55 hp	
Top ring	0.004-0.014
Bottom ring	0.006-0.016
70-115 hp	0.006-0.016
125-140 hp	
Top ring	0.004-0.014
Bottom ring	0.006-0.016

Chapter Nine

Gearcase

Torque is transferred from the engine crankshaft to the gearcase by a drive shaft. A pinion gear on the drive shaft meshes with a drive gear in the gearcase to change the vertical power flow into a horizontal flow through the propeller shaft. The power head drive shaft rotates clockwise continuously when the engine is running, but propeller rotation is controlled by the gear train shifting mechanism.

On Chrysler outboards with a reverse gear, a sliding clutch engages the appropriate gear in the gearcase. This creates a direct coupling that transfers the power flow from the pinion to the propeller shaft. **Figure 1** shows the operation of the gear train.

Smaller outboards with a NEUTRAL but no REVERSE gear utilize a spring-loaded clutch to shift between NEUTRAL and FORWARD gear. Gear train operation is shown in **Figure 2** (disengaged) and **Figure 3** (engaged).

The gearcase can be removed without removing the entire outboard from the boat. This chapter contains removal, overhaul and installation procedures for the propeller, gearcase and water pump. **Table 1** is at the end of the chapter.

The gearcases covered in this chapter differ somewhat in design and construction over the years covered and thus require slightly different service procedures. The chapter is arranged in a normal disassembly/assembly sequence. When only a partial repair is required, follow the procedure(s) for your gearcase to the point where the faulty parts can be replaced, then assemble the unit.

Since this chapter covers a wide range of models from 1966-on, the gearcases shown in the accompanying illustrations are the most common ones. While it is possible that the components shown in the pictures may not be identical with those being serviced, the step-by-step procedures may be used with all models covered in this manual.

PROPELLER

The outboards covered in this manual use variations of 2 propeller attachment designs.

Smaller gearcases use a drive pin that engages a slot in the propeller hub, which is retained by a cotter pin.

On some models, the cotter pin passes through a cone-type nut that is separate from the propeller. See **Figure 4**. In this design, a metal pin installed in the propeller shaft engages a recessed slot in the propeller hub. As the shaft rotates, the pin rotates the propeller. The drive pin is designed to break if the propeller hits an obstruction in the water and has 2 advantages. The pin absorbs the impact to prevent possible propeller damage. It also alerts the user to the fact that something is wrong, since the engine speed will increase immediately if the pin breaks.

Propellers on the larger gearcases ride on thrust bearings and are retained by a castellated nut and cotter pin or a nut and screw arrangement. See **Figure 5**. Any underwater impact is absorbed by the propeller hub.

Removal/Installation

1. To remove the propeller on smaller units:
   a. Remove and discard the cotter pin.
   b. Unscrew the propeller nut from the propeller shaft.
   c. Remove drive pin from propeller with an appropriate punch. Remove propeller and O-ring. Discard the O-ring.

   NOTE
   Propeller shaft on 4.5 and 6 hp Special models is not splined but should be cleaned in Step d.

   d. Clean propeller shaft splines thoroughly. Inspect the pin engagement slot in the propeller hub and shaft for wear or damage.
   e. Installation is the reverse of removal. Lubricate the propeller shaft with

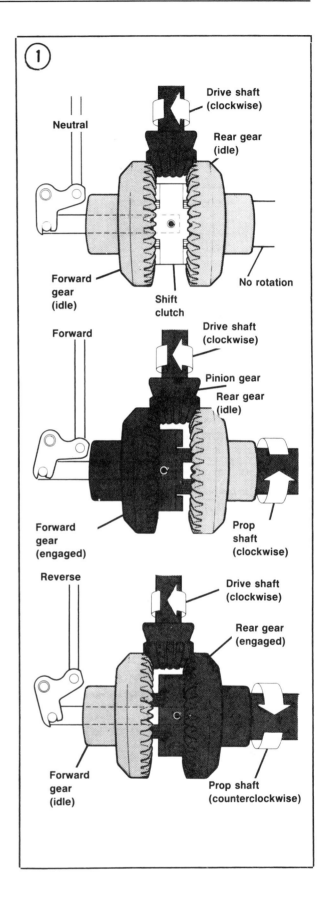

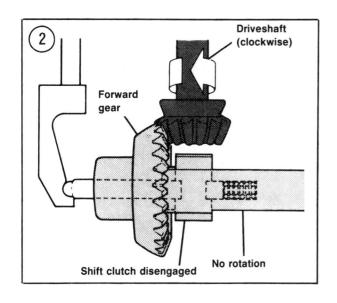

Driveshaft (clockwise)

Forward gear

Shift clutch disengaged

No rotation

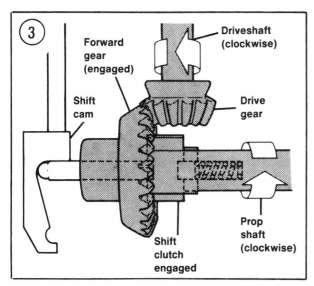

Forward gear (engaged)

Driveshaft (clockwise)

Shift cam

Drive gear

Prop shaft (clockwise)

Shift clutch engaged

anti-seize lubricant. Use a new drive pin and cotter pin.

2. To remove the propeller on larger units:
   a. Remove the cotter pin from the propeller nut. Discard the cotter pin.
   b. Remove the nut from the shaft.
   c. Remove the stop nut, plain washer and hub or flare washer from the shaft.
   d. Remove the propeller and clean the propeller shaft splines thoroughly.
   e. Installation is the reverse of removal. Lubricate propeller shaft with anti-seize lubricant and use a new cotter pin.

GEARCASE REMOVAL/INSTALLATION

1966-1969 3.5 hp; 1970-1977 3.6 hp

1. Disconnect the spark plug lead as a safety precaution to prevent any accidental starting of the engine during lower unit removal.

2. Remove the propeller as described in this chapter.

3. Pry the clip from the front of the motor leg with a screwdriver. See **Figure 6**.

4. Remove the nut under the clip at the upper front of the gearcase.

5. Remove the hex head screw from the upper rear edge of the gearcase.

9

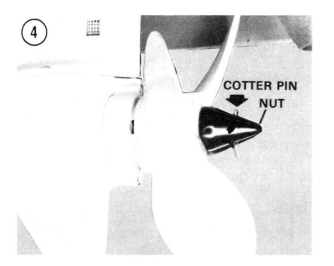

COTTER PIN

NUT

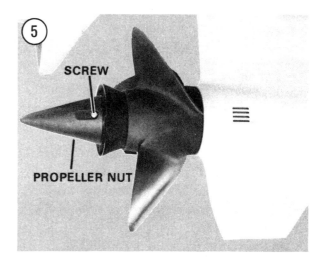

SCREW

PROPELLER NUT

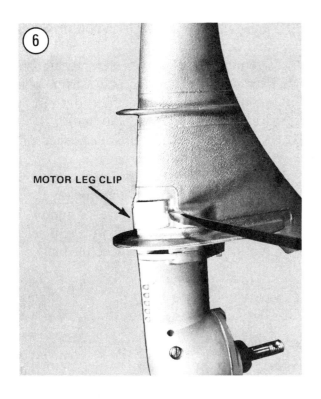

MOTOR LEG CLIP

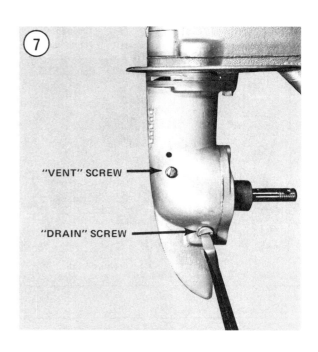

"VENT" SCREW

"DRAIN" SCREW

the lubricant. Check the drain container for signs of water separation from the lubricant.

CAUTION
Do not grease the top of the drive shaft in Step 10. This may excessively preload the drive shaft and crankshaft when the mounting bolts are tightened and cause a premature failure of the power head or gearcase.

6. Tilt the motor leg up and carefully separate the gearcase. Remove the gearcase from the motor leg.

7. Install the gearcase in an appropriate holding fixture.

8. Place a suitable container under the gearcase. Remove the vent and drain screws (**Figure 7**). Drain the lubricant from the unit.

NOTE
If the lubricant is creamy in color or metallic particles are found in Step 9, the gearcase must be completely disassembled to determine and correct the cause of the problem.

9. Wipe a small amount of lubricant on a finger and rub the finger and thumb together. Check for the presence of metallic particles in the lubricant. Note the color of the lubricant. A white or creamy color indicates water in

10. To reinstall the gearcase, lightly lubricate the drive shaft splines with anti-seize compound.

11. Position gearcase under motor leg and align drive shaft splines with the crankshaft.

12. 3.6 hp—Slip water tube end into rubber seal of water pump.

CAUTION
Do not rotate the flywheel counterclockwise in Step 13. This can damage the water pump impeller.

13. Push the gearcase into place, rotating the flywheel clockwise as required to let the drive shaft and crankshaft engage.

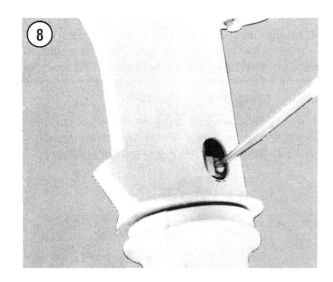

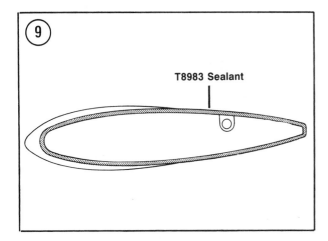

T8983 Sealant

14. Install hex head screw at rear of gearcase and tighten to specifications (**Table 1**).

15. Install nut on stud at front of gearcase. Tighten nut to specifications (**Table 1**).

16. Position motor leg clip and tap into place with a soft mallet.

17. Install the propeller as described in this chapter.

18. Reconnect the spark plug lead and refill the gearcase with proper type and quantity of lubricant. See Chapter Four.

1980-on 3.5 hp; 1976-on 4 hp

1. Disconnect the spark plug lead as a safety precaution to prevent any accidental starting of the engine during lower unit removal.

2. Remove the propeller as described in this chapter.

3. Pry the shift coupler cover from the side of the motor leg with a screwdriver. See **Figure 8**.

4. Loosen the shift coupler screw (**Figure 8**).

5. Remove the 2 screws holding the gearcase to the motor leg.

6. Tilt the motor leg up and carefully separate the gearcase. Remove the gearcase from the motor leg.

7. Install the gearcase in an appropriate holding fixture.

8. Place a suitable container under the gearcase. Remove the vent and drain screws. Drain the lubricant from the unit.

> *NOTE*
> *If the lubricant is creamy in color or metallic particles are found in Step 9, the gearcase must be completely disassembled to determine and correct the cause of the problem.*

9. Wipe a small amount of lubricant on a finger and rub the finger and thumb together. Check for the presence of metallic particles in the lubricant. Note the color of the lubricant. A white or creamy color indicates water in the lubricant. Check the drain container for signs of water separation from the lubricant.

> *CAUTION*
> *Do not grease the top of the drive shaft in Step 10. This may excessively preload the drive shaft and crankshaft when the mounting bolts are tightened and cause a premature failure of the power head or gearcase.*

10. To reinstall the gearcase, lightly lubricate the drive shaft splines with anti-seize compound.

11. Pull lower shift rod upward as far as possible.

12. Run a bead of RTV sealant along the gearcase mating surface as shown in **Figure 9**.

9

13. Position gearcase under motor leg and align drive shaft splines with the crankshaft.

CAUTION
Do not rotate the flywheel counter-clockwise in Step 14. This can damage the water pump impeller.

14. Start the gearcase into place, rotating the flywheel clockwise as required to let the drive shaft and crankshaft engage.

15. Align water tube with water tube seal. Align the upper shift rod with the lower shift rod coupler. See **Figure 10**.

16. Seat gearcase against motor leg.

17. Coat motor leg screw threads with RTV sealant. Install both screws and tighten to specifications (**Table 1**).

18. Place shift lever in NEUTRAL. Make sure the lower shift rod is all the way up and tighten the coupler screw.

19. Position shift coupler cover on motor leg and tap into place with a soft mallet.

20. Install the propeller as described in this chapter.

21. Reconnect the spark plug lead and refill the gearcase with proper type and quantity of lubricant. See Chapter Four.

4.9 hp; 1974-1976 5 hp

1. Disconnect the spark plug leads as a safety precaution to prevent any accidental starting of the engine during lower unit removal.

2. Remove the propeller as described in this chapter.

3. Remove the screw at the front of the exhaust housing.

4. Remove the screw near the exhaust hole.

5. Tilt the motor leg up and carefully separate the gearcase and anti-cavitation plate from the motor leg.

6. Install the gearcase in an appropriate holding fixture.

7. Place a suitable container under the gearcase. Remove the vent and drain screws

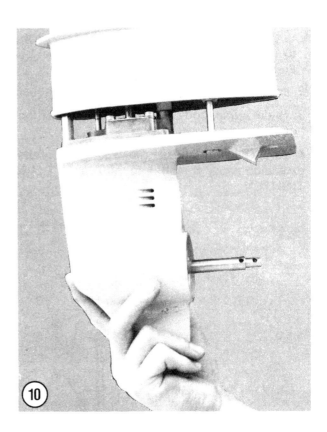

(**Figure 11**). Drain the lubricant from the unit.

NOTE
If the lubricant is creamy in color or metallic particles are found in Step 8, the gearcase must be completely disassembled to determine and correct the cause of the problem.

8. Wipe a small amount of lubricant on a finger and rub the finger and thumb together. Check for the presence of metallic particles in the lubricant. Note the color of the lubricant. A white or creamy color indicates water in the lubricant. Check the drain container for signs of water separation from the lubricant.

CAUTION
Do not grease the top of the drive shaft in Step 9. This may excessively preload the drive shaft and crankshaft when the mounting bolts are tightened and cause a premature failure of the power head or gearcase.

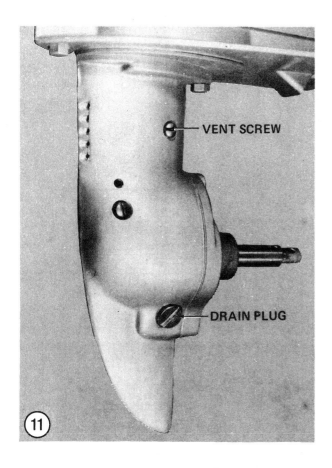

VENT SCREW

DRAIN PLUG

⑪

9. To reinstall the gearcase, lightly lubricate the drive shaft splines with anti-seize compound.

10. Position gearcase under motor leg and align drive shaft splines with the crankshaft.

CAUTION
Do not rotate the flywheel counterclockwise in Step 11. This can damage the water pump impeller.

11. Push the gearcase into place, rotating the flywheel clockwise as required to let the drive shaft and crankshaft engage.

12. Align water tube with water tube seal and seat gearcase against motor leg.

13. Coat screw threads with anti-seize compound. Install screws and tighten to specifications (**Table 1**).

14. Install the propeller as described in this chapter.

15. Reconnect the spark plug leads and refill the gearcase with proper type and quantity of lubricant. See Chapter Four.

4.4 hp; 4.5 hp; 1969-1970 5 hp; 6-15 hp

1. Disconnect the spark plug leads as a safety precaution to prevent any accidental starting of the engine during lower unit removal.

2. Place a container under the gearcase. Remove the vent and fill plugs. Drain the lubricant from the unit.

NOTE
If the lubricant is creamy in color or metallic particles are found in Step 3, the gearcase must be completely disassembled to determine and correct the cause of the problem.

3. Wipe a small amount of lubricant on a finger and rub the finger and thumb together. Check for the presence of metallic particles in the lubricant. Note the color of the lubricant. A white or creamy color indicates water in the lubricant. Check the drain container for signs of water separation from the lubricant.

4. Remove the propeller as described in this chapter.

5. Move the shift lever into FORWARD (7.5 and 9.2-15 hp) or REVERSE (4.4-8 hp). If necessary, rotate the propeller shaft slightly to help unit engage.

6. Remove the 4 hex head screws at the base of the motor leg.

7. Separate the gearcase from the motor leg enough to expose the screw holding the gear shift rods together. Remove the screw (**Figure 12**).

8. Remove the gearcase from the motor leg.

9. Mount the gearcase in a suitable holding fixture.

10. To reinstall the gearcase, check the lower shift rod adjustment. The centerline of the hole in the rod must be 3/16 ± 3/64 in. below

9

the motor leg-to-gearcase mounting surface (motor leg-to-extension on long shaft models) with the gearcase in NEUTRAL. See **Figure 13**.

11. Pull up on gearcase lower shift rod to place gearcase in FORWARD (7.5 and 9.2-15 hp) or REVERSE (4.4-8 hp).

> *CAUTION*
> *Do not grease the top of the drive shaft in Step 12. This may excessively preload the drive shaft and crankshaft when the mounting bolts are tightened and cause a premature failure of the power head or gearcase.*

12. Lightly lubricate the drive shaft splines with anti-seize compound.

> *CAUTION*
> *Do not rotate the flywheel counter-clockwise in Step 13. This can damage the water pump impeller.*

13. Position gearcase under motor leg. Align water tube in water pump seal and drive shaft with crankshaft splines.

14. Push gearcase toward the motor leg, rotating the flywheel clockwise as required to let the drive shaft and crankshaft engage. Align the lower and upper shift rods as shown in **Figure 14**.

15. Install the shift rod screw and tighten securely.

16. Make sure the water tube is seated in the water pump seal, then push gearcase against motor leg and install the gearcase screws. Tighten screws to specifications (**Table 1**).

17. Install the propeller as described in this chapter.

18. Reconnect the spark plug leads and refill the gearcase with proper type and quantity of lubricant. See Chapter Four.

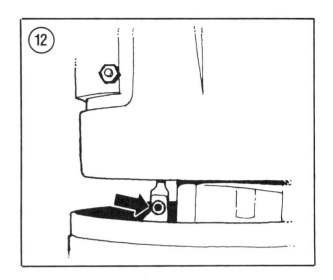

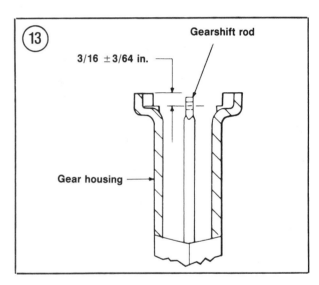

1966-1975 20 hp; 1973-1978 25 hp; 1973-1975 30 hp

1. Disconnect the spark plug leads as a safety precaution to prevent any accidental starting of the engine during lower unit removal.

2. Place a container under the gearcase. Remove the vent and drain plugs. Drain the lubricant from the unit.

> *NOTE*
> *If the lubricant is creamy in color or metallic particles are found in Step 3, the gearcase must be completely disassembled to determine and correct the cause of the problem.*

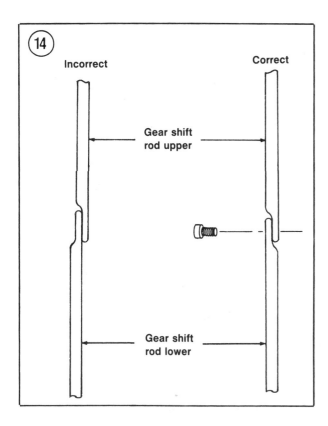

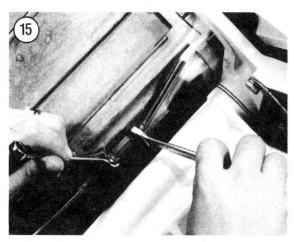

3. Wipe a small amount of lubricant on a finger and rub the finger and thumb together. Check for the presence of metallic particles in the lubricant. Note the color of the lubricant. A white or creamy color indicates water in the lubricant. Check the drain container for signs of water separation from the lubricant.

4. Remove the propeller as described in this chapter.

5. Remove the motor leg covers.

6. Loosen the jam nut above the shift rod coupling. See **Figure 15**.

7. Shift the engine into REVERSE. If necessary, rotate the propeller shaft slightly to help unit engage.

8. Rotate coupling clockwise until the lower and upper shift rods separate.

9. Remove the 4 hex head bolts holding the gearcase to the motor leg (motor leg extension on long shaft models).

10. Remove the gearcase from the motor leg or extension.

11. Mount the gearcase in a suitable holding fixture.

12. To reinstall the gearcase, thread the lower shift rod in the shift cam until it bottoms. Back shift rod out until the bend in the rod is centered over the front starboard water pump mounting screw.

CAUTION
Do not grease the top of the drive shaft in Step 13. This may excessively preload the drive shaft and crankshaft when the mounting bolts are tightened and cause a premature failure of the power head or gearcase.

13. Lightly lubricate the drive shaft splines with anti-seize compound.

14. Slide gearcase into place on the motor leg. Make sure the water pick-up tube seats in the water pump seal and the lower shift rod aligns with the hole in the front of the motor leg.

CAUTION
Do not rotate the flywheel counterclockwise in Step 15. This can damage the water pump impeller.

15. Push gearcase upward until drive shaft contacts crankshaft. Align drive shaft and crankshaft splines, rotating the flywheel clockwise as required to let the drive shaft and crankshaft engage.

9

16. Install and tighten the gearcase-to-motor leg bolts to specifications (**Table 1**).

17. Shift gearcase into NEUTRAL.

18. Thread coupler on upper shift rod until lower end of coupler engages lower shift rod, then thread coupler on shift rods until shift knob on support plate is vertical (NEUTRAL position).

19. Tighten coupler jam nut and install motor leg covers.

20. Install the propeller as described in this chapter.

21. Reconnect the spark plug leads and refill the gearcase with proper type and quantity of lubricant. See Chapter Four.

All Other 20-140 hp

1. Disconnect the spark plug leads as a safety precaution to prevent any accidental starting of the engine during lower unit removal.

2. Place a container under the gearcase. Remove the vent and drain plugs. Drain the lubricant from the unit.

> *NOTE*
> *If the lubricant is creamy in color or metallic particles are found in Step 3, the gearcase must be completely disassembled to determine and correct the cause of the problem.*

3. Wipe a small amount of lubricant on a finger and rub the finger and thumb together. Check for the presence of metallic particles in the lubricant. Note the color of the lubricant. A white or creamy color indicates water in the lubricant. Check the drain container for signs of water separation from the lubricant.

4. Remove the propeller as described in this chapter.

5. Remove the motor leg covers and upper side shock mounts, if so equipped.

6. Remove the fasteners holding the gearcase to the motor leg (motor leg extension on long shaft models).

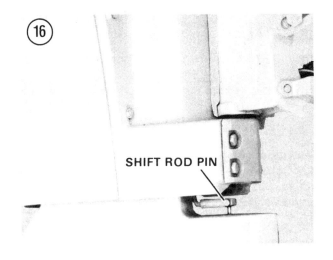
SHIFT ROD PIN

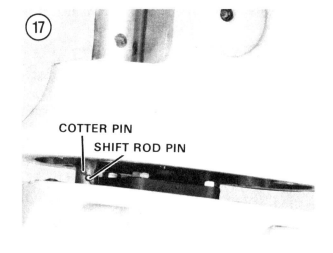
COTTER PIN
SHIFT ROD PIN

7. Remove exhaust snout on one-piece gearcases. Reach into the exhaust snout cavity and remove the screw and lockwasher.

8A. On models with an exposed shift rod pin (**Figure 16**), remove the cotter pin holding the shift rod pin to the upper shift rod coupler.

8B. On models with an internal shift rod pin (**Figure 17**), separate gearcase from motor leg enough to expose shift rod clevis or coupler. Remove cotter pin and shift rod pin. Remove the clevis, if so equipped.

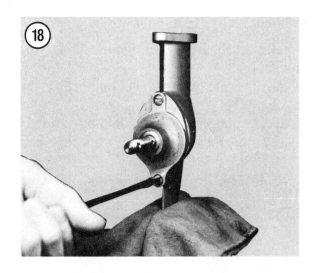

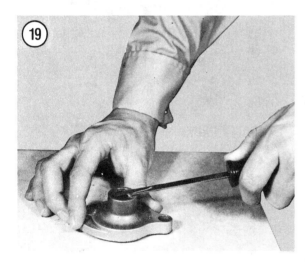

9. Remove the gearcase from the motor leg and mount in a suitable holding fixture.

> *CAUTION*
> *Do not grease the top of the drive shaft in Step 10. This may excessively preload the drive shaft and crankshaft when the mounting bolts are tightened and cause a premature failure of the power head or gearcase.*

10. Lightly lubricate the drive shaft splines with anti-seize compound.

> *CAUTION*
> *Do not rotate the flywheel counter-clockwise in Step 11. This can damage the water pump impeller.*

11. Position gearcase under motor leg. Align water tube with water pump grommet, drive shaft with crankshaft splines and the shift rod with the shift rod connector. Rotate the flywheel clockwise as required to let the drive shaft and crankshaft engage.

12. Wipe gearcase screw threads with RTV sealant. Install screws with lockwashers, but do not tighten.

13. Connect shift rods and install shift rod clevis (if used) and clevis or shift rod pin. Install a new cotter pin to retain the clevis or shift rod pin.

14. Tighten mounting screws to specifications (**Table 1**).

15. Install the propeller as described in this chapter.

16. Reconnect the spark plug leads and refill the gearcase with proper type and quantity of lubricant. See Chapter Four.

GEARCASE DISASSEMBLY/ASSEMBLY

1966-1969 3.5 hp;
1970-1977 3.6 hp;
4.9 hp; 5 hp

1. Remove the gearcase as described in this chapter.

2. 3.6-5 hp—Remove the water pump as described in this chapter.

3. Remove the 2 slotted head screws holding the cap to the gearcase (**Figure 18**).

4. Gently tap the cap edge with a plastic mallet to break cap loose from gasket.

5. Remove cap from propeller shaft.

6. Remove and discard the gearcase cap gasket.

7. Pry seal retainer from cap. See **Figure 19**. Discard retainer.

8. Remove cap bore seal.

9. Pull propeller shaft from gearcase.

10. Remove the drive shaft retaining clip as shown in **Figure 20**.

11. Remove drive shaft and pinion gear from gearcase.

12. Pry gearcase seal retainer out as shown in **Figure 21**. Discard retainer.

13. Remove the 2 gearcase bore seals.

14. Install 2 new seals in gearcase bore with their lips facing toward the top of the gearcase.

15. Position a new seal retainer over the gearcase seals. Use a suitable installer and drive seal retainer down until it bottoms.

16. Position the pinion gear in the gearcase. Insert the drive shaft with its retaining clip slot end down until it protrudes through the gear, then install the retaining clip.

17. Install propeller shaft in gearcase, rotating shaft until forward gear teeth engage pinion gear teeth.

18. Install a new propeller shaft seal in the gearcase cap bore. Lip side of seal must face down toward bottom of core.

19. Install a new seal retainer over the seal. The bowed-out side should face up toward the top of the bore. Use a suitable installer and drive seal retainer down until it bottoms.

20. Install a new cap gasket on the gearcase. Slide cap on the propeller shaft and seat against the gearcase.

21. Install cap screws and tighten to specifications (**Table 1**).

22. 3.6-5 hp—Install the water pump as described in this chapter.

23. Pressure test the gearcase as described in this chapter.

24. Install the gearcase as described in this chapter. Fill with recommended type and quantity of lubricant. See Chapter Four.

25. Check gearcase lubricant level after engine has been run. Change the lubricant after 10 hours of operation (break-in period). See Chapter Four.

1980-on 3.5 hp; 1976-on 4 hp

Refer to **Figure 22** or **Figure 23** for this procedure.

1. Remove the gearcase as described in this chapter.

2. Secure the gearcase in a holding fixture or a vise with protective jaws. If protective jaws

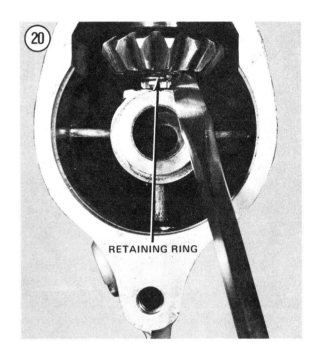

RETAINING RING

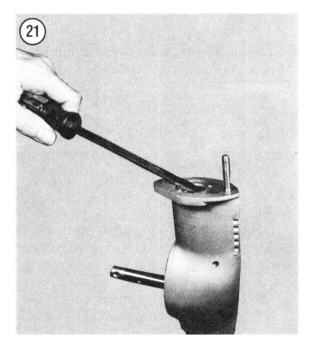

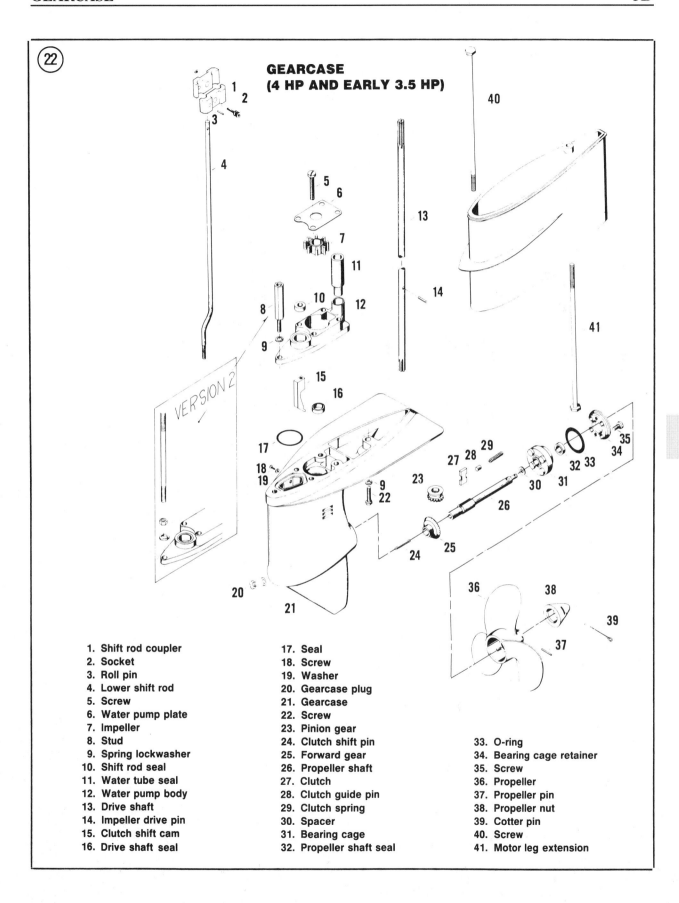

**GEARCASE
(4 HP AND EARLY 3.5 HP)**

1. Shift rod coupler
2. Socket
3. Roll pin
4. Lower shift rod
5. Screw
6. Water pump plate
7. Impeller
8. Stud
9. Spring lockwasher
10. Shift rod seal
11. Water tube seal
12. Water pump body
13. Drive shaft
14. Impeller drive pin
15. Clutch shift cam
16. Drive shaft seal
17. Seal
18. Screw
19. Washer
20. Gearcase plug
21. Gearcase
22. Screw
23. Pinion gear
24. Clutch shift pin
25. Forward gear
26. Propeller shaft
27. Clutch
28. Clutch guide pin
29. Clutch spring
30. Spacer
31. Bearing cage
32. Propeller shaft seal
33. O-ring
34. Bearing cage retainer
35. Screw
36. Propeller
37. Propeller pin
38. Propeller nut
39. Cotter pin
40. Screw
41. Motor leg extension

9

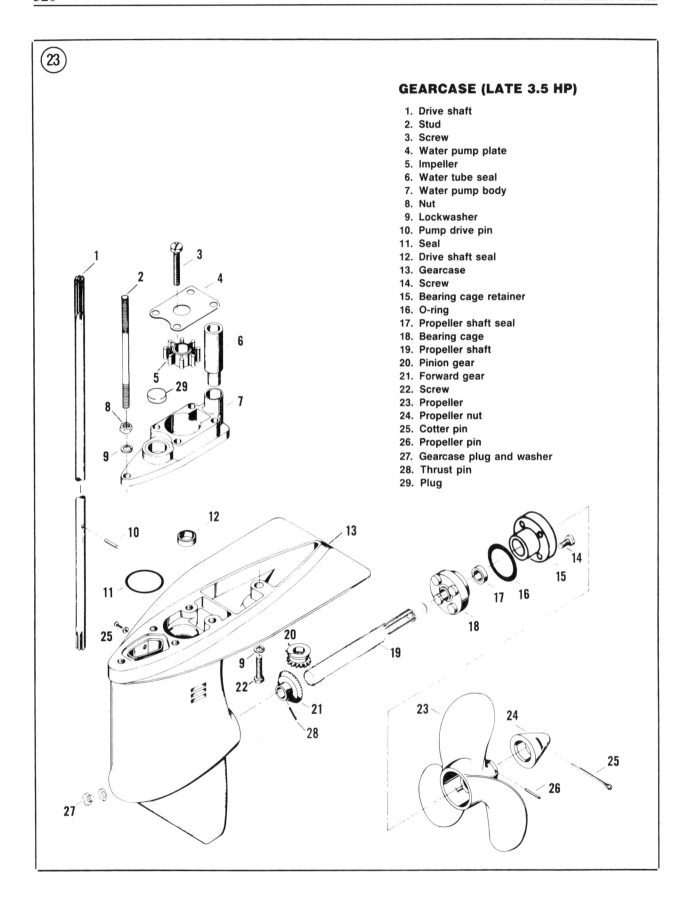

㉓

GEARCASE (LATE 3.5 HP)

1. Drive shaft
2. Stud
3. Screw
4. Water pump plate
5. Impeller
6. Water tube seal
7. Water pump body
8. Nut
9. Lockwasher
10. Pump drive pin
11. Seal
12. Drive shaft seal
13. Gearcase
14. Screw
15. Bearing cage retainer
16. O-ring
17. Propeller shaft seal
18. Bearing cage
19. Propeller shaft
20. Pinion gear
21. Forward gear
22. Screw
23. Propeller
24. Propeller nut
25. Cotter pin
26. Propeller pin
27. Gearcase plug and washer
28. Thrust pin
29. Plug

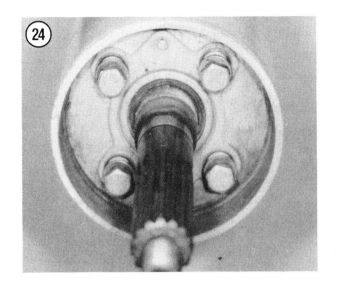

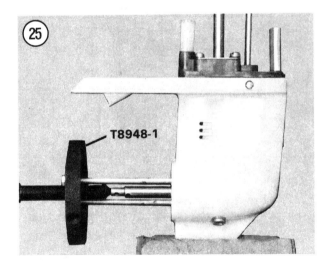

T8948-1

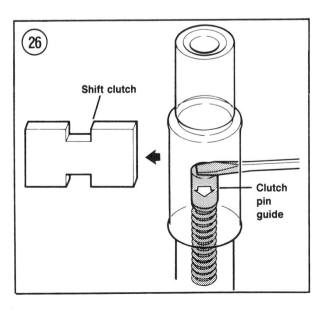

Shift clutch

Clutch
pin
guide

are not available, position the gearcase upright with the skeg between wooden blocks in a vise.

3. Remove the water pump as described in this chapter.

4. Remove the 4 screws holding the propeller retainer to the bearing cage. See **Figure 24**.

5. Install flywheel puller (part No. T 8948-1) as shown in **Figure 25** and remove propeller retainer.

6. Remove the propeller shaft bearing cage with O-ring and spacer.

7. Withdraw the propeller shaft from the gearcase.

8. Remove the clutch shift rod and cam.

CAUTION
Cover shaft end with a shop cloth in Step 9. If spring pressure is released suddenly, the clutch guide pin may shoot out rapidly.

9. Clamp the propeller shaft (prop end down) in a vise with protective jaws or wrapped in shop cloths. Use a small screwdriver to compress the clutch spring and remove the clutch from the shaft slot (**Figure 26**). Remove the clutch guide pin and spring.

10. Check front gear assembly on propeller shaft. If fitted with a roll pin, drive pin out and remove the gear. If there is no roll pin used, separate the gear from the shaft.

11. Pry the drive shaft seal from the gearcase bore with a screwdriver. See **Figure 27**.

12. Position bearing cage face down on open vise jaws or two blocks of wood. Drive seal out with a suitable punch.

13. Clean and inspect all parts as described in this chapter. Check drive shaft bearing in gearcase. If worn or damaged, replace the gearcase housing.

14. Install a new seal in the bearing cage with a suitable installer. Spring side should face out and seal case should be flush to 0.030 in. below bearing cage surface. See **Figure 28**.

9

15. Install clutch spring in propeller shaft cavity, then insert clutch pin guide. Place a shop cloth over the shaft end and use a small screwdriver to compress the spring, then slide the clutch in the prop shaft slot.

16. Install a new drive shaft seal in gearcase housing with a suitable installer. Spring side should face out.

17. Position a new O-ring in the shift rod cavity groove.

18. Install propeller shaft with forward gear in gearcase.

19. Install water pump as described in this chapter.

20. Install spacer and bearing cage. Push bearing cage into gearcase until the O-ring groove is exposed. Install a new O-ring.

21. Install the bearing cage retainer and align the bolt holes. Wipe retainer bolt threads with RTV sealant. Install bolts and tighten to specifications (**Table 1**).

22. Pressure test the gearcase as described in this chapter.

23. Install the gearcase as described in this chapter. Fill with recommended type and quantity of lubricant. See Chapter Four.

24. Check gearcase lubricant level after engine has been run. Change the lubricant after 10 hours of operation (break-in period). See Chapter Four.

4.4-15 hp (One-piece Gearcase)

Refer to **Figure 29** for this procedure.

1. Remove the gearcase as described in this chapter.

2. Secure the gearcase in a suitable holding fixture or a vise with protective jaws. If protective jaws are not available, position the gearcase upright with the skeg between wooden blocks in a vise.

3. Remove the water pump as described in this chapter.

4. Remove the 4 screws holding the propeller retainer to the bearing cage. See **Figure 24**.

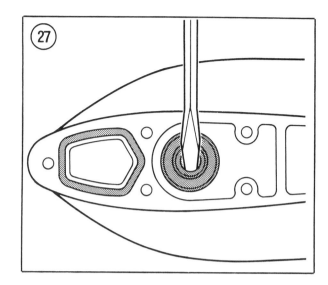

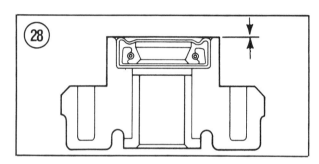

5. Install flywheel puller (part No. T 8948-1) as shown in **Figure 25** and remove propeller retainer.

6. Remove the propeller shaft bearing cage and O-ring.

7. Remove the reverse gear and thrust washer from the propeller shaft.

8. Withdraw the propeller shaft from the gearcase.

9. Tilt gearcase opening downward and catch the clutch shift cam as it drops out.

10. Remove the forward and pinion gears from the gearcase.

11. Remove the clutch shift pin from the front of the propeller shaft.

CAUTION
Cover shaft end with a shop cloth in Step 12. If spring pressure is released suddenly, the clutch guide pin may shoot out rapidly.

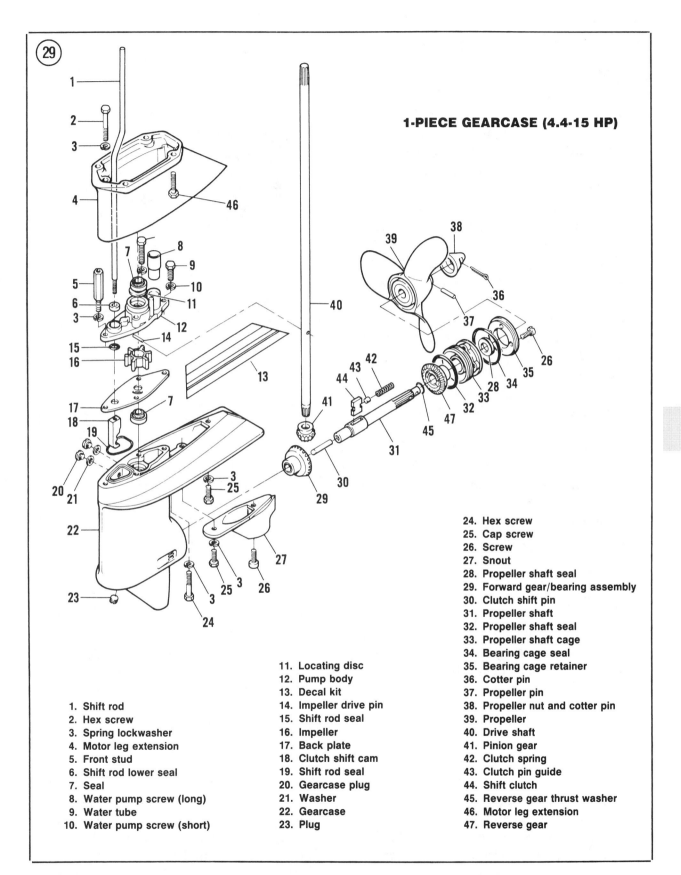

1-PIECE GEARCASE (4.4-15 HP)

1. Shift rod
2. Hex screw
3. Spring lockwasher
4. Motor leg extension
5. Front stud
6. Shift rod lower seal
7. Seal
8. Water pump screw (long)
9. Water tube
10. Water pump screw (short)

11. Locating disc
12. Pump body
13. Decal kit
14. Impeller drive pin
15. Shift rod seal
16. Impeller
17. Back plate
18. Clutch shift cam
19. Shift rod seal
20. Gearcase plug
21. Washer
22. Gearcase
23. Plug

24. Hex screw
25. Cap screw
26. Screw
27. Snout
28. Propeller shaft seal
29. Forward gear/bearing assembly
30. Clutch shift pin
31. Propeller shaft
32. Propeller shaft seal
33. Propeller shaft cage
34. Bearing cage seal
35. Bearing cage retainer
36. Cotter pin
37. Propeller pin
38. Propeller nut and cotter pin
39. Propeller
40. Drive shaft
41. Pinion gear
42. Clutch spring
43. Clutch pin guide
44. Shift clutch
45. Reverse gear thrust washer
46. Motor leg extension
47. Reverse gear

9

12. Clamp the propeller shaft (prop end down) in a vise with protective jaws or wrapped in shop cloths. Use a small screwdriver to compress the clutch spring and remove the clutch from the shaft slot (**Figure 26**). Remove the clutch guide pin and spring.

13. Pry the drive shaft seal from the gearcase bore with a screwdriver at the point shown in **Figure 30**.

14. Position bearing cage face down on open vise jaws or two blocks of wood. Drive seal out with a punch.

15. Clean and inspect all parts as described in this chapter. Check drive shaft bearing in gearcase. If worn or damaged, replace the gearcase housing.

16. Install a new seal in the bearing cage with an installer. Spring side should face out and seal case should be flush to 0.030 in. below bearing cage surface. See **Figure 31**.

17. Install clutch spring in propeller shaft cavity, then insert clutch pin guide. Place a shop cloth over the shaft end and use a small screwdriver to compress the spring, then slide the clutch in the prop shaft slot. See **Figure 32**.

18. Install a new drive shaft seal in gearcase housing with an installer. Spring side should face out.

19. Position a new O-ring in the shift rod cavity groove.

20. Install the water pump as described in this chapter. Do not install the pump body fasteners at this time.

21. Drop clutch shift cam into the gearcase shift rod cavity.

22. Pull the drive shaft up about 5/8 in. and install the forward gear and bearing assembly.

23. Position pinion gear in gearcase under drive shaft. Slowly rotate drive shaft while holding pinion gear until their splines align, then seat the drive shaft and water pump assembly to engage the pinion gear.

24. Wipe the water pump fastener threads with RTV sealant. Install and tighten

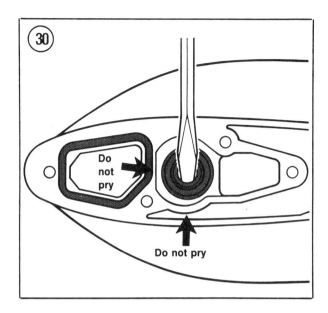

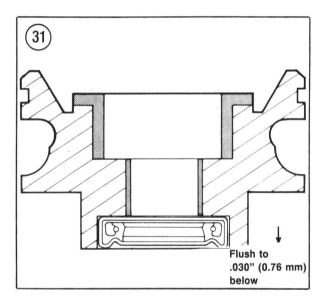

Flush to .030" (0.76 mm) below

fasteners and extension stud to specifications (**Table 1**).

25. Install the gear shift rod, engage the shift cam and push down on rod to position shift cam in REVERSE.

26. Install thrust washer and reverse gear in that order on the propeller shaft, then insert the propeller shaft in the gearcase. Shaft should engage forward gear with clutch shift pin sliding into reverse detent on shift cam.

27. Lubricate propeller shaft splines with anti-seize compound and install the bearing

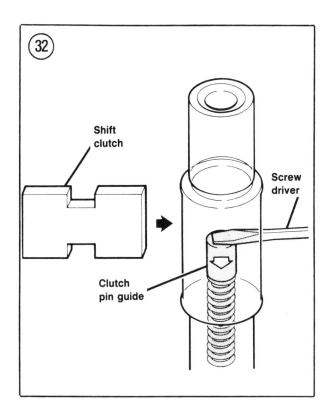

(32)

Shift
clutch

Screw
driver

Clutch
pin guide

cage into gearcase while rotating propeller shaft to engage reverse gear with pinion gear.

28. Seat bearing cage until gearcase O-ring groove is exposed. Install a new O-ring and lubricate with Rykon No. 2EP grease.

29. Install the bearing cage retainer and align the bolt holes. Wipe retainer bolt threads with RTV sealant. Install bolts and tighten to specifications (**Table 1**).

30. Pressure test the gearcase as described in this chapter.

31. Install the gearcase as described in this chapter. Fill with recommended type and quantity of lubricant. See Chapter Four.

32. Check gearcase lubricant level after engine has been run. Change the lubricant after 10 hours of operation (break-in period). See Chapter Four.

4.4-15 hp (Two-piece Gearcase)

Refer to **Figure 33** for this procedure.

1. Remove the gearcase as described in this chapter.

2. Secure the gearcase in a suitable holding fixture or a vise with protective jaws. If protective jaws are not available, position the gearcase upright with the skeg between wooden blocks in a vise.

3. Remove the water pump as described in this chapter.

4. Remove the 2 screws holding the bearing cage assembly to the lower gearcase.

5. Install flywheel puller (part No. T 8948-1) as shown in **Figure 25** and remove bearing cage assembly.

6. Remove the long hex bolt in front of the water pump cavity.

7. Slowly rotate lower gear shift rod counterclockwise to avoid cutting the shift rod seal and unthread it from the shift cam.

8. Remove the nut and lockwasher inside the gearcase, then separate the lower and upper gearcase housings. Remove and discard the gasket.

9. Remove the propeller shaft assembly from the lower gearcase. Slide reverse gear and thrust washer off propeller shaft.

10. Remove the pinion gear from the gearcase.

11. Remove the shift cam detent spring and slide cam from gearcase slot.

12. Remove forward gear with thrust bearing and race from gearcase bore.

13. Remove the propeller shaft clutch shift pin. Slide the clutch against the spring and insert a 1/8 in. pin punch through the propeller shaft clutch slot.

14. Secure propeller shaft in a vise with protective jaws or wrap in shop cloths. Drive spring pin from clutch with a suitable drift. See **Figure 34**. Remove clutch dog from propeller shaft.

15. Thread a 5/16 in. lag screw into the upper gearcase shift rod seal. Insert a 1/4 in. diameter rod from the other side and drive screw and seal from housing.

16. Pry drive shaft seal from upper gearcase housing.

9

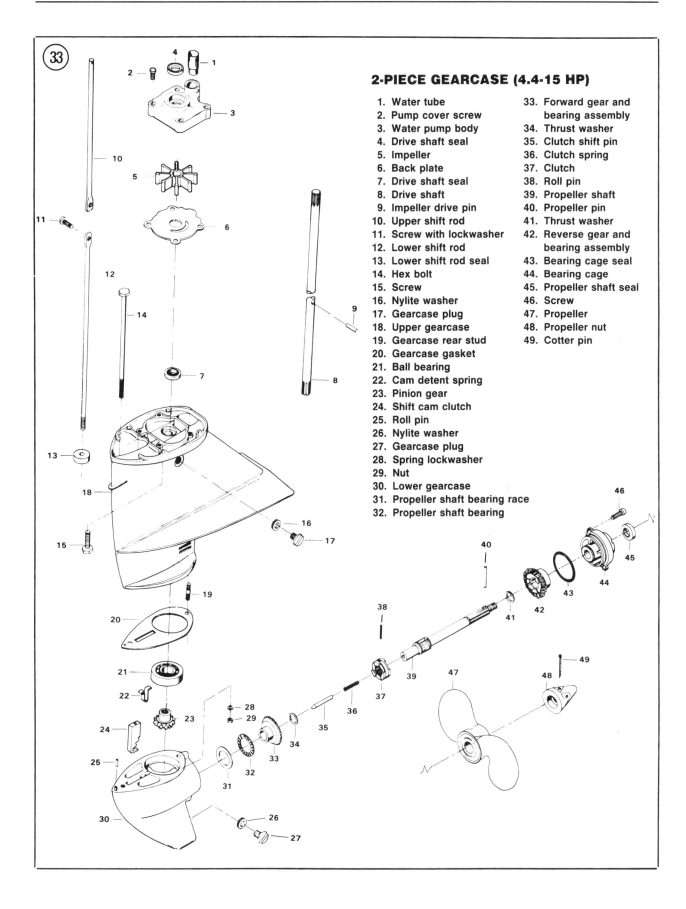

2-PIECE GEARCASE (4.4-15 HP)

1. Water tube
2. Pump cover screw
3. Water pump body
4. Drive shaft seal
5. Impeller
6. Back plate
7. Drive shaft seal
8. Drive shaft
9. Impeller drive pin
10. Upper shift rod
11. Screw with lockwasher
12. Lower shift rod
13. Lower shift rod seal
14. Hex bolt
15. Screw
16. Nylite washer
17. Gearcase plug
18. Upper gearcase
19. Gearcase rear stud
20. Gearcase gasket
21. Ball bearing
22. Cam detent spring
23. Pinion gear
24. Shift cam clutch
25. Roll pin
26. Nylite washer
27. Gearcase plug
28. Spring lockwasher
29. Nut
30. Lower gearcase
31. Propeller shaft bearing race
32. Propeller shaft bearing
33. Forward gear and bearing assembly
34. Thrust washer
35. Clutch shift pin
36. Clutch spring
37. Clutch
38. Roll pin
39. Propeller shaft
40. Propeller pin
41. Thrust washer
42. Reverse gear and bearing assembly
43. Bearing cage seal
44. Bearing cage
45. Propeller shaft seal
46. Screw
47. Propeller
48. Propeller nut
49. Cotter pin

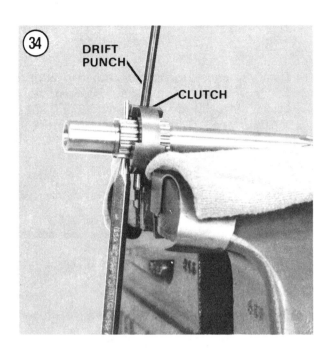

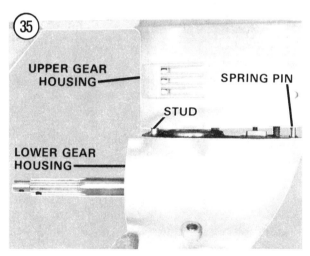

17. Remove and discard bearing cage O-ring. Drive seal from bearing cage with a suitable punch.

18. Clean and inspect all parts as described in this chapter. Check drive shaft bearing in upper gearcase housing. If worn or damaged, replace the upper gearcase housing.

19. Install a new shift rod seal. Drive into place until top is flush with bottom of seal bore chamfer.

20. Lubricate the lips of a new drive shaft seal with anti-seize compound. Install seal in bore of upper gearcase housing with a suitable installer until it seats at the bottom of the bore.

21. Install a new bearing cage seal. Install metal-cased seals with the garter spring facing toward the lower gearcase housing gears. Install rubber-cased seals with the garter spring facing the propeller.

22. Position clutch dog with its chamfered edges facing the propeller end of shaft. Drive the spring pin partially into the clutch dog and install on propeller shaft.

23. Secure propeller shaft in a vise with protective jaws or wrap in shop cloths. Insert a screwdriver blade in end of propeller shaft and compress the clutch spring. Position clutch dog over shaft splines and carefully tap drive pin into the shaft until flush.

24. Remove screwdriver blade and install shift pin with flat end facing the clutch spring pin.

25. Install thrust bearing and race on forward gear. Install thrust washer on the forward gear face, then install forward gear assembly on propeller shaft.

26. Insert clutch shift cam in lower gearcase housing. Hold top of cam about 3/16 in. out of housing and install propeller shaft assembly until forward gear engages the shift pin.

27. Install the pinion gear.

28. Install reverse gear and thrust washer on propeller shaft.

29. Install the shift cam detent spring in the hole in front of the pinion gear.

30. Coat both sides of a new gearcase housing gasket with RTV sealant and install on lower housing.

31. Align the spring pin at the front of the lower gearcase housing with the bore in the upper housing (**Figure 35**) and mate the housings.

32. Install the spring lockwasher and hex nut on the upper housing stud that protrudes into the lower housing.

9

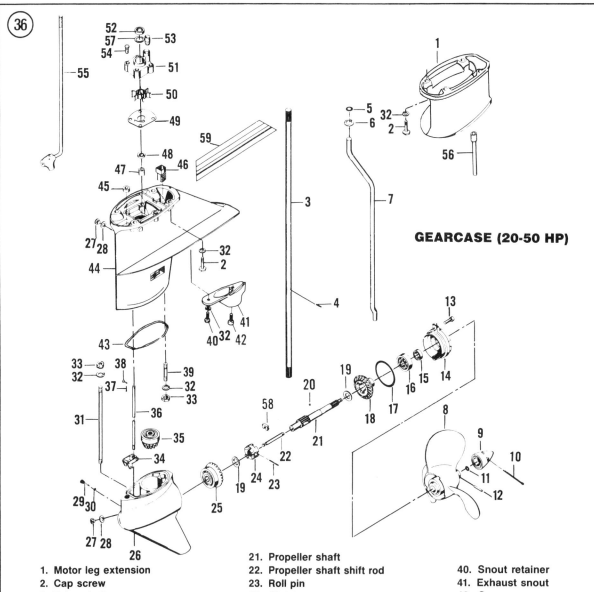

GEARCASE (20-50 HP)

1. Motor leg extension
2. Cap screw
3. Drive shaft
4. Impeller drive pin
5. Water tube seal
6. Washer
7. Upper water tube
8. Propeller
9. Propeller nut
10. Cotter pin
11. Propeller nut seal
12. Propeller drive pin
13. Hex socket cap screw
14. Bearing cage
15. Propeller shaft seal
16. Propeller shaft ball bearing
17. Bearing cage seal
18. Reverse gear
19. Thrust washer
20. Propeller shaft ball
21. Propeller shaft
22. Propeller shaft shift rod
23. Roll pin
24. Clutch
25. Forward gear and bearing assembly
26. Lower gearcase
27. Gearcase plug
28. Plug washer
29. Gear shift arm pin
30. O-ring
31. Long stud
32. Spring lockwasher
33. Hex nut
34. Gear shift arm
35. Pinion gear and bearing assembly
36. Lower shift rod
37. Cotter pin
38. Lower shift rod pin
39. Short stud
40. Snout retainer
41. Exhaust snout
42. Screw
43. Gasket
44. Upper gearcase
45. Shift rod lower seal
46. Water inlet screen
47. Upper drive shaft bearing
48. Drive shaft seal
49. Back plate
50. Impeller
51. Water pump body
52. Pump seal
53. Lower water tube
54. Pump screw
55. Intermediate shift rod
56. Water tube extension
57. Locating disc
58. Gear shift arm yoke
59. Decal set

33. Coat the long hex bolt threads with RTV sealant. Install bolt in upper gearcase housing and tighten to specifications (**Table 1**).

34. Lubricate the bearing cage O-ring seal with Rykon No. 2EP grease. Coat the propeller shaft seal lips with anti-seize compound.

35. Slide the bearing cage on the propeller shaft and press into the gearcase. Install and tighten the bearing cage screws to specifications (**Table 1**).

36. Carefully rotate lower shift rod through the seal and thread into the shift cam.

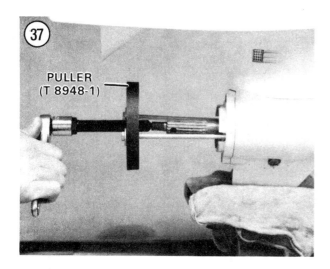

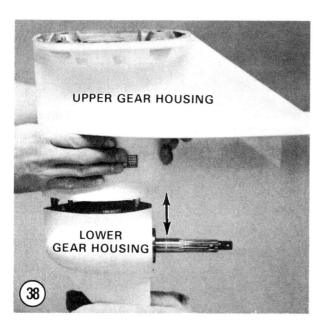

37. Install the water pump as described in this chapter.

38. Fill the gearcase with recommended type and quantity of lubricant. See Chapter Four.

39. Pressure test the gearcase as described in this chapter.

40. Install the gearcase as described in this chapter.

41. Check gearcase lubricant level after engine has been run. Change the lubricant after 10 hours of operation (break-in period). See Chapter Four.

20-50 hp

Refer to **Figure 36** for this procedure.

1. Remove the gearcase as described in this chapter.

2. Secure the gearcase in a holding fixture or a vise with protective jaws. If protective jaws are not available, position the gearcase upright with the skeg between wooden blocks.

3. Remove the water pump as described in this chapter.

4. Remove the 2 screws holding the bearing cage assembly to the lower gearcase.

5. Install flywheel puller (part No. T 8948-1) as shown in **Figure 37** and remove bearing cage assembly.

6. Remove the nut in the lower gearcase housing and the nut on the upper gearcase housing stud. Separate the housings (**Figure 38**).

7. Remove the pinion gear, bearing and cup assembly (**Figure 39**).

8. Remove the propeller shaft assembly from the lower gearcase housing (**Figure 40**).

9. Remove the forward gear, bearing, thrust washer and shift yoke from the propeller shaft.

10. Use a pencil-type magnet to remove the propeller shaft ball.

11. Press the reverse gear from the shaft.

9

12. Open the vise jaws about 1/2 in. Position propeller shaft and clutch assembly on shop cloths as shown in **Figure 41**. Clutch pin hole should rest over opening in vise jaws. Drive clutch pin out with a suitable punch. Remove spring pin, clutch, shift pin and spring.

13. Unscrew and remove the gear shift rod.

14. Remove the gearcase pivot screw from outside the lower gearcase housing. Discard the O-ring. Remove the gear shift arm from the housing.

15. Assemble special tool part No. T 8964 as shown in **Figure 42** and press drive shaft bearing and seal from upper gearcase housing.

16. Use a slide hammer with a hooked end to remove the shift rod seal. See **Figure 43**.

17. If the forward gear bearing is to be replaced, remove the bearing cup with cup remover part No. T 8921 and bearing guide set part No. T 8918 as shown in **Figure 44**.

18. Remove and discard the bearing cage O-ring.

19. Pad the vise with shop cloths and rap bearing cage against it to remove the bearing.

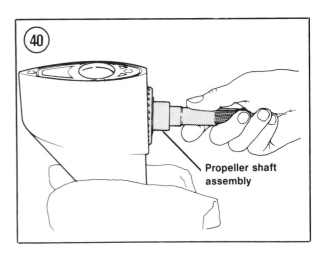

Propeller shaft assembly

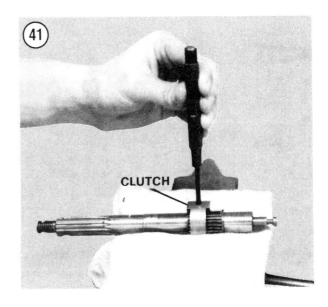

CLUTCH

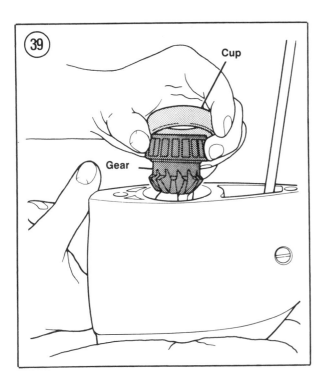

Cup

Gear

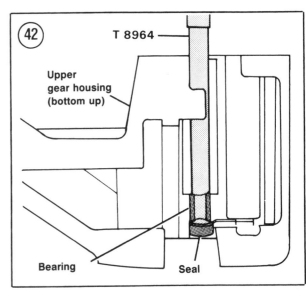

T 8964

Upper gear housing (bottom up)

Bearing

Seal

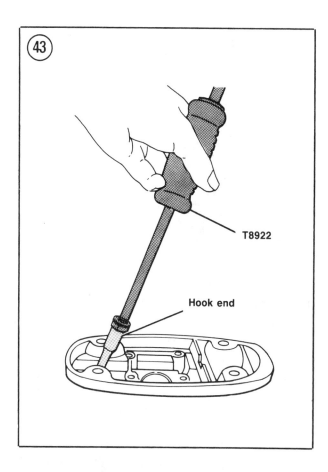

43

T8922

Hook end

20. Pry the bearing cage seal out with a screwdriver. Discard the seal.

21. Clean and inspect all parts as described in this chapter.

22. Press a new drive shaft bearing in the upper gearcase housing with special tool part No. T 8962 or a mandrel. See **Figure 45**.

23. Coat the lips of a new drive shaft bearing seal with anti-seize compound and install (rubber side facing up) with a drift.

24. Install a new shift rod seal (raised bead facing up) until seal edge is flush with bore surface.

25. If forward gear bearing cup was removed, press a new cup in lower gearcase housing with cup installer part No. T 8904, driver handle part No. T 8907 and guide plate from bearing guide set part No. T 8918. See **Figure 46**.

26. Install a new bearing cage seal with installer part No. T 3431. Garter spring should face outward. Lubricate seal lips with anti-seize compound.

9

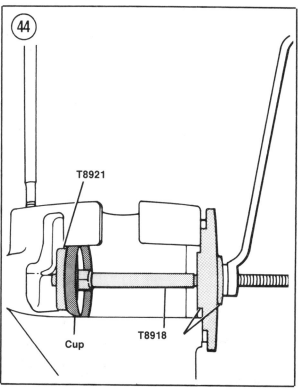

44

T8921

T8918

Cup

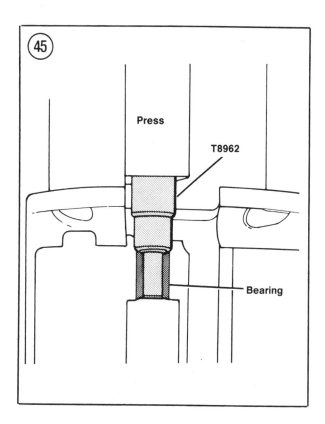

45

Press

T8962

Bearing

27. Install a new cage bearing. Tap gently on the outer bearing race to seat it fully.

28. Lubricate a new O-ring with anti-seize compound and install on the bearing cage.

29. Install shift arm (fork ends facing down) in lower gearcase housing. Install the gearcase pivot screw with a new O-ring.

30. Install the shift rod in the shift arm.

31. Install the propeller shaft rear thrust washer. Position clutch dog and shift pin on shaft. Align pin holes in clutch dog and shaft and install the spring pin. See **Figure 47**.

32. Install reverse gear to propeller shaft.

33. Coat the shift pin groove with anti-seize compound and install shift yoke in groove. Lubricate the shaft ball and install in propeller shaft.

34. Slide thrust washer, forward gear and bearing assembly on propeller shaft.

35. Install propeller shaft assembly in lower gearcase. Yoke pins must engage shift arm slots.

36. Install a new seal in the lower gearcase housing groove. Fit upper and lower housings together. Install screws and lockwashers and tighten to specifications (**Table 1**).

37. Rotate the propeller shaft until the shaft ball faces straight up. Position the slot in the bearing cage bearing inner race facing straight up. Slide bearing cage onto propeller shaft and engage shaft ball with bearing slot. Push bearing cage fully into housing bore.

38. Wipe the bearing cage screw threads with anti-seize compound. Install and tighten to specifications.

39. Install the water pump as described in this chapter.

40. Fill gearcase with recommended type and quantity of lubricant. See Chapter Four.

41. Pressure test the gearcase as described in this chapter.

42. Install the gearcase as described in this chapter.

43. Check gearcase lubricant level after engine has been run. Change the lubricant

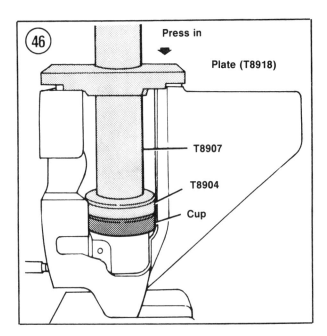

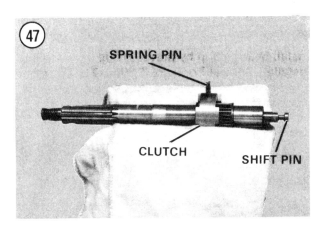

after 10 hours of operation (break-in period). See Chapter Four.

55-85 hp (Two-piece Gearcase)

Refer to **Figure 48** (55-60 hp) or **Figure 49** (70-85 hp) for this procedure.

1. Remove the gearcase as described in this chapter.

2. Secure the gearcase in a holding fixture or a vise with protective jaws. If protective jaws are not available, position the gearcase upright with the skeg between wooden blocks.

3. Remove the water pump as described in this chapter.

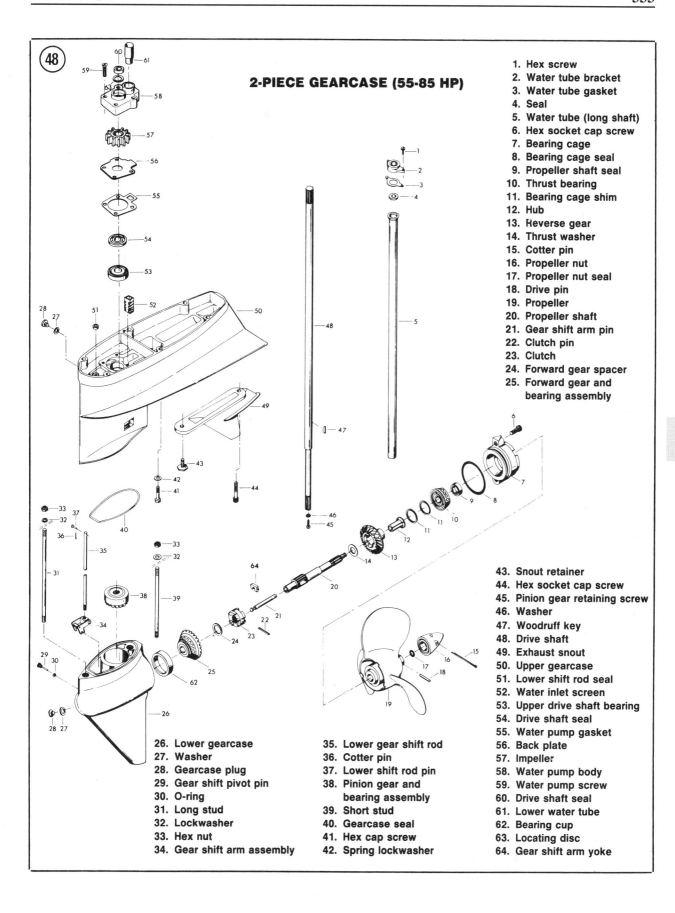

2-PIECE GEARCASE (55-85 HP)

1. Hex screw
2. Water tube bracket
3. Water tube gasket
4. Seal
5. Water tube (long shaft)
6. Hex socket cap screw
7. Bearing cage
8. Bearing cage seal
9. Propeller shaft seal
10. Thrust bearing
11. Bearing cage shim
12. Hub
13. Reverse gear
14. Thrust washer
15. Cotter pin
16. Propeller nut
17. Propeller nut seal
18. Drive pin
19. Propeller
20. Propeller shaft
21. Gear shift arm pin
22. Clutch pin
23. Clutch
24. Forward gear spacer
25. Forward gear and
    bearing assembly

9

43. Snout retainer
44. Hex socket cap screw
45. Pinion gear retaining screw
46. Washer
47. Woodruff key
48. Drive shaft
49. Exhaust snout
50. Upper gearcase
51. Lower shift rod seal
52. Water inlet screen
53. Upper drive shaft bearing
54. Drive shaft seal
55. Water pump gasket
56. Back plate
57. Impeller
58. Water pump body
59. Water pump screw
60. Drive shaft seal
61. Lower water tube
62. Bearing cup
63. Locating disc
64. Gear shift arm yoke

26. Lower gearcase
27. Washer
28. Gearcase plug
29. Gear shift pivot pin
30. O-ring
31. Long stud
32. Lockwasher
33. Hex nut
34. Gear shift arm assembly

35. Lower gear shift rod
36. Cotter pin
37. Lower shift rod pin
38. Pinion gear and
    bearing assembly
39. Short stud
40. Gearcase seal
41. Hex cap screw
42. Spring lockwasher

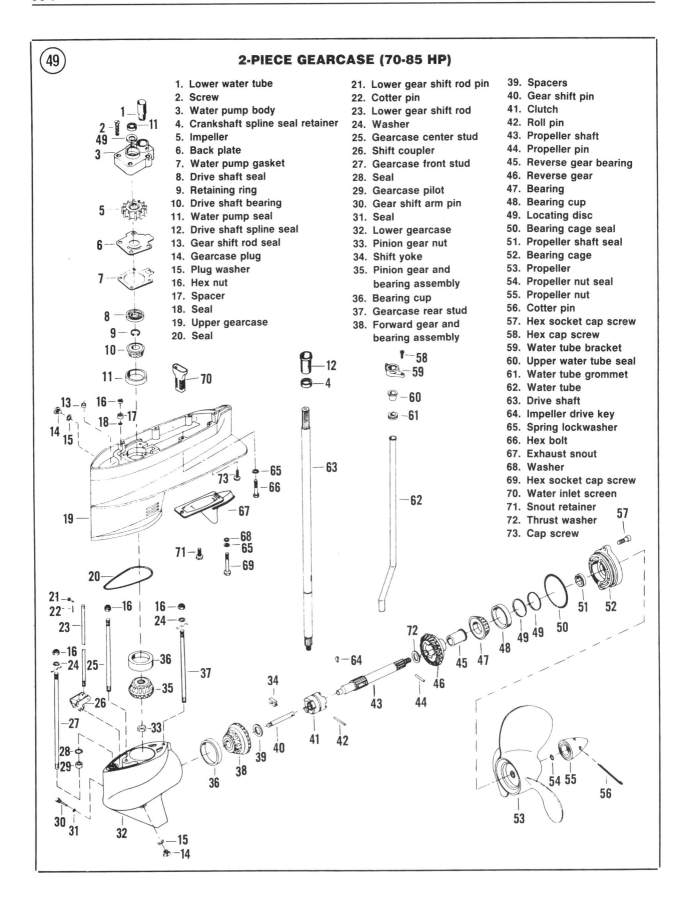

2-PIECE GEARCASE (70-85 HP)

1. Lower water tube
2. Screw
3. Water pump body
4. Crankshaft spline seal retainer
5. Impeller
6. Back plate
7. Water pump gasket
8. Drive shaft seal
9. Retaining ring
10. Drive shaft bearing
11. Water pump seal
12. Drive shaft spline seal
13. Gear shift rod seal
14. Gearcase plug
15. Plug washer
16. Hex nut
17. Spacer
18. Seal
19. Upper gearcase
20. Seal
21. Lower gear shift rod pin
22. Cotter pin
23. Lower gear shift rod
24. Washer
25. Gearcase center stud
26. Shift coupler
27. Gearcase front stud
28. Seal
29. Gearcase pilot
30. Gear shift arm pin
31. Seal
32. Lower gearcase
33. Pinion gear nut
34. Shift yoke
35. Pinion gear and bearing assembly
36. Bearing cup
37. Gearcase rear stud
38. Forward gear and bearing assembly
39. Spacers
40. Gear shift pin
41. Clutch
42. Roll pin
43. Propeller shaft
44. Propeller pin
45. Reverse gear bearing
46. Reverse gear
47. Bearing
48. Bearing cup
49. Locating disc
50. Bearing cage seal
51. Propeller shaft seal
52. Bearing cage
53. Propeller
54. Propeller nut seal
55. Propeller nut
56. Cotter pin
57. Hex socket cap screw
58. Hex cap screw
59. Water tube bracket
60. Upper water tube seal
61. Water tube grommet
62. Water tube
63. Drive shaft
64. Impeller drive key
65. Spring lockwasher
66. Hex bolt
67. Exhaust snout
68. Washer
69. Hex socket cap screw
70. Water inlet screen
71. Snout retainer
72. Thrust washer
73. Cap screw

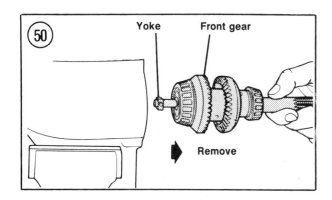

Yoke   Front gear

⑤⓪

Remove

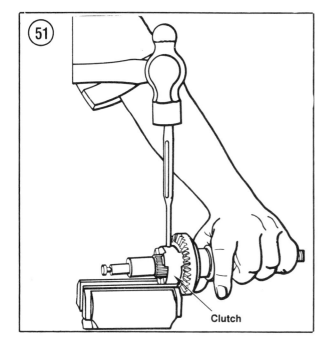

⑤①

Clutch

4. Remove the 2 screws holding the bearing cage assembly to the lower gearcase.

5. Install flywheel puller (part No. T 8948-1) as shown in **Figure 37** and remove bearing cage assembly.

6A. 55-60 hp—Remove the nut in the lower gearcase housing and the nut on the upper gearcase housing stud. Separate the housings (**Figure 38**).

6B. 70-85 hp—Remove the 3 upper gearcase stud nuts holding the lower gearcase housing in place. Separate the housings.

7A. 55-60 hp—Unscrew and remove the shift rod from the shift arm. Remove the shift arm pin and O-ring. Discard the O-ring.

7B. 70-85 hp—Remove shift arm pin and O-ring. Discard the O-ring. Lift gear shift rod with coupler from housing. Unscrew coupler from shift rod.

8. Remove the propeller shaft assembly from the lower gearcase housing (**Figure 50**).

9. Remove shift yoke and spacer from front of propeller shaft.

10. Remove forward gear and bearing assembly from propeller shaft.

11. Open the vise jaws about 1/2 in. Position propeller shaft and clutch assembly on shop cloths with clutch pin hole resting over opening between vise jaws. Drive clutch pin out with a punch. See **Figure 51**. Remove shift pin and clutch from shaft.

12. Remove reverse gear and bearing from propeller shaft with an arbor press.

13. Wrap drive shaft with shop cloths and secure in a vise with protective jaws.

   a. 55-60 hp—Remove the screw and washer holding the pinion gear/bearing assembly to the drive shaft. Discard the screw. Remove the gear, bearing and cup.

   b. 70-85 hp—Remove the retaining nut and washer holding the pinion gear/bearing assembly to the drive shaft. Discard the nut. Remove the gear and bearing. If cup requires removal, use tool part No. T 8919.

14. Remove the drive shaft and seal from the upper gearcase housing with an arbor press.

15. If upper drive shaft bearing requires replacement, remove bearing snap ring (70-85 hp only) and press bearing off shaft.

16. Thread a 5/16 in. lag screw into the shift rod seal. Insert a 1/4 in. diameter rod from the other end and press the seal and screw from the gearcase. See **Figure 52**.

NOTE
Replace the forward gear, bearing and cup as a set.

9

17. If forward gear and bearing assembly requires replacement, remove the bearing cup with cup remover part No. T 8921 (55-60 hp) or part No. T 2995 (70-85 hp) and bearing guide plate part No. T 8918 as shown in **Figure 53**.

18. If upper drive shaft bearing is to be replaced, remove bearing cup from upper gearcase housing with a punch.

19A. 55-60 hp—Remove and discard the bearing cage O-ring. Drive bearing cage seal out with an appropriate punch.

19B. 70-85 hp—Remove bearing cup from bearing cage with a slide hammer and tool part No. T 8917. Remove and retain the shims under the cup. Remove and discard the bearing cage O-ring. Press seal out with remover part No. T 8914.

20. Clean and inspect all parts as described in this chapter.

21. If forward gear bearing cup was removed, press a new cup in lower gearcase housing with cup installer part No. T 8904, driver handle part No. T 8907 and guide plate from bearing guide set part No. T 8918. See **Figure 46**.

22. If upper drive shaft bearing cup was removed, install a new cup with its larger diameter facing up. Tap lightly around the cup edges with a punch and hammer until it is fully seated.

23. If upper drive shaft bearing was removed, install a new bearing with an arbor press. Install a new snap ring on 70-85 hp models.

24. 55-60 hp—Install pinion gear to drive shaft as follows:

a. Coat drive shaft internal threads and splines with sealant primer. Apply Loctite D to threads of a new pinion gear screw.

b. Install drive shaft in upper gearcase housing. Slide pinion gear and bearing assembly on drive shaft spline. Install washer and pinion gear screw finger-tight.

c. Secure the drive shaft horizontally in a vise with protective jaws. If protective jaws are not available, wrap drive shaft in shop cloths.

d. Hold the pinion screw with a screwdriver and rotate gearcase to tighten screw until it just starts to bind. See **Figure 54**.

e. Position a drift against the pinion gear screw. Tap screw twice to seat the pinion and drive shaft bearings properly. See **Figure 55**.

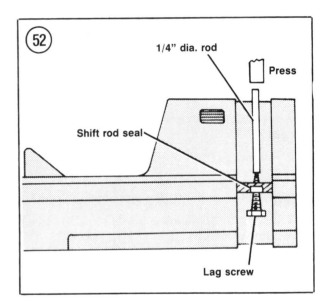

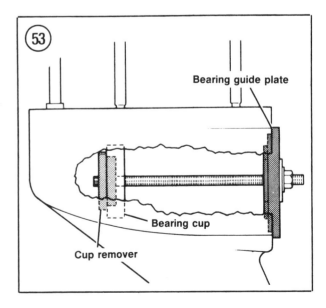

25. 70-85 hp—Install pinion gear to drive shaft as follows:

a. Install drive shaft in upper gearcase housing. Slide pinion gear and bearing assembly on drive shaft spline. Install washer and a new pinion gear nut finger-tight.

b. Secure the drive shaft horizontally in a vise with protective jaws. If protective jaws are not available, wrap drive shaft in shop cloths.

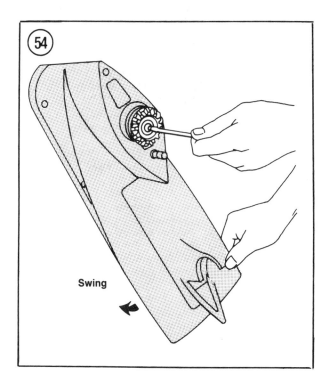

Swing

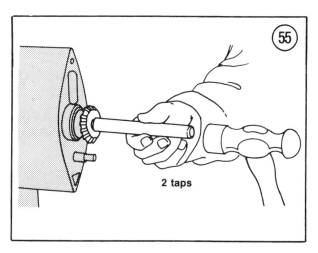

2 taps

c. Tighten pinion gear nut to specifications (**Table 1**).

26. Slide a new drive shaft seal over the drive shaft (spring side up) and install with tool part No. T 8985 until it seats in the bore.

27. Wipe outer diameter of a new shift rod seal with Loctite H. Position seal with raised bend facing upward and install in gearcase cavity until it bottoms on the bore.

28. Shim the pinion gear as described in this chapter.

29. Check the propeller shaft end float as described in this chapter.

30. Insert the gear shift arm in the lower gearcase housing. Install the gear shift arm pin with a new O-ring.

31. Thread the shift rod into the shift arm.

32. Install the front gear and bearing assembly in the lower gearcase.

33. Install the propeller shaft assembly.

34. Install a new bearing cage seal with an installer. Garter spring should face toward rear of cage.

35. Install a new bearing cage O-ring and lubricate with Rykon No. 2EP grease.

36. Install bearing cage in gearcase bore. Wipe screw threads with RTV sealant. Install screws and tighten to specifications (**Table 1**).

37. Install a new seal in the lower gearcase housing groove. Fit upper and lower housings together. Install fasteners and tighten to specifications (**Table 1**).

38. Install the water pump as described in this chapter.

39. Fill gearcase with recommended type and quantity of lubricant. See Chapter Four.

40. Pressure test the gearcase as described in this chapter.

41. Install the gearcase as described in this chapter.

42. Check gearcase lubricant level after engine has been run. Change the lubricant after 10 hours of operation (break-in period). See Chapter Four.

9

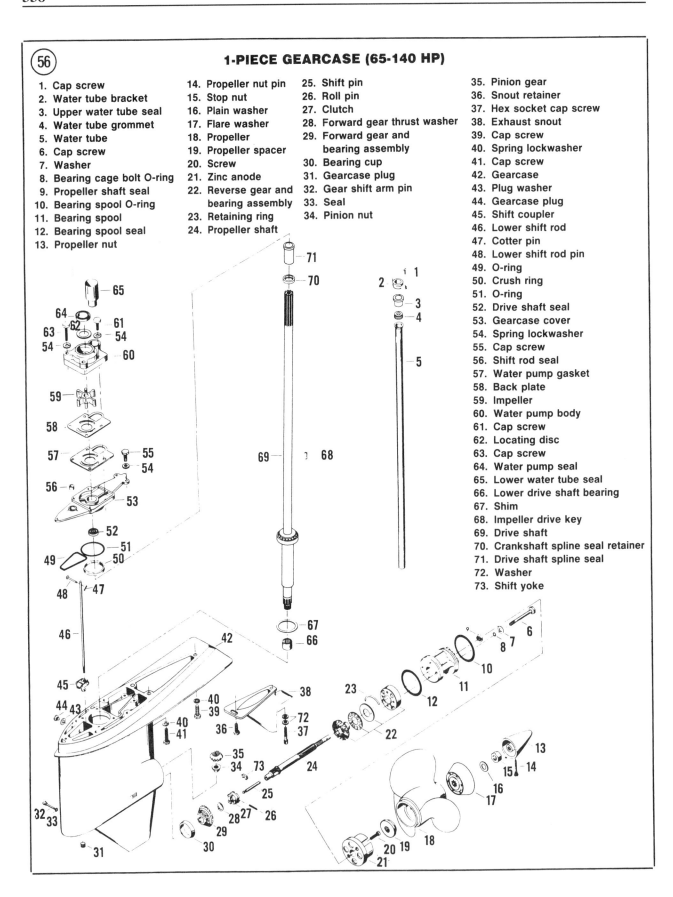

56 **1-PIECE GEARCASE (65-140 HP)**

1. Cap screw
2. Water tube bracket
3. Upper water tube seal
4. Water tube grommet
5. Water tube
6. Cap screw
7. Washer
8. Bearing cage bolt O-ring
9. Propeller shaft seal
10. Bearing spool O-ring
11. Bearing spool
12. Bearing spool seal
13. Propeller nut
14. Propeller nut pin
15. Stop nut
16. Plain washer
17. Flare washer
18. Propeller
19. Propeller spacer
20. Screw
21. Zinc anode
22. Reverse gear and bearing assembly
23. Retaining ring
24. Propeller shaft
25. Shift pin
26. Roll pin
27. Clutch
28. Forward gear thrust washer
29. Forward gear and bearing assembly
30. Bearing cup
31. Gearcase plug
32. Gear shift arm pin
33. Seal
34. Pinion nut
35. Pinion gear
36. Snout retainer
37. Hex socket cap screw
38. Exhaust snout
39. Cap screw
40. Spring lockwasher
41. Cap screw
42. Gearcase
43. Plug washer
44. Gearcase plug
45. Shift coupler
46. Lower shift rod
47. Cotter pin
48. Lower shift rod pin
49. O-ring
50. Crush ring
51. O-ring
52. Drive shaft seal
53. Gearcase cover
54. Spring lockwasher
55. Cap screw
56. Shift rod seal
57. Water pump gasket
58. Back plate
59. Impeller
60. Water pump body
61. Cap screw
62. Locating disc
63. Cap screw
64. Water pump seal
65. Lower water tube seal
66. Lower drive shaft bearing
67. Shim
68. Impeller drive key
69. Drive shaft
70. Crankshaft spline seal retainer
71. Drive shaft spline seal
72. Washer
73. Shift yoke

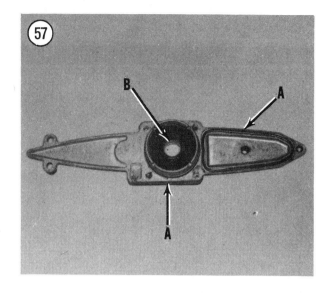

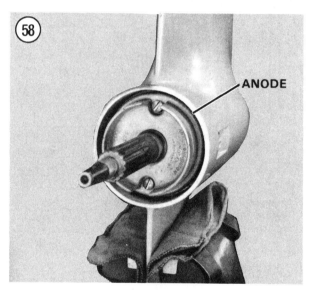

65-140 hp (One-piece Gearcase)

Refer to **Figure 56** for this procedure.

1. Remove the gearcase as described in this chapter.

2. Secure the gearcase in a suitable holding fixture or a vise with protective jaws. If protective jaws are not available, position the gearcase upright with the skeg between wooden blocks.

3. Remove the water pump as described in this chapter.

4. Remove the spline seal and retainer from the end of the drive shaft.

5. Remove the screws and lockwashers holding the gearcase cover. Pull upward on cover and move back and forth until it comes free of the gearcase. Remove cover from drive shaft.

6. Remove the crush ring from the gearcase bore.

7. Invert the gearcase cover and remove the 2 O-ring seals (A, **Figure 57**). Pry the drive shaft seal from the cover bore (B, **Figure 57**).

8. Thread a 5/16 in. lag screw into the shift rod seal. Insert a 1/4 in. diameter rod from the other end and press the seal and screw from the gearcase. See **Figure 52**.

9. Rotate lower shift rod counterclockwise and unscrew from shift arm coupler. Remove the shift rod.

10. Remove the 2 screws holding the anode to the propeller shaft spool. See **Figure 58**. Remove the anode.

11. Remove the 4 screws and O-rings holding the propeller shaft bearing spool to the bearing cage. Discard O-rings.

12. Install flywheel puller (part No. T 8948-1) as shown in **Figure 59**. Shift gearcase into gear and rotate drive shaft. Puller will unthread itself from the propeller shaft, drawing the spool from the gearcase.

13. Remove and discard the 2 shaft spool O-rings. See **Figure 60**. Drive the seal from

9

the spool with a suitable pin punch. See **Figure 61**.

14. Remove the 2 retaining rings in the gearcase bore.

15. Reinstall shaft spool (without O-rings) and secure with 2 screws. Repeat Step 12 and puller will remove the shaft spool and bearing cage.

16. Remove reverse gear from the propeller shaft.

17. Withdraw the propeller shaft from the gearcase housing. Remove the spacer from the propeller shaft. Remove the yoke from the shift pin.

18. Open the vise jaws about 1/2 in. Position propeller shaft and clutch assembly on shop cloths with clutch pin hole positioned over the opening in the vise jaws. Drive spring pin out with a punch. Remove clutch and shift pin from shaft.

19. Install spline adapter (part No. T 7848), socket and flex handle on drive shaft splines.

20. Hold pinion nut with a 3/4 in. socket and flex handle. Pad that part of the gearcase where the flex handle will hit with shop cloths to prevent housing damage.

21. Holding the pinion nut from moving, break the nut loose by turning the drive shaft. Remove pinion nut and drive shaft adapter wrenches.

22. Holding pinion nut with one hand, rotate drive shaft with the other and unscrew it from the nut. Remove the nut.

23. Secure drive shaft in a vise with protective jaws or wooden blocks, clamping as close as possible to the gearcase.

24. Pull gearcase toward you to remove any drive shaft play, then place a wooden block against the drive shaft counterbore area and rap block sharply with a hammer until the housing comes off the drive shaft.

25. Remove gearcase from drive shaft. Upper drive shaft bearing should remain on the drive shaft.

26. Remove the shims installed between the bearing and the gearcase bore.

27. Reach into gearcase bore and remove the forward gear and bearing assembly.

28. Remove the shift arm pin and O-ring from the gearcase. Discard the O-ring. Remove the gear shift arm coupler.

NOTE
Replace the forward gear, bearing and cup as a set.

29. If forward gear bearing requires replacement, remove the bearing cup with cup remover part No. T 11207 and bearing

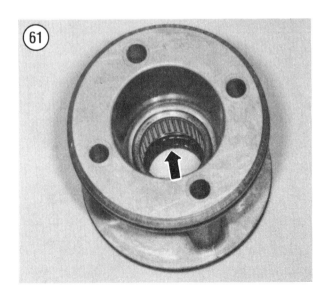

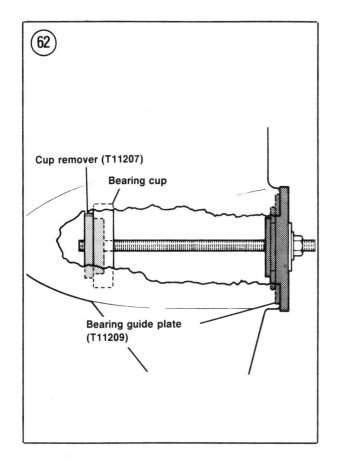

Cup remover (T11207)

Bearing cup

Bearing guide plate
(T11209)

guide plate part No. T 11209 as shown in **Figure 62**.

30. If lower gearcase drive shaft bearing requires replacement, remove the bearing with remover part No. T 11205 and bearing guide plate part No. T 11209 as shown in **Figure 63**.

31. Clean and inspect all parts as described in this chapter.

32. If forward gear bearing cup was removed, position gearcase as shown in **Figure 64** and install a new bearing cup with installer part No. T 11204, driver handle T 8907 and guide plate part No. T 11209.

33. Shim the pinion gear as described in this chapter.

34. Check the propeller shaft end float as described in this chapter.

35. Insert the shift arm coupler in the lower gearcase housing. Install the shift arm pin with a new O-ring.

36. Install the shift yoke and thrust washer on the propeller shaft assembly. Insert shaft

9

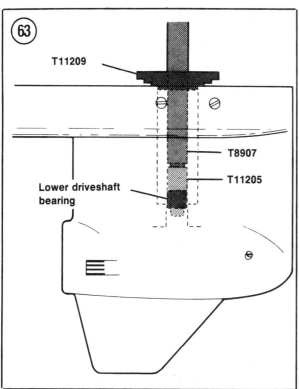

T11209

T8907

T11205

Lower driveshaft
bearing

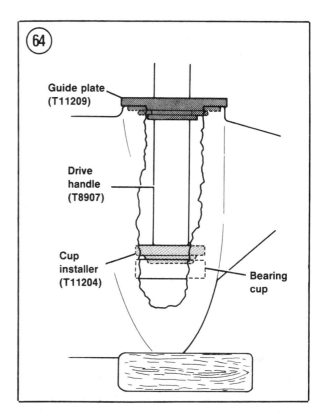

Guide plate
(T11209)

Drive
handle
(T8907)

Cup
installer
(T11204)

Bearing
cup

assembly in gearcase bore and engage yoke with shift coupler arms.

37. Install the retaining rings in the gearcase bore groove.

38. Install new prop shaft spool O-rings. Note that the smaller ring goes on the front of the spool and the larger ring goes on the rear. Light coat O-rings with Rykon No. 2EP grease.

39. Install prop shaft spool in gearcase bore until O-rings touch their counterbore. Install spool bolts with new O-rings and hand-tighten, then torque to specifications (**Table 1**).

40. Install anode on gearcase. Wipe anode screw threads with Loctite D. Install and tighten screws to specifications (**Table 1**).

41. Coat drive shaft splines with anti-seize compound and install a new spline seal and retainer.

42. Install new gearcase cover O-rings and seal. Fit cover over drive shaft and position on gearcase. Wipe screw threads with RTV sealant. Install screws finger-tight.

43. Install water pump as described in this chapter.

44. Fill gearcase with recommended type and quantity of lubricant. See Chapter Four.

45. Pressure test the gearcase as described in this chapter.

46. Install the gearcase as described in this chapter.

47. Check gearcase lubricant level after engine has been run. Change the lubricant after 10 hours of operation (break-in period). See Chapter Four.

GEARCASE CLEANING AND INSPECTION

1. Clean all parts in fresh solvent. Blow dry with compressed air, if available.

2. Clean all nut and screw threads thoroughly if RTV sealant or Loctite has been used. Soak nuts and screws in solvent and use a fine wire brush to remove residue.

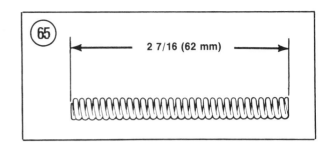

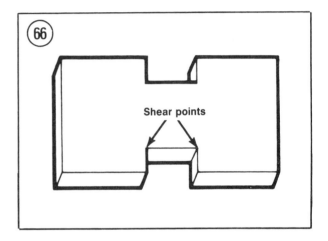

3. Remove and discard all O-rings, gaskets and seals. Clean all gasket or RTV sealant residue from mating surfaces.

4. Check drive shaft splines for wear or damage. If gearcase has struck a submerged object, the drive shaft and propeller shaft may suffer severe damage. Replace drive shaft as required and check crankshaft splines for similar wear or damage.

5. Check propeller shaft splines and threads for wear, rust or corrosion damage. Replace shaft as necessary.

6. Install V-blocks under the drive shaft bearing surfaces at each end of the shaft. Slowly rotate the shaft while watching the crankshaft end. Replace the shaft if any signs of wobble are noted.

7. Repeat Step 6 with the propeller shaft. Also check the shaft surfaces where oil seal lips make contact. Replace the shaft as required.

8. Check the propeller shaft bearing cage or spool and needle bearing for wear or damage.

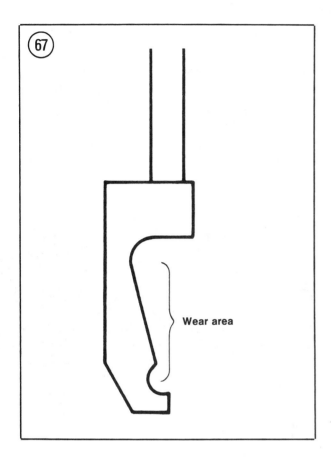

Wear area

Replace bearing as required. If bearing wear is excessive, replace cage or spool.

9. Check bearing cage or spool contact points on the propeller shaft. If shaft shows signs of pitting, grooving, scoring, heat discoloration or embedded metallic particles, replace shaft and bearings.

10. Check water pump as described in this chapter.

11. Check all shift components for wear or damage:

a. Look for excessive wear on the clutch dog and forward/reverse gear engagement surfaces. If clutch dogs are pitted, chipped, broken or excessively worn, replace the gear(s) and clutch dog.

b. Measure the clutch spring free length (**Figure 65**) and roll on a flat surface to check straightness. If spring is not 2 7/16 in. in length or if it does not roll freely, replace the spring.

c. Check clutch for cracks at shear point (**Figure 66**) and rounded areas that contact forward gear clutch dogs. Replace as required.

d. Check shift pin and cam for wear. See **Figure 67** (typical) for wear points. Excessive wear on shift cam crests can let the engine drop out of gear. Replace as required.

12. Clean all roller bearings with solvent and lubricate with Chrysler/U.S. Marine Gear Lube to prevent rusting. Check all bearings for rust, corrosion, flat spots or excessive wear. Replace as required.

13. Check the forward, reverse and pinion gears for wear or damage. If teeth are pitted, chipped, broken or excessively worn, replace the gear. Check pinion gear splines for excessive wear or damage. Replace as required.

14. Check the propeller for nicks, cracks or damaged blades. Minor nicks can be removed with a file, taking care to retain the shape of the propeller. Replace any propeller with bent, cracked or badly chipped blades.

PINION GEAR SHIMMING AND PROPELLER END FLOAT

The pinion gear must be correctly shimmed or the gears will not engage properly. If the clearance is not within specifications, excessive or insufficient, gear and bearing wear will occur, leading to premature failure.

Always check propeller shaft end float once shimming is correct.

55-60 hp (1967-1979)

1. Install thrust washer, reverse gear, reverse gear bearing and propeller shaft bearing on propeller shaft.

2. With installer part No. T 8920 on end of shaft, insert assembly in an arbor press

(**Figure 68**). Make sure installer is centered on the bearing inner race and press components on shaft until the bearing bottoms out.

3. Secure checking gauge part No. T 8924-1 in a vise. Attach tool part No. T 8982 on checking gauge and tighten the 2 socket head screws. Insert tool part No. T 8982A in the bore at the bottom of the checking gauge and install the propeller shaft. See **Figure 69**.

4. Install propeller shaft bearing cage with 2 shim spacers (part No. T 8982B). Tighten socket head screws to 25 in.-lb. while rotating spacers so they will seat evenly.

5. Install a dial indicator as shown in **Figure 69**. Indicator plunger must rest on machined end surface of propeller shaft.

6. Depress propeller shaft to seat the gear against the surface of tool part No. T 8982.

7. Set the dial indicator gauge to zero, then grasp the gear on each side and snap it upward with a quick, even pressure. See **Figure 70**. Do not rotate gear during this step.

8. Note the indicator reading, then allow the gear assembly to drop back down. The indicator should return to a zero reading.

9. Repeat Step 7 and Step 8 to recheck the reading. A reading of 0.004-0.006 in. indicates that the gear is correctly shimmed.

10. To determine the amount of shimming required, subtract 0.005 in. from the dial indicator reading. As an example, suppose the dial indicator reading in Step 7 was 0.012 in. Subtracting 0.005 in. from the reading leaves a difference of 0.007 in.; this is the thickness of the required shim pack.

11. Once the shim pack thickness has been established, remove the dial indicator,

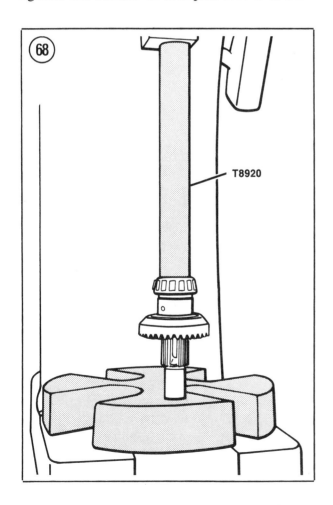

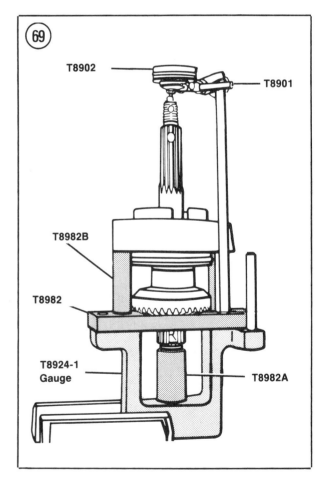

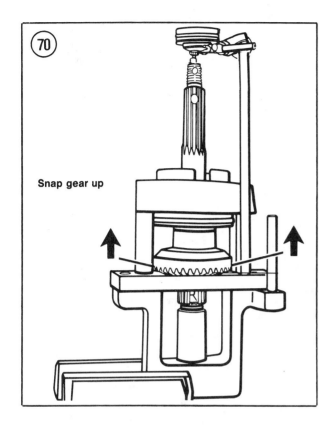

Snap gear up

SHIM(S) HERE

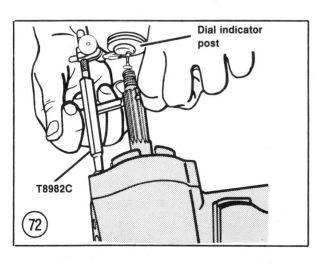

Dial indicator post

T8982C

bearing cage and propeller shaft from the checking gauge.

12. Remove the reverse gear and bearing assembly. Install the required shims between the reverse gear and propeller shaft bearings. See **Figure 71**.

13. Reassemble the propeller shaft. Install propeller shaft, bearing cage and dial indicator to checking gauge.

14. Repeat Steps 7-9 to assure that the shim pack thickness is correct. The dial indicator should read 0.004-0.006 in. If it does not, recheck your calculations in Step 10.

15. Disassemble the pinion gear shimming tools and clamp the lower gearcase in the vise (use protective jaws or wooden blocks) with the prop shaft bore horizontal.

16. Install the clutch dog and shift pin on the propeller shaft. Secure in place with the spring pin.

17. Install a 0.056 in. forward gear thrust washer (part No. 2087) on the front end of the propeller shaft, then install the forward gear. Insert the assembled propeller shaft into the gearcase bore.

18. Install the bearing cage (without O-ring) on the propeller shaft. Secure cage in place with a bearing cage screw in the lower hole and dial indicator post (part No. T 8982C) in the upper hole.

19. Reposition gearcase in vise so propeller shaft is vertical and install the dial indicator (**Figure 72**). Indicator plunger must rest on machined end surface of propeller shaft.

20. Depress and rotate propeller shaft 90° to seat the forward gear bearing in its cup.

21. Set the dial indicator gauge to zero, pull the propeller shaft upward sharply and note indicator reading. Repeat this step several times. The reading should be the same each time. The required end float is 0.004-0.006 in. If the indicator reading in this step falls within this range, end float is correct. If it does not, perform Step 22.

9

22. To determine the required thrust washer thickness, subtract 0.005 in. from the dial indicator reading and add the remainder to the thickness of the thrust washer installed. As an example, suppose the dial indicator reading in Step 21 was 0.011 in. Subtracting 0.005 in. from the reading leaves a difference of 0.006 in. Add this 0.006 in. to the 0.056 in. thickness of the thrust washer originally installed. The sum of these 2 figures (0.062 in.) is the required thrust washer thickness.

23. Remove the dial indicator and bearing cage from the propeller shaft bore.

24. Remove the propeller shaft assembly. Exchange the 0.056 in. thrust washer originally installed with the one determined in Step 22.

25. Reinstall the forward gear and insert the propeller shaft assembly into the gearcase bore. Install bearing cage O-ring and continue gearcase assembly as described in this chapter.

70-85 hp (Two-piece Gearcase)

1. Install the reverse gear bearing with its flange facing away from the gear teeth.

2. Install thrust washer, reverse gear and propeller shaft bearing on propeller shaft with an arbor press. If a press is not available, drive shaft installer (part No. T 8920) and bearing installer (part No. T 8906) can be used. See **Figure 73**.

3. Install the clutch dog and shift pin on the propeller shaft. Secure in place with the spring pin.

4. Install the bearing cup in bearing cage with installer part No. T 8910 and drive handle part No. T 8907.

5. Secure checking gauge part No. T 8924-1 in a vise. Insert the propeller shaft assembly in the gauge.

6. Install the bearing cage with 2 spacers (part No. T 8981C) and tighten cage screws to 25 in.-lb.

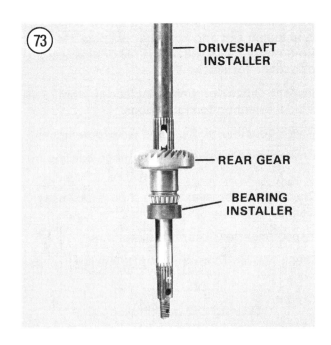

7. Install a dial indicator as shown in **Figure 74**. Indicator plunger must rest on machined surface of reverse gear.

8. Depress reverse gear to assure good contact with dial indicator plunger.

9. Set the dial indicator gauge to zero, then grasp the bearing cage and gear on each side and snap it upward with a quick, even pressure. Do not rotate gear during this step.

10. Note the indicator reading, then allow the bearing cage and gear assembly to drop back down. The indicator should return to a zero reading.

11. Repeat Step 9 and Step 10 to recheck the reading. The clearance required for proper gear mesh is 0.004-0.006 in.; the final indicator reading must fall within this range.

12. To determine the amount of shimming required, subtract 0.005 in. from the dial indicator reading. As an example, suppose the dial indicator reading in Step 9 was 0.024 in. Subtracting 0.005 in. from the reading leaves a difference of 0.019 in.; this is the thickness of the required shim pack.

13. Once the shim pack thickness has been established, remove the dial indicator and bearing cage from the checking gauge.

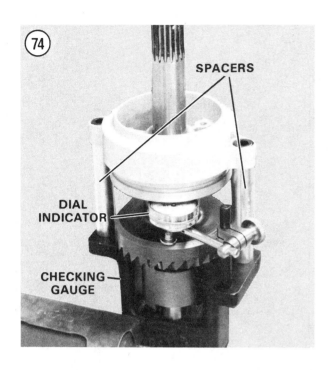

SPACERS

DIAL
INDICATOR

CHECKING
GAUGE

14. Remove the bearing cup from the bearing cage. Install the proper shim pack in the bearing cage, then reinstall the bearing cup.

15. Reinstall the bearing cage and dial indicator on the propeller shaft/checking gauge assembly and repeat Steps 9-11 to make sure the indicator reading is between 0.004-0.006 in. If it is not, recheck your calculations in Step 12.

16. Disassemble the pinion gear shimming tools and clamp the lower gearcase in the vise (use protective jaws or wooden blocks) with the prop shaft bore horizontal.

17. Install the forward gear in the gearcase housing.

18. Install a 0.054 in. thrust washer in the forward gear bore and insert the assembled propeller shaft into the gearcase bore.

19. Install the bearing cage (without O-ring) on the propeller shaft. Secure cage in place with a bearing cage screw in the lower hole and dial indicator post (part No. T 8982C) in the upper hole.

20. Reposition gearcase in vise so propeller shaft is vertical and install the dial indicator.

Indicator plunger must rest on machined end surface of propeller shaft.

21. Depress propeller shaft and rotate it 90°. This will seat the forward gear and remove all bearing clearance.

22. Set the dial indicator to zero and pull upward sharply on the propeller shaft. Note indicator reading. Repeat this step several times. The reading should be the same each time. The required end float is 0.004-0.006 in. If the indicator reading in this step falls within this range, end float is correct. If it does not, perform Step 23.

23. To determine the required thrust washer thickness, subtract 0.005 in. from the dial indicator reading and add the remainder to the thickness of the thrust washer installed. As an example, suppose the dial indicator reading in Step 22 was 0.026 in. Subtracting 0.005 in. from the reading leaves a difference of 0.021 in. Add this 0.021 in. to the 0.054 in. thickness of the thrust washer originally installed. The sum of these 2 figures (0.075 in.) is the required thrust washer thickness.

24. Remove the dial indicator and bearing cage from the propeller shaft bore.

25. Remove the propeller shaft assembly. Exchange the 0.054 in. thrust washer originally installed with the one determined in Step 23.

26. Reinstall the propeller shaft assembly in the gearcase bore. Install the shift yoke and move shift rod up and down to make sure the yoke and shift arm engage properly.

27. Install bearing cage O-ring and continue gearcase assembly as described in this chapter.

65-140 hp (One-piece Gearcase)

1. Insert the master shim from shimming tool set (part No. T 8997) in gearcase counterbore.

2. Install the drive shaft assembly in the gearcase. Install the pinion gear on the drive shaft splines with an old pinion nut.

9

3. Install drive shaft spline adapter (part No. T 7848) over the drive shaft end. Attach a torque wrench and suitable socket to the adapter. Hold pinion nut with a 3/4 in. socket and flex handle. Pad that part of the gearcase where the handle will hit with shop cloths to prevent housing damage. Torque the pinion gear nut to specifications (**Table 1**).

4. Insert shimming plug (part No. T 8997A) in forward gear. Rotate plug until flat on plug faces pinion gear. Install plug cradle between plug and gearcase bottom. See **Figure 75**.

5. Pull up on the drive shaft to provide maximum clearance between the pinion gear face and shimming plug ring.

NOTE
Do not insert feeler gauge blade more than 1/4 in. in Step 6.

6. Insert a 0.035 in. or thicker flat feeler gauge blade between the bottom of the pinion gear and the larger outer diameter of the shimming plug.

7. Holding feeler gauge in place, depress drive shaft sharply and try to remove the gauge. There should be a slight drag when gauge is removed if clearance is correct. If feeler gauge does not come out easily or if it comes out too easily, repeat Step 6 with different gauge thicknesses until the correct one is determined.

8. When the correct feeler gauge thickness is determined, subtract 0.005 in. (final clearance required) from the blade thickness. Subtract this figure from the thickness of the master shim installed in Step 1. The result is the shim pack thickness required to provide the final clearance necessary. As an example, suppose the feeler gauge thickness is 0.029 in. Subtracting the final clearance of 0.005 in. leaves 0.024 in. When 0.024 in. is subtracted from the master shim thickness of 0.050 in., the result is 0.026 in.; this is the required shim pack thickness.

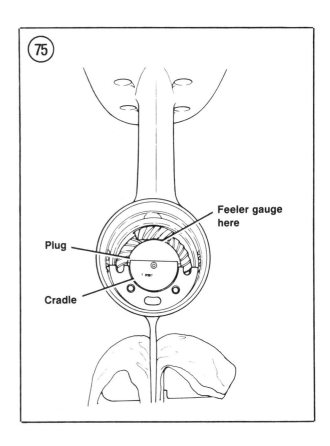

9. After the correct shim pack thickness is determined, remove the shimming plug and cradle from the gearcase bore.

10. Perform Steps 19-25 of *Gearcase Disassembly/Assembly, 65-140 hp (One-piece Gearcase)* in this chapter to remove the pinion gear and drive shaft.

11. Remove the master shim from the gearcase. Install the shim pack determined in Step 8.

12. Repeat Step 2 and Step 3 to reinstall the drive shaft and pinion gear. Install a new crush ring in the drive shaft bearing bore and use a new pinion gear nut.

13. Install the clutch dog and shift pin on the propeller shaft. Secure in place with the spring pin.

14. Install a 0.054 in. thrust washer in the forward gear bore.

15. Lightly grease the bearing cage spacer with Rykon No. 2EP and assembly bearing cage, spacer, bearing and reverse gear as

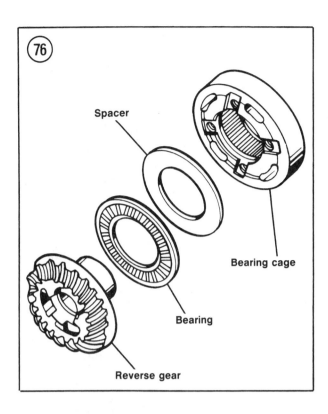

(76)

Spacer

Bearing cage

Bearing

Reverse gear

shown in **Figure 76**. Install reverse gear assembly on propeller shaft.

CAUTION
Make sure the bearing cage spacer does not move out of place in Step 16. If it does, remove the propeller shaft assembly and reposition the spacer to prevent serious gearcase damage.

16. Install the propeller shaft assembly in the gearcase housing. When properly seated, the gearcase retaining ring grooves will be exposed.

17. Install and seat the 2 retaining rings in the gearcase grooves.

18. Install propeller shaft spool (without O-rings) in gearcase. Install and tighten the spool bolts to 160 in.-lb.

19. Install a dial indicator so that the plunger will contact the machined surface on the end of the propeller shaft.

20. Depress and rotate the propeller shaft back and forth to remove all forward gear bearing clearance.

21. Set the dial indicator to zero and pull up sharply on the propeller shaft. Note the indicator reading. Repeat this step several times to make sure the reading is consistent. The required end float is 0.009-0.011 in. If the indicator reading in this step falls within this range, the end float is correct. If it does not, perform Step 22.

22. To determine the required thrust washer thickness, subtract 0.010 in. from the dial indicator reading and add the remainder to the thickness of the thrust washer installed. As an example, suppose the dial indicator reading in Step 21 was 0.029 in. Subtracting 0.010 in. from the reading leaves a difference of 0.019 in. Add this 0.019 in. to the 0.054 in. thickness of the thrust washer originally installed. The sum of these 2 figures (0.073 in.) is the required thrust washer thickness.

23. Remove the dial indicator, propeller shaft spool, retaining rings and propeller shaft assembly from the gearcase bore.

24. Exchange the 0.054 in. thrust washer originally installed with the one determined in Step 22.

25. Reinstall the propeller shaft assembly, retaining rings and propeller shaft spool in the gearcase.

26. Install bearing cage O-ring and continue gearcase assembly as described in this chapter.

9

PRESSURE TESTING

Whenever a gearcase is overhauled, it should be pressure tested after refilling it with lubricant. If the gearcase fails the pressure test, it must be disassembled and the source of the problem located and corrected. Failure to perform a pressure test or ignoring the results and running a gearcase which failed the test will result in major gearcase damage.

1. Remove the vent plug from the gearcase housing.

2. Thread pressure tester adapter (part No. T 8950) into vent hole. Tighten adapter securely.

3. Pump the pressure to 10 psi and note gauge for 5 minutes. If pressure does not hold, submerge the gearcase in water and repressurize to 10 psi. Check for the presence of air bubbles to indicate the source of the leak.

4. If the source of a pressure leak cannot be determined visually, disassemble the gearcase and locate it.

ZINC ANODE

Some gearcases are fitted with a sacrificial zinc alloy anode to reduce galvanic corrosion. On some models, the anode is a circular cover installed over the propeller shaft spool (**Figure 58**). On other gearcases, the exhaust snout (**Figure 77**) is the sacrificial anode.

The anode should not be painted, as this destroys its protective value. Check an anode periodically for erosion. If badly eroded, it should be replaced and the source of the

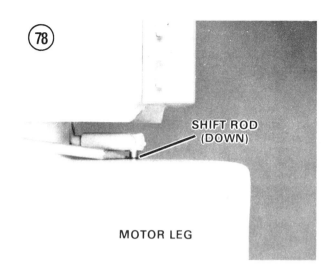

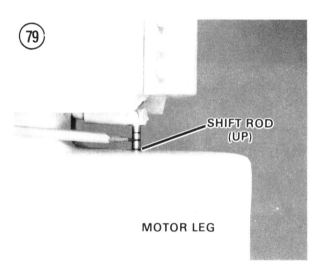

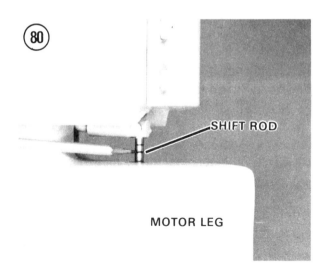

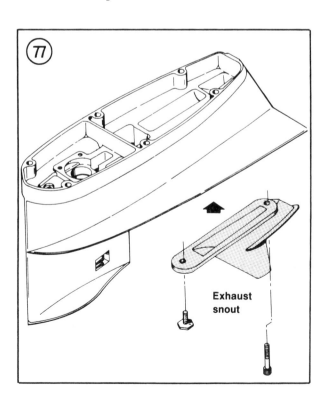

erosion located and corrected if possible. Erosion may be caused by faulty wiring in the boat or in a boat moored nearby, as well as by the water.

GEAR SHIFT LINKAGE ADJUSTMENT

70-140 hp

1. Shift the motor into FORWARD. The lower shift rod should travel downward as far as it will go.
2. Mark the shift rod at the point shown in **Figure 78**.
3. Shift the motor into REVERSE. The lower shift rod should travel upward as far as it will go.

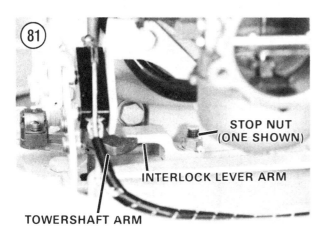

STOP NUT (ONE SHOWN)

INTERLOCK LEVER ARM

TOWERSHAFT ARM

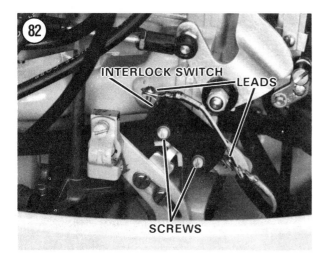

INTERLOCK SWITCH

LEADS

SCREWS

4. Mark the shift rod at the point shown in **Figure 79**.
5. Make a third mark at a point midway between those made in Step 2 and Step 4. See **Figure 80**.
6. Move the shift lever until the mark made in Step 5 is flush with the top of the motor leg. The interlock lever arm connected to the upper shift rod should align with the lower tower shaft as shown in **Figure 81**.
7. If the interlock lever arm and tower shaft do not align properly, loosen the 2 stop nuts on the upper shift rod. Align the interlock lever arm and tower shaft and reposition the stop nuts to hold the mechanism in alignment.
8. Check the neutral interlock switch (**Figure 82**). The button on the switch should be depressed with the shift lever in NEUTRAL. If it is not, reposition the switch.

55-65 hp

1. Remove the starboard support plate.
2. Remove and discard the cotter pin from the gear shift pin. Remove the gear shift pin.
3. Shift the motor into FORWARD. The lower shift rod should travel downward as far as it will go.
4. Loosen the locknut under the shift rod coupler. Adjust coupler until the shift pin hole is 1.38 in. (35 mm) above the steering arm. See **Figure 83**.
5. Reinstall the gear shift pin and secure with a new cotter pin.
6. Shift the motor into NEUTRAL. Check the position of the shift arm. It should be centered on the neutral interlock switch. See **Figure 84**. If it is not, loosen the shift cable connector locknut and adjust the connector as required, then tighten the locknut.
7. Disconnect the shift arm link. Shift the motor into FORWARD. Loosen the locknut on the shift arm link and adjust the link

9

connector until it aligns with the shift arm pin. See **Figure 85**.

20-50 hp

1. Make sure the shift rod coupling is centered on the shift rods. Each rod should thread into the coupling at least 1/4 in. (6.35 mm). If coupling position is not correct, loosen the coupling locknut.

2A. Electric start models—Move the shift lever slowly toward forward gear while rotating the propeller manually until you hear the clutch dogs engaging forward gear. Mark this position on the shift lever opposite the center of the interlock switch button.

2B. Manual start models—Place shift lever roller in NEUTRAL position. Mark the cylinder drain cover at the top of the shift lever apex for reference purposes, then move the shift lever slowly toward forward gear while rotating the propeller manually until you feel the clutch dogs engaging the gear.

3. Repeat Step 2A or Step 2B while shifting toward REVERSE to locate the start of reverse gear engagement.

4. Move the shift lever back to NEUTRAL. If the distance between the FORWARD and REVERSE marks is not equal, adjust the shift rod coupling until forward and reverse gear travel is equal, then tighten the locknut.

WATER PUMP

The outboards covered in this manual use a volume-type water pump on top of the gear housing. The pump body impeller on small gearcases is secured to the drive shaft by a pin that fits into the drive shaft and a similar cutout in the impeller hub. On larger gearcases, a drive shaft key engages a flat on the drive shaft and a cutout in the impeller hub. As the drive shaft rotates, the impeller rotates with it. Water between the impeller

blades and pump housing is pumped up to the power head through the water tube.

The offset center of the pump housing causes the impeller vanes to flex during rotation. At low speeds, the pump acts as a displacement type; at high speeds, water resistance forces the vanes to flex inward and the pump becomes a centrifugal type. See **Figure 86**.

All seals and gaskets should be replaced whenever the water pump is removed. Since

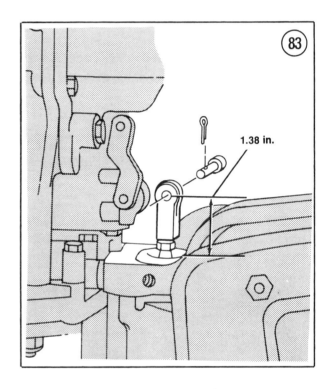

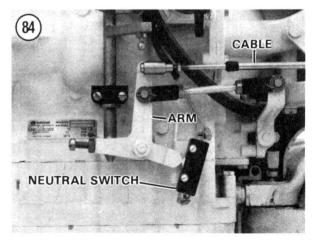

proper water pump operation is critical to outboard operation, it is also a good idea to install a new impeller at the same time.

Do not turn a used impeller over and reuse it. The impeller rotates in a clockwise direction with the drive shaft and the vanes gradually take a "set" in one direction. Turning the impeller over will cause the vanes to move in a direction opposite to that which caused the "set." This will result in premature impeller failure and can damage a power head extensively.

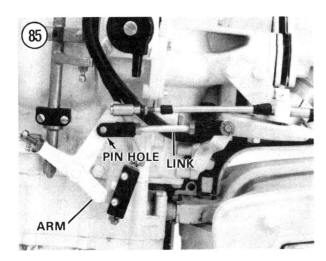

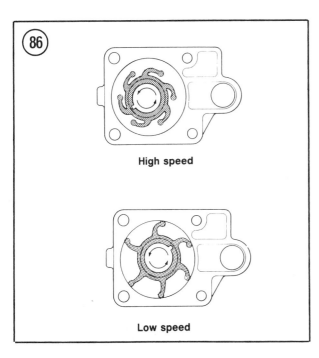

Removal and Disassembly (1970-1977 3.6 hp)

Although this engine is air-cooled, it has a water-cooled motor leg (unlike the air-cooled 3.5 hp).

1. Remove the single nut holding the water pump body to the front gearcase stud.
2. Pry the pump body loose. Slide it up and off the drive shaft. Remove the impeller from the pump body.
3. Remove the impeller drive pin from the drive shaft.
4. Remove the pump back plate from the gearcase.
5. Lubricate the pump body bore with Rykon No. 2EP grease.
6. Reinstall the impeller in the pump body with a counterclockwise rotating motion.
7. Install back plate to base of pump body, aligning back plate hole with pump body locating pin.
8. Insert impeller drive pin in drive shaft slot.
9. Install pump body and base plate on the drive shaft and slide down to drive pin position.
10. Rotate drive shaft until drive pin slips into impeller slot. Seat pump assembly on gearcase.
11. Install gearcase stud nut finger-tight.
12. Insert a 1/4-20 bolt through the rear gearcase mounting hole. Align gearcase and pump body holes.
13. Tighten stud nut to specifications (**Table 1**) and remove alignment bolt.

Removal and Disassembly (1980-on 3.5 hp; 4 hp 1979-on)

Refer to **Figure 87** for this procedure.
1. Secure the gearcase in a holding fixture or a vise with protective jaws. If protective jaws are not available, position the gearcase upright in the vise with the skeg between wooden blocks.

9

2. Remove the spring pin holding the shift coupler to the shift rod. Remove the shift coupler.

3. Remove the water pump plate from the pump body.

4. Pull upward on the drive shaft and remove it from the gearcase with impeller attached.

5. Slide impeller from drive shaft and remove impeller drive pin.

6. Remove the extension stud holding the pump body to the gearcase.

7. Carefully pry pump body free and remove from gearcase.

8. Remove and discard the gearcase O-ring underneath the pump body.

Removal and Disassembly (4.9 and 5 hp)

1. Secure the gearcase in a holding fixture or a vise with protective jaws. If protective jaws are not available, position the gearcase upright in the vise with the skeg between wooden blocks.

2. Remove the screw on each side of the anti-cavitation plate. Separate the gear housing from the plate.

3. Remove the impeller drive pin from the drive shaft. Slide water pump plate from drive shaft.

4. Remove the impeller from the anti-cavitation plate.

5. Pry water tube seal from anti-cavitation plate with a small screwdriver.

Removal and Disassembly (Two-piece Gearcase)

Refer to **Figure 88** for this procedure.

1. Secure the gearcase in a holding fixture or a vise with protective jaws. If protective jaws are not available, position the gearcase upright in the vise with the skeg between wooden blocks.

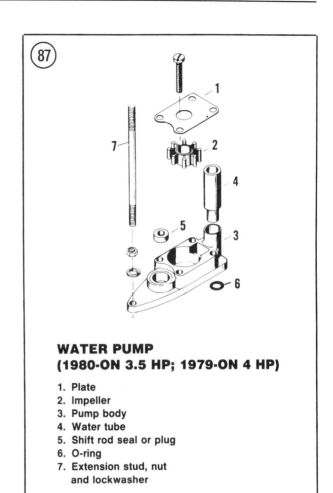

WATER PUMP (1980-ON 3.5 HP; 1979-ON 4 HP)

1. Plate
2. Impeller
3. Pump body
4. Water tube
5. Shift rod seal or plug
6. O-ring
7. Extension stud, nut and lockwasher

2. Remove the water tube from the pump body.

3. Remove the screws holding the pump body to the gearcase. Slide the pump body up and off the drive shaft.

4. Remove the impeller drive key from the drive shaft flat.

5. Carefully pry the back plate and gasket from the gearcase housing. Remove the back plate and discard the gasket.

6. Remove the impeller from the pump body.

Removal and Disassembly (One-piece Gearcase)

Refer to **Figure 89** for this procedure.

1. Secure the gearcase in a holding fixture or a vise with protective jaws. If protective jaws

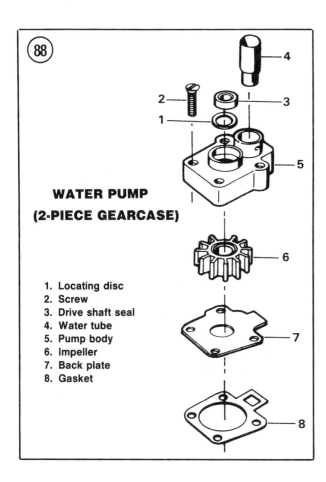

**WATER PUMP
(2-PIECE GEARCASE)**

1. Locating disc
2. Screw
3. Drive shaft seal
4. Water tube
5. Pump body
6. Impeller
7. Back plate
8. Gasket

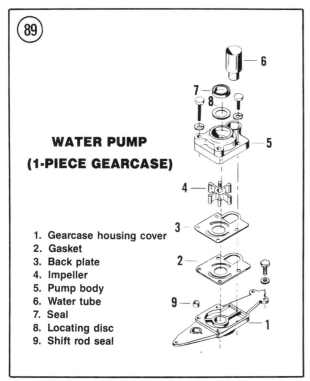

**WATER PUMP
(1-PIECE GEARCASE)**

1. Gearcase housing cover
2. Gasket
3. Back plate
4. Impeller
5. Pump body
6. Water tube
7. Seal
8. Locating disc
9. Shift rod seal

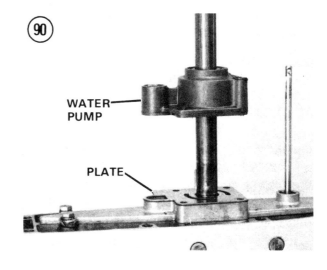

are not available, position the gearcase upright in the vise with the skeg between wooden blocks.

2. Remove the water tube from the pump body.

3. Remove the screws and lockwashers holding the pump body to the gearcase cover. Carefully pry the pump body loose and slide it off the drive shaft (**Figure 90**).

4. Remove the impeller key from the drive shaft flat.

5. Carefully pry water pump plate and gasket (if used) from gearcase cover. Discard the gasket.

NOTE
It is not necessary to remove the gearcase cover for water pump service.

6. Remove impeller from pump body.

Cleaning and Inspection

When removing seals from impeller housing, note and record the direction in which the lip of each seal faces for proper reinstallation.

1. Remove the shift rod seal from the pump body, if so equipped.

9

2. Remove and discard the drive shaft seal and centering disc from the pump body. See **Figure 91** (typical).

3. Check the pump body for cracks, distortion or melting. Replace as required.

4. Clean the pump body and back plate in solvent and blow dry with compressed air, if available.

5. Carefully remove all gasket residue from the mating surfaces.

6. If original impeller is to be reused, check bonding to hub. Check side seal surfaces and vane ends for cracks, tears, wear or a glazed or melted appearance. If any of these defects are noted, do *not* reuse impeller.

Assembly and Installation (1980-on 3.5 hp; 1979-on 4 hp)

1. Install a new shift rod seal in the pump body. The metal capped end should face up, with the top of the seal flush to 0.010 in. below the seal bore chamfer. See **Figure 92**.

2. Install a new O-ring seal in the gearcase groove.

3. Install pump body over shift rod and seat on gearcase.

4. Coat extension stud threads with RTV sealant. Install stud but do not tighten the nut at this time.

5. Insert drive pin in drive shaft slot. Install impeller on drive shaft and slide down over the drive pin.

6. Lightly coat end of drive shaft with Chrysler/U.S. Marine Gear Lube. Install drive shaft end through the drive shaft seal with a rotating motion.

7. Hold impeller in place on drive shaft. Depress and turn drive shaft clockwise to engage the bevel pinion gear splines.

8. Install water pump plate on pump body and tighten screws securely. Tighten extension stud nut at this time.

9. Install shift rod coupler to shift rod with spring pin.

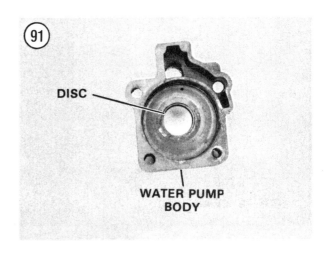

DISC

WATER PUMP BODY

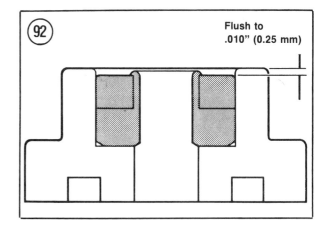

Flush to .010" (0.25 mm)

Assembly and Installation (4.9 and 5 hp)

1. To assemble, coat water tube bore in anti-cavitation plate with a water-resistant adhesive. Install a new seal in the bore until its large flange rests against the plate.

2. Install the water pump impeller in the anti-cavitation plate with slot in hub facing gearcase.

3. Position water pump plate on drive shaft with rounded hole facing pump impeller.

4. Install impeller drive pin in drive shaft hole. Slide anti-cavitation plate down drive shaft and align impeller slot with drive shaft pin.

5. Position rounded end of water pump plate toward the front of the gearcase and install anti-cavitation plate screws.

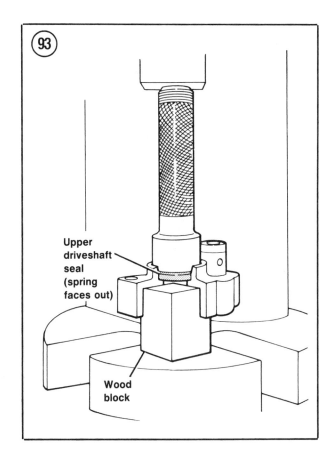

Upper
driveshaft
seal
(spring
faces out)

Wood
block

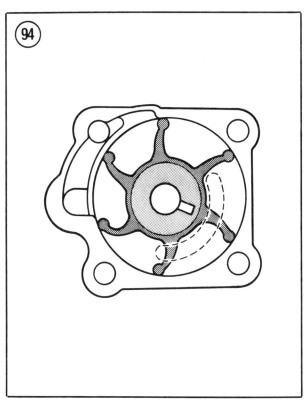

Assembly and Installation
(All Others)

Refer to **Figure 88** (2-piece gearcase) or **Figure 89** (1-piece gearcase) for this procedure.

1. Support the water pump body and insert the centering disc in the pump body seal bore.

2. Lubricate the lips of a new drive shaft seal with Chrysler Gear Lube. Install 2-piece gearcase seal with spring side up. Install 1-piece gearcase seal with its major sealing lip facing up. Drive seal in place with an installer until the seal bottoms out. See **Figure 93**.

> *CAUTION*
> *If the original impeller is to be reused, install it in the same rotational direction as removed to avoid premature failure. The drive pin groove should be visible and the curl of the blades positioned as shown in **Figure 94**.*

3. Install the impeller in the water pump body.

4. Install a new back plate gasket (if used), then install the back plate on the gearcase cover. Make sure the back plate holes align with the gearcase cover holes.

5. Slide the water pump body down the drive shaft. Align the impeller slot with the drive shaft flat. Install impeller key on drive shaft slot and seat pump body.

> *CAUTION*
> *Correct housing fastener torque is important in Step 6. Excessive torque can cause the pump to crack during operation; insufficient torque may result in leakage and exhaust induction which will cause overheating.*

6. Wipe pump body screw threads with RTV sealant. Install screws with lockwashers and tighten to specifications (**Table 1**).

7. Install the water tube in the pump body.

9

Table 1 GEARCASE TIGHTENING TORQUES*

Fastener	in.-lb.
Bearing cage screws	
(4.4-8 hp)	70
Drive shaft nut	85 ft.-lb.
Gearcase cover screws	70
Lower gearcase-to-upper gearcase	260-270
Upper gearcase-to-motor leg	
3.5-4 hp	110
20-85 hp	
1-piece gearcase	160
2-piece gearcase	260-270
Standard torque values	
(screw or nut size)	
6-32	9
8-32	20
10-24	30
10-32	35
12-24	45
1/4-20	70
5/16-18	160
3/8-16	270

* Use standard torque values if specific fastener is not listed.

Chapter Ten

Automatic Rewind Starters

All 3.5-35 hp manual start (and 20-30 hp electric start) models are equipped with a rope-operated rewind starter. On models with an integral fuel tank, the starter assembly is mounted on the power head and attached to the fuel tank. Models without an integral fuel tank may have the starter bracket-mounted to the front of the power head or mounted in a housing above the flywheel.

Pulling the rope handle causes the starter spindle or spool shaft to rotate against spring tension, moving the drive pawl or pinion gear to engage the flywheel and turn the engine over. When the rope handle is released, the spring inside the assembly reverses direction of the spindle or spool shaft and winds the rope around the pulley or spool.

A starter interlock feature is used on all models without an integral fuel tank. This prevents operation of the rewind starter unless the shift lever is in the NEUTRAL position.

Automatic rewind starters are relatively trouble-free, with a broken or frayed rope the most common malfunction. This chapter covers rewind starter and rope/spring service.

ENGINE COVER STARTER

This starter type is used on all models with an integral fuel tank. See **Figure 1** (starter removed).

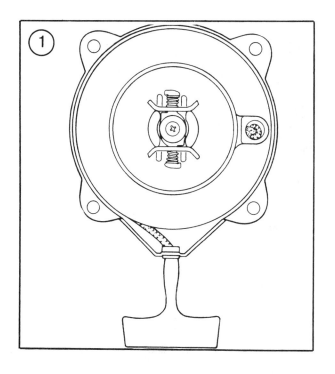

Removal/Installation

1. Disconnect the spark plug lead to prevent the engine from accidentally starting.

2. Pry retainer from starter rope handle. Remove the rope from retainer and pull the rope through the engine cover hole, then tie a large slip knot in the rope to prevent it from winding into the starter.

3. Remove the fuel tank cap. Remove the upper engine cover. Reinstall the cap on the fuel tank.

4. Remove the 4 nuts and lockwashers holding the starter to the fuel tank (**Figure 2**).

5. Installation is the reverse of removal. Pull starter rope to engage friction shoe plates against starter cup before tightening the nuts.

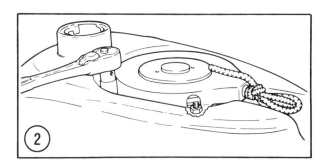

Starter Rope/Rewind
Spring Replacement

WARNING
Disassembling this starter mechanism without holding the spring in place can result in the spring unwinding violently and can cause serious personal injury. Wear safety glasses and gloves during this procedure.

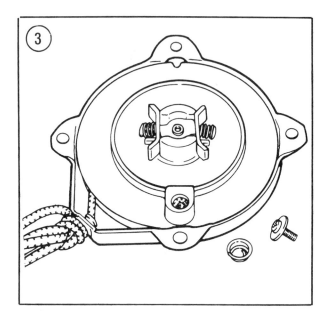

1. Invert the starter cover housing on a flat surface.

2. Remove the screw and washer holding the friction shoe assembly (**Figure 3**).

3. Remove the 2 shoe plates, the brake cover and 2 springs, the conical brake spring, washer and bearing.

4. Untie the slip knot in the rope and let pulley unwind slowly.

5A. If the spring is good but the rope needs replacement, carefully lift one edge of the pulley, insert a thin screwdriver blade under it and disengage the spring loop from the pulley slot. Remove the pulley and rope.

5B. If the spring requires replacement, place starter assembly on the floor (right side up) and gently tap on the top of the cover housing. The spring will drop out and unwind

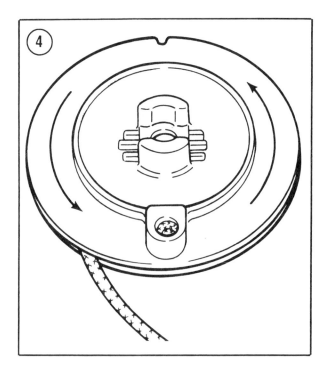

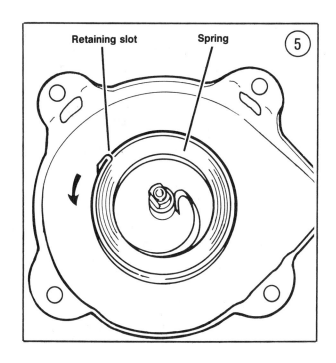

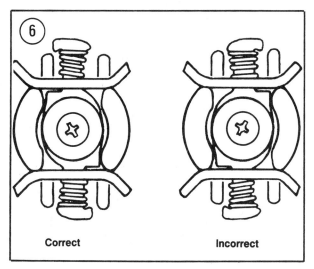

inside the housing. Remove the cover and discard the spring.

6. Check pulley bore bearing, brake spring and friction shoe plates for wear. Replace as required. If shoe plate edges are not sharp, dress them with a file.

7. Tie a knot in one end of the new starter rope. Insert rope through pulley hole, then fold rope end over and pull into pulley recess to lock rope in place. Wind rope counterclockwise (**Figure 4**).

8. If a new spring is being installed, place it in the cover housing. Align outer loop with cover retaining slot (**Figure 5**) and remove the spring retainer carefully to prevent it from flying out of the cover.

9. Lubricate the pulley shaft bore with Rykon No. 2EP grease and carefully insert pulley in cover, engaging looped end of starter spring with pulley slot.

10. Insert loose end of rope in the slot located on the outer edge of the pulley. Turn pulley counterclockwise 3 full turns to tension the spring.

11. Holding pulley to maintain spring tension, insert rope end through cover guide hole and tie a large slip knot to keep the rope in place.

12. Pull rope out as far as possible and hold in that position, then try turning the pulley counterclockwise. If it will not turn at least another 1/8-3/8 of a turn, the spring is bottomed out. In this case, carefully disassemble the starter and begin reassembly again.

13. Insert pulley bearing in bore, then place the washer and brake spring on the pulley shaft. Install the shoe plates, springs and lever on the pulley as shown in **Figure 6** and tighten screw and washer.

14. Install the starter as described in this chapter.

SPOOL STARTER

(4.4-8 hp)

This starter is bracket-mounted to the power head. The starter rope is wound around a spring-loaded spool. A pinion gear at the top of the spool (**Figure 7**) engages the flywheel when the rope is pulled. A neutral interlock shaft and tab at the front of the starter bracket prevent the engine from being started in gear.

Figure 8 shows the starter components.

10

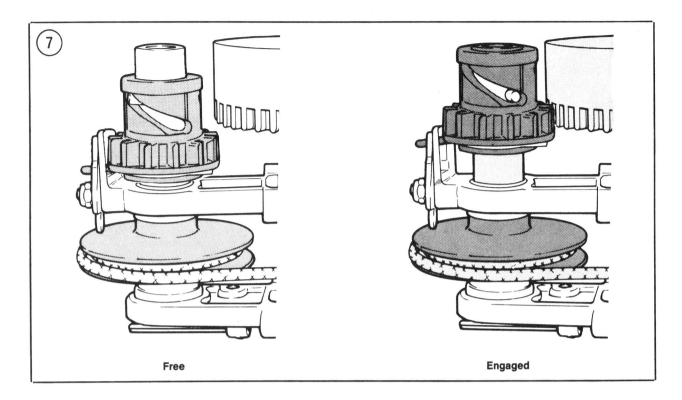

Free   Engaged

Starter Pinion Gear and Rewind Spring Removal/Installation

1. Remove the engine cover.
2. Remove the flywheel. See Chapter Eight.
3. Remove the starter spool cap screw located on top of the starter spring rewind drive in the spool shaft.
4. Install rewind key part No. T 2985 (1968-1978) or part No. T 3139-1 (1979-on) in place of the screw removed in Step 3.
5. Hold the rewind key to prevent the starter spring from unwinding abruptly and remove the pinion gear pin. Use the rewind key to let the rewind spring unwind slowly, then remove the key, pinion spring and pinion gear from the spool shaft. Remove the rewind drive from the key.
6. Remove the rewind spring guide post, spring and retainer from the spool.
7. Lightly lubricate the inner and outer diameter of the rewind spring and the inner

diameter and groove of pinion gear with Rykon No. 2EP grease.
8. Install pinion spring in pinion gear groove.
9. Place pinion gear on spool shaft with spring loop over the starter bracket boss.
10. Fit spring retainer in open end of spring until only the retainer prongs protrude and spring loop is anchored to retainer base.
11. Insert guide post in spring. Install spring drive (tapered end first) until spring loop engages the drive notch. See **Figure 9**.

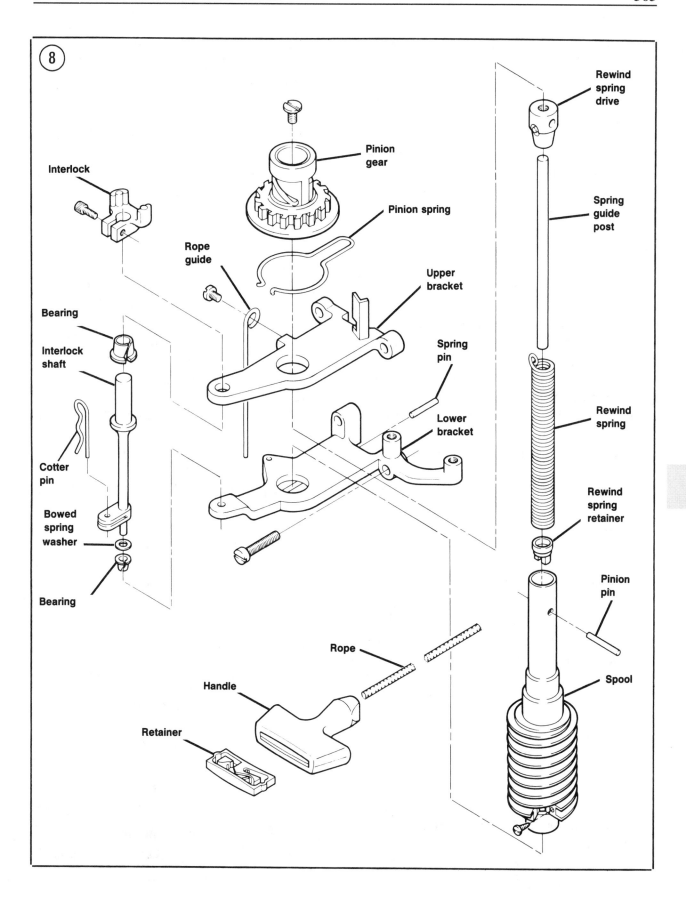

Interlock

Pinion gear

Pinion spring

Rope guide

Bearing

Interlock shaft

Upper bracket

Spring pin

Lower bracket

Cotter pin

Bowed spring washer

Bearing

Rope

Handle

Retainer

Rewind spring drive

Spring guide post

Rewind spring

Rewind spring retainer

Pinion pin

Spool

10

12. Install spring assembly in spool shaft. Align spring retainer notch with pin in lower starter bracket.

13. Install rewind key part No. T 2985 (1968-1978) or part No. T 3139-1 (1979-on) in tapped hole of starter spring arbor.

14. Align pin slot in gear with pin holes in spool shaft and spring arbor. Insert pinion gear pin through gear slot and partially into spool shaft and arbor pin holes. Hold end of pinion pin and tighten rewind key on end of arbor.

15. Remove the pin and turn the rope spool to remove all slack. Turn rewind key counterclockwise until spring tension starts to be felt. At this point, continue rewinding the spring counterclockwise another 3 1/2-4 turns (1968-1978) or 7 1/2-8 turns (1979-on).

16. Align gear slot and pin holes. Insert pin partially as in Step 14. Hold end of pin and remove rewind key. Install pin completely and center in gear to make equal contact on both sides of the gear drive groove.

17. Install cap screw to retain pin.

18. Install the flywheel. See Chapter Eight.

19. Install the engine cover.

Starter Spool and Rope
Removal/Installation

1. Remove the engine cover.

2. Pry the rope handle retainer free and remove the rope from the retainer.

3. Grasp the rope from the inside support plate and let the spring unwind slowly.

4. Remove the pinion gear and rewind spring as described in this chapter.

5. Loosen the screw holding the neutral interlock link. Remove the link.

6. Remove the screws holding the upper starter bracket. Lift the starter spool and upper bracket from the power head.

7. Remove the rope guide from the upper bracket.

8. Unwind the rope from the spool. Remove the screw holding the rope to the spool. Remove the rope.

9. Install the end of a new rope through the spool hole and pull it through until about 3/16 in. of the end is exposed in the spool. Install and tighten the screw.

10. Wind the rope counterclockwise (as seen from the top of the spool) around the spool grooves.

11. Lightly lubricate the base of the spool with Rykon No. 2EP grease. Install spool in base of lower bracket.

12. Route the rope to the starboard side of the interlock shaft and through the support plate grommet.

13. Install the rope in the handle retainer. Install retainer in handle.

14. Reinstall interlock shaft if removed from lower bracket.

15. Lightly lubricate the upper end of starter spool with Rykon No. 2EP grease.

16. Attach rope guide to upper bracket. Fit bracket over starter spool and interlock shaft bearing, guiding end of rope guide into lower bracket hole. Install and tighten bracket screws.

17. Install the neutral interlock link.

18. Install the starter spring and pinion gear as described in this chapter.

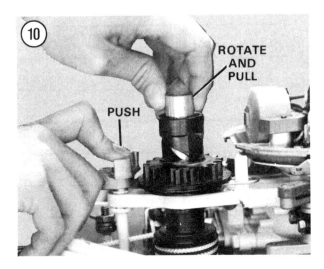

19. Adjust the neutral interlock as described in this chapter.

Neutral Interlock Adjustment

1. Make sure the engine is in NEUTRAL.
2. Depress and hold interlock shaft in lower starter bracket.
3. Rotate starter pinion gear downward as far as possible. Pull up on starter spool shaft to eliminate end play. See **Figure 10**.
4. Loosen interlock lever screw and adjust lever height to provide 0.020 in. clearance between top of pinion gear flange and bottom of lever.
5. Position interlock lever with curved edge parallel to outer diameter of pinion gear flange. See **Figure 11**. Tighten lever screw.
6. Shift engine into FORWARD and then REVERSE gear to check interlock operation. When starter rope is pulled, interlock should hold pinion gear down to prevent engagement with flywheel.
7. Shift engine into NEUTRAL. Interlock should clear pinion gear flange and allow

engagement with flywheel when starter rope is pulled.
8. If interlock does not function as described in Step 6 and Step 7, repeat adjustment procedure.

SPOOL STARTER

(9.2-15 hp)

This starter is bracket-mounted to the power head. The starter rope is wound around a spring-loaded spool. A pinion gear at the top of the spool (**Figure 7**) engages the flywheel when the rope is pulled. A neutral interlock bracket and link connected to the pinion gear prevent the engine from being started in gear.

Figure 12 shows the starter components.

**Starter Pinion Gear and
Rewind Spring
Removal/Installation**

1. Remove the engine cover.
2. Remove the flywheel. See Chapter Eight.
3. Hold a punch against the pinion gear pin and remove the starter spring arbor screw.
4. Push pinion gear pin half-way out of the gear.
5. Install rewind key part No. T 2985 in place of the screw removed in Step 3.
6. Hold the rewind key to prevent the starter spring from unwinding abruptly and remove the pinion gear pin. Use the rewind key to let the rewind spring unwind slowly, then remove the key and starter spring arbor from the spool shaft. Remove the key from the arbor.
7. Remove the pinion gear and slide the pinion spring off the gear.
8. Remove the rewind spring, spring retainer and spring end from the spool.
9. Lightly lubricate the inner and outer diameter of the rewind spring and the inner

10

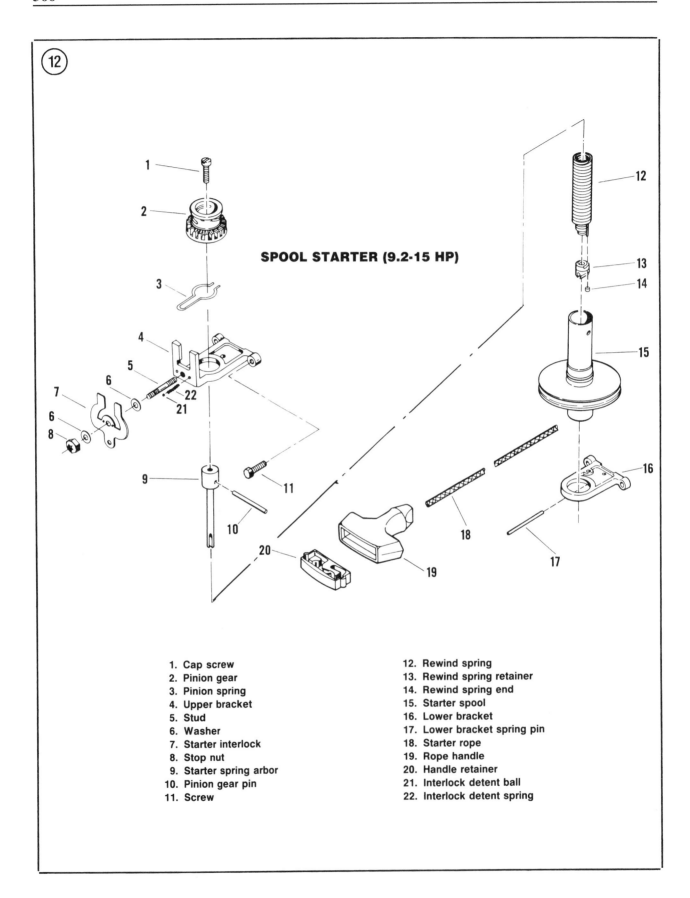

SPOOL STARTER (9.2-15 HP)

1. Cap screw
2. Pinion gear
3. Pinion spring
4. Upper bracket
5. Stud
6. Washer
7. Starter interlock
8. Stop nut
9. Starter spring arbor
10. Pinion gear pin
11. Screw
12. Rewind spring
13. Rewind spring retainer
14. Rewind spring end
15. Starter spool
16. Lower bracket
17. Lower bracket spring pin
18. Starter rope
19. Rope handle
20. Handle retainer
21. Interlock detent ball
22. Interlock detent spring

diameter and groove of pinion gear with Rykon No. 2EP grease.

10. Insert arbor shaft in rewind spring. Shaft groove should catch the spring end.

11. Install arbor/spring assembly in spool shaft. Align spring end notch with pin in lower starter bracket.

12. Install pinion spring in pinion gear groove.

13. Place pinion gear on spool shaft with spring loop protruding through the starter bracket slot.

14. Install rewind key part No. T 2985 in tapped hole of rewind spring arbor.

15. Align pin slot in gear with pin holes in spool shaft and spring arbor. Insert pinion gear pin through gear slot and partially into spool shaft and arbor pin holes. Hold end of pinion pin and tighten rewind key on end of arbor.

16. Remove the pin and turn the rope spool to remove all slack. Turn rewind key counterclockwise until spring tension starts to be felt. At this point, continue rewinding the spring counterclockwise another 3 1/2-4 turns.

17. Align gear slot and pin holes. Insert pin partially as in Step 15. Hold end of pin and remove rewind key. Install pin completely and center in gear to make equal contact on both sides of the gear drive groove.

18. Install cap screw to retain pin.

19. Install the flywheel. See Chapter Eight.

20. Install the engine cover.

Starter Spool and Rope Removal/Installation

1. Remove the engine cover.

2. Pry the rope handle retainer free and remove the rope from the retainer.

3. Remove the pinion gear and rewind spring as described in this chapter.

4. Remove the stop nut, washer and 2 O-rings holding the interlock connecting link to the gear shift shaft lever. See **Figure 13**.

5. Remove the upper bracket screws. Remove the upper bracket and interlock assembly.

6. Hold the bracket end upright and remove the stop nut. Remove the interlock, 2 washers, interlock link, 2 detent balls and 2 springs from the bracket.

7. Remove the starter spool from the power head.

8. Unwind and remove the rope from the spool.

9. Install the end of a new rope through the spool hole and pull it through until about 3/16 in. of the end is exposed in the spool. Install and tighten the screw.

10. Wind the rope counterclockwise (as seen from the top of the spool) around the spool grooves.

11. Lightly lubricate the spool bore with Rykon No. 2EP grease. Install spool in lower bracket.

12. Route the rope around the rope pulley on the manifold and through the support plate grommet.

10

13. Install the rope in the handle retainer. Install retainer in handle.

14. Position upper bracket on spool shaft. Install assembly and tighten screws securely.

15. Reassemble and install interlock assembly to bracket. Connect interlock link to shift lever (**Figure 13**).

16. Install the starter spring and pinion gear as described in this chapter.

17. Check the neutral interlock operation. If the interlock does not prevent starting in gear, disconnect the gear shift lever and shift rod. Rotate shift rod connector on rod as required. Reconnect rod to lever and recheck interlock. Repeat this step until interlock functions correctly.

SPOOL STARTER

(35 hp)

This starter is bracket-mounted to the power head. The starter rope is wound around a spring-loaded spool. A pinion gear at the top of the spool (**Figure 14**) engages the flywheel when the rope is pulled.

A neutral interlock device and interlock rod connected to the gear shift lever prevent the engine from being started in gear. When the engine is in NEUTRAL, the rod engages a slot in the interlock to allow the starter to function. If the engine is in gear, the rod cannot engage the interlock slot.

Figure 15 shows the starter components.

Starter Removal/Installation

1. Remove the engine cover.

2. Remove the flywheel. See Chapter Eight.

3. Remove the starter spool cap screw located on top of the starter spring rewind drive in the spool shaft.

4. Install rewind key part No. T 3139-1 in place of the screw removed in Step 2.

5. Hold the rewind key to prevent the starter spring from unwinding abruptly and remove

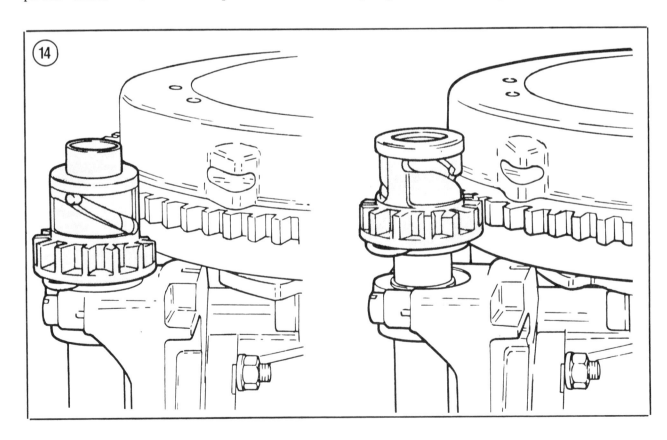

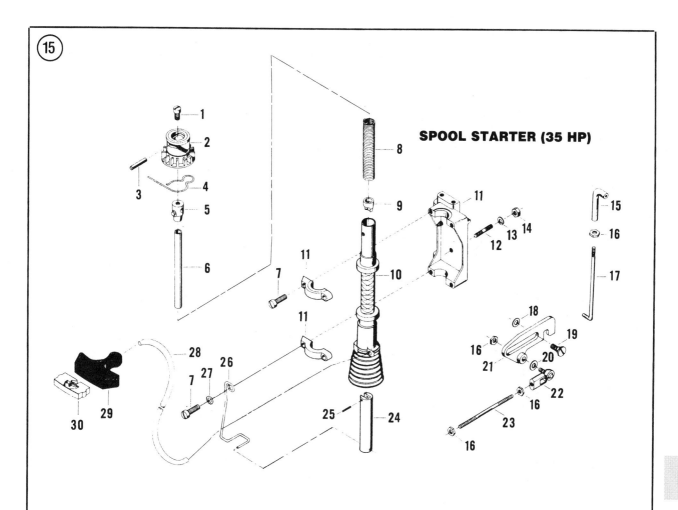

SPOOL STARTER (35 HP)

1. Cap screw
2. Pinion gear
3. Pinion gear pin
4. Pinion spring
5. Rewind spring drive
6. Rope guide post
7. Screw
8. Rewind spring
9. Spring retainer
10. Starter spool
11. Mounting bracket/cap assembly
12. Stud
13. Spring lockwasher
14. Hex nut
15. Spring extension
16. Hex nut
17. Interlock rod
18. Plain washer
19. Interlock screw
20. Internal tooth lockwasher
21. Interlock link
22. Ball joint with bearing
23. Link
24. Retainer extension
25. Roll pin
26. Rope guide
27. External tool lockwasher
28. Starter rope
29. Rope handle
30. Handle retainer

10

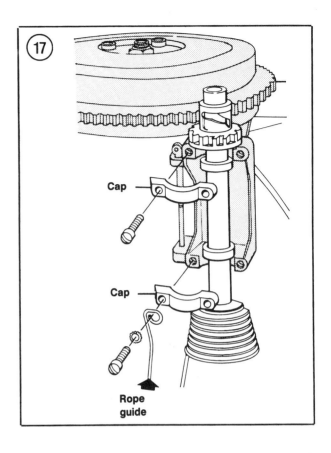

the pinion gear pin. Use the rewind key to let the rewind spring unwind slowly, then remove the key.

6. Grasp the end of the starter pinion spring with needlenose pliers and remove from the base of the pinion gear. See **Figure 16**.

7. Pry the retainer from the starter handle. Remove the rope from the retainer.

8. Remove the tie wrap holding the shorting switch lead to the starter rope guide.

9. Remove the screw and lockwasher holding the rope guide to the racket. Push guide end from retainer extension. See **Figure 17**.

10. Mark bracket caps for reinstallation reference and remove the remaining screws. Remove bracket caps and spool assembly (**Figure 17**).

11. Remove the rewind spring assembly from the spool. If spring replacement is required, refer to **Figure 18**:

    a. Remove spring pin holding extension to retainer.

    b. Remove retainer from spring.

    c. Remove guide post and spring drive from spring.

12. Unwind and remove rope from spool.

13. Insert the free end of a new rope through the spool bore and pull rope through spool until it seats on the spool. Wind the rope onto the spool counterclockwise. See **Figure 19**.

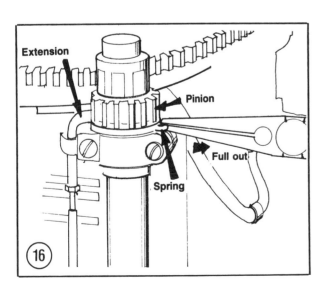

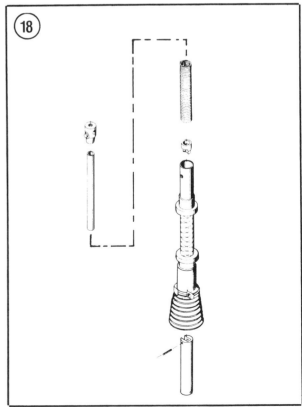

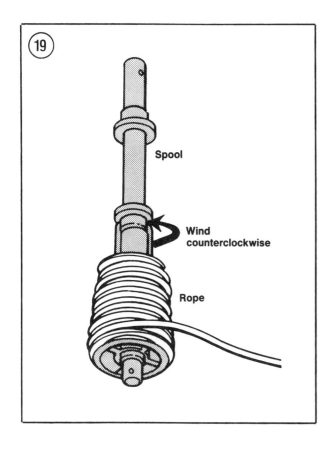

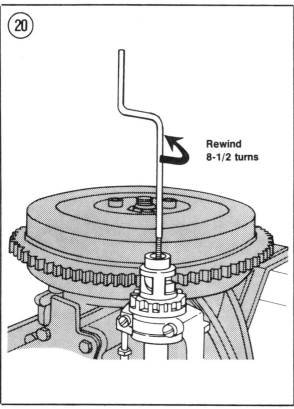

14. If rewind spring assembly was disassembled, install guide post and spring drive to rewind spring, then attach extension to retainer with spring pin. See **Figure 18**.

15. Install rewind spring assembly in starter spool.

NOTE
There should be at least 0.06 in. clearance between the rope guide and spool edge when installed in Step 16. Carefully bend guide as required to establish this clearance.

16. Position spool assembly on bracket. Install bracket caps and 3 screws. Insert rope guide into extension and install remaining bracket screw. See **Figure 17**.

17. Position shorting switch wire to guide and install a tie wrap to hold it in place.

18. Pull the rope through the support plate grommet.

19. Install the rope in the handle retainer. Install retainer in handle.

20. Lightly lubricate the inner and outer diameter of the rewind spring and the inner diameter and groove of pinion gear with Rykon No. 2EP grease.

21. Install pinion gear on spool shaft.

22. Install pinion spring in pinion gear groove and push into place with end in starter spring extension.

23. Install rewind key part No. T 2985 in tapped hole of rewind spring drive.

24. Align pin slot in gear with pin holes in spool shaft and spring drive. Insert pinion gear pin through gear slot and partially into spool shaft and drive pin holes. Hold end of pinion pin and tighten rewind key on end of arbor.

25. Remove the pin and turn the rope spool to remove all slack. Turn rewind key counterclockwise until spring tension starts to be felt. At this point, continue rewinding the spring counterclockwise another 8-8 1/2 turns. See **Figure 20**.

10

26. Align gear slot and pin holes. Insert pin partially as in Step 24. Hold end of pin and remove rewind key. Install pin completely and center in gear to make equal contact on both sides of the gear drive groove.

27. Install cap screw to retain pin.

28. Install the flywheel. See Chapter Eight.

29. Check the neutral interlock operation. If the interlock does not prevent starting in gear, adjust the interlock device and interlock rod (**Figure 15**) as required.

30. Install the engine cover.

FLYWHEEL MOUNTED STARTER

This starter is mounted in a housing on top of the power head. The starter rope is wound around a spring-loaded pulley. A pawl plate attached to the pulley (**Figure 21**) engages the flywheel when the rope is pulled. A neutral interlock lever at the front of the starter housing prevents the engine from being started in gear.

Figure 22 shows the starter components.

Removal/Installation

1. Remove the engine cover.

2. Disconnect the spark plug wires to prevent accidental starting of the engine.

3. Electric start—Remove ignition component bracket.

4. Remove 3 screws and lockwashers holding starter housing to power head (**Figure 23**).

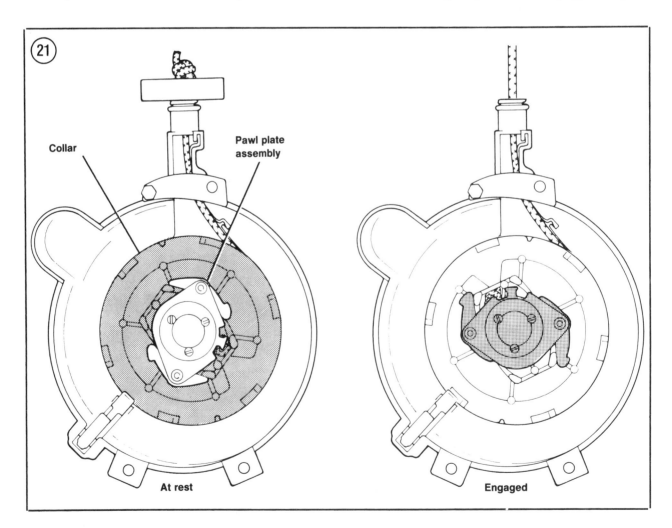

Collar

Pawl plate assembly

At rest

Engaged

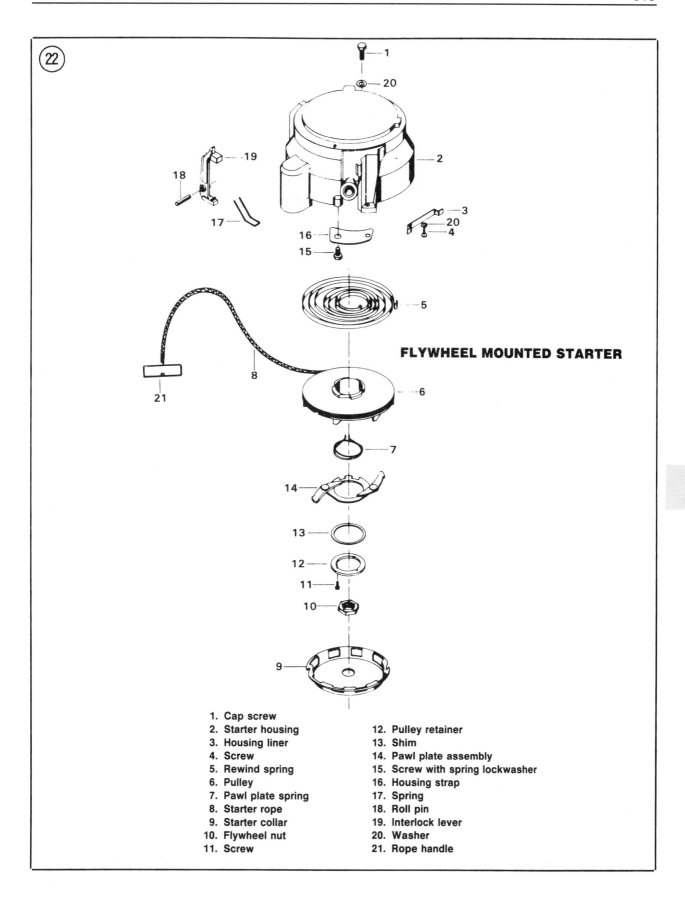

FLYWHEEL MOUNTED STARTER

1. Cap screw
2. Starter housing
3. Housing liner
4. Screw
5. Rewind spring
6. Pulley
7. Pawl plate spring
8. Starter rope
9. Starter collar
10. Flywheel nut
11. Screw
12. Pulley retainer
13. Shim
14. Pawl plate assembly
15. Screw with spring lockwasher
16. Housing strap
17. Spring
18. Roll pin
19. Interlock lever
20. Washer
21. Rope handle

10

5. Lift starter housing and tilt to the starboard side until interlock lever clears the flywheel, then remove the housing from the power head.

6. Installation is the reverse of removal. Pull starter rope to engage the pawl assembly with the collar before tightening the screws.

Starter Rope Replacement

WARNING
Disassembling this starter mechanism without holding the spring in place can result in the spring unwinding violently and can cause serious personal injury. Wear safety glasses and gloves during this procedure.

1. Remove the 3 screws and retainer holding the pawl plate to the starter housing. See **Figure 24**.

2. Remove the pawl plate, spring and shims (if used).

3. Hold the pulley to prevent it from unwinding and remove the starter rope handle.

4. Depress the interlock lever to move it out of the way and let the rewind spring slowly wind up the starter rope.

5. Rotate the pulley in each direction to make sure that all spring tension is gone.

6. With interlock lever depressed, carefully lift pulley and slide a thin blade screwdriver under it to disengage the spring, then remove the pulley and rope assembly from the housing.

7. Grasp the rope knot with needlenose pliers and remove the rope from the pulley.

8. Tie a knot in one end of a new rope and melt the end of the rope with a match to fuse the strands. Insert the other end of the rope through the hole between the pulley flanges.

9. Wrap the rope around the pulley in a counterclockwise direction. Route the rope through the pulley slot. Leave about 9 inches protruding from the slot.

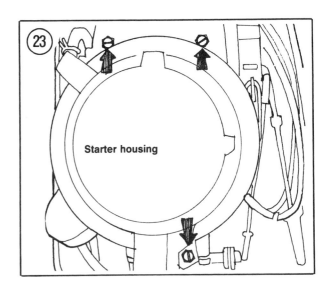

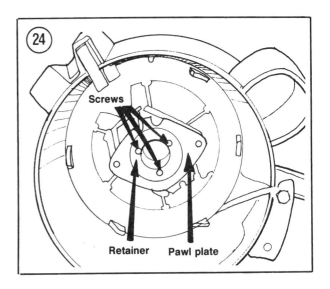

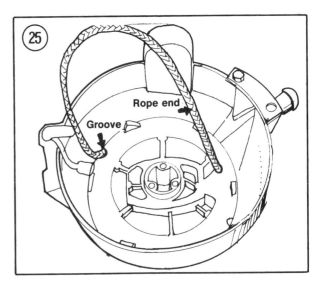

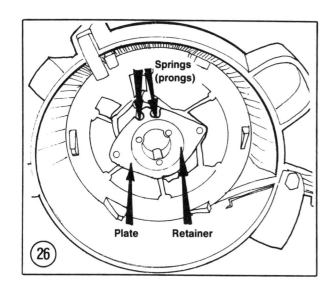

26

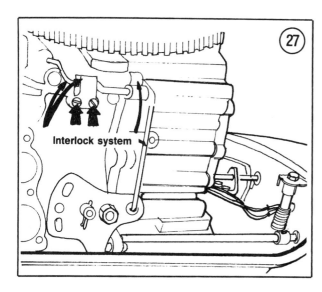

27

NOTE
Shims are used to eliminate excessive up-and-down play in starter assembly. If play is present but no shims were removed during disassembly, obtain the proper shim(s) from your Chrysler/U.S. Marine dealer before reassembling the housing.

14. Reinstall the shims (if used) and install the retainer. Tighten the 3 screws securely.

15. Install the starter assembly as described in this chapter.

16. Check the neutral interlock operation. If the interlock does not prevent starting in gear, adjust the interlock device and interlock rod (**Figure 27**) as required.

Rewind Spring Replacement

WARNING
Disassembling this starter mechanism without holding the spring in place can result in the spring unwinding violently and can cause serious personal injury. Wear safety glasses and gloves during this procedure.

1. Remove the starter pulley and rope as described in this chapter.

2. Place starter assembly on the floor (right side up) and gently tap on the top of the cover housing. The spring will drop out and unwind inside the housing. Remove the cover and discard the spring.

3. Install new spring (with retainer) in starter housing. Outer spring loop should fit over spring retaining boss. When properly installed, the spring will run counterclockwise from the retaining boss.

4. Install the starter pulley and rope as described in this chapter.

5. Check the neutral interlock operation. If the interlock does not prevent starting in gear, adjust the interlock device and interlock rod (**Figure 27**) as required.

10. Install the pulley in the housing with the rope groove positioned at about the 9 o'clock position and the rope outlet boss at about the 12 o'clock position. See **Figure 25**.

11. Rotate the pulley 1 3/4 turns counterclockwise. Hold pulley in that position and install rope handle.

12. Position the pawl plate spring over the housing with its ends facing upward and pointing towards the gap between the pawl guides.

13. Install the pawl plate. Pawl lever side should face pulley with slots over the ends of the pawl plate spring. See **Figure 26**.

10

6. Install the engine cover.

Interlock Lever Replacement

If the interlock lever or spring requires replacement, remove the starter housing as described in this chapter. Invert housing on a solid surface and drive out the roll pin holding the interlock lever with a suitable pin punch. Remove the old interlock lever and spring. Insert a new spring or lever as required in the housing cutout. Secure interlock lever in place with a new roll pin.

Chapter Eleven

Power Trim and Tilt and Remote Control Systems

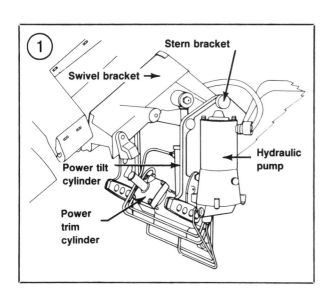

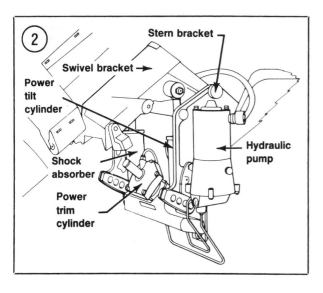

POWER TRIM AND TILT

The usual method of raising and lowering the outboard gearcase is a mechanical one, consisting of a series of holes in the transom mounting bracket. To trim the engine, an adjustment stud is removed from the bracket, the outboard is repositioned and the stud reinserted in the proper set of holes to retain the unit in place.

A power trim and tilt system is available as standard equipment on 3- and 4-cylinder models. It is also offered as an option on some 2-cylinder models. The use of power trim and tilt provides low-effort control whether the boat is underway or at rest.

This chapter includes maintenance, troubleshooting procedures and pump and trim/tilt cylinder replacement.

Components

The power trim and tilt system consists of a hydraulic pump (containing an electric motor, oil reservoir, oil pump and valve body), a hydraulic trim cylinder, a hydraulic tilt cylinder, shock absorber (4-cylinder only), a control switch, sender switch and the necessary hydraulic and electrical lines. See **Figure 1** (3-cylinder) or **Figure 2** (4-cylinder) for major components.

Operation

Moving the trim switch to LIFT closes the pump motor circuit. The motor drives the oil pump, forcing oil into the base of the trim and tilt cylinders. The larger trim cylinder diameter produces more force and the trim cylinder extends first. Once it has fully extended, the tilt cylinder starts its upward travel. When it reaches its highest position, the drive motor stalls. If the trim switch is not released at this time, a thermal relay opens to shut off power to the motor. **Figure 3** shows the operational sequence.

Moving the trim switch to DOWN also closes the pump motor circuit. The reversible motor runs in the opposite direction, driving the oil pump to force oil to the rod end of the trim and tilt cylinders. The weight of the outboard causes the tilt cylinder to retract first. As the outboard lowers, it contacts the trim cylinder rod and both cylinders continue retracting until they are fully collapsed. If the switch is not released when the engine reaches the limit of its downward travel, the pump motor will continue running and cause the oil to foam. **Figure 4** shows the operational sequence.

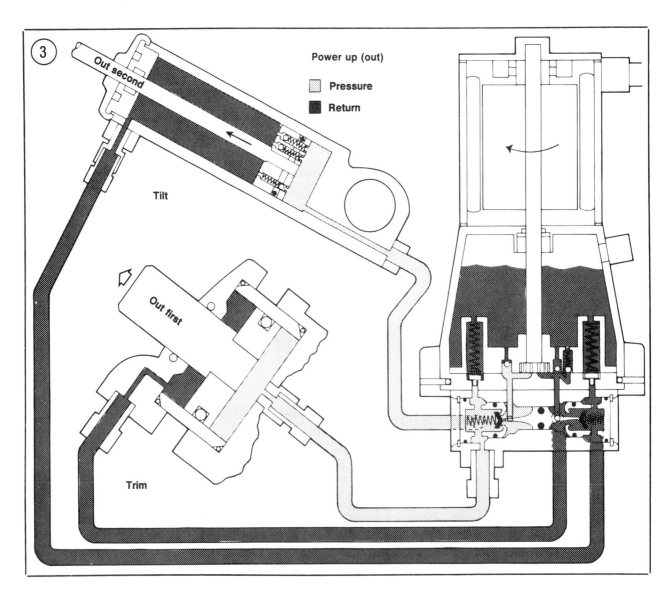

The power tilt will temporarily maintain the engine at any angle within its range to allow shallow water operation at slow speed, launching, beaching or trailering.

To prevent damage from striking an underwater object, spring-loaded check valves in the tilt cylinder and hydraulic pump allow the outboard to pivot upward quickly and return slowly.

Hydraulic Pump
Fluid Check

The outboard should be fully retracted for this procedure.

1. Clean area around pump fill plug. See **Figure 5**. Remove the plug and visually check the fluid level in the pump reservoir. It should be at the bottom of the fill hole threads.

CAUTION
Do not overfill reservoir in Step 2. Excessive oil will be forced into the motor and cause a premature failure.

2. Top up if necessary with a non-detergent SAE 30 engine oil.

3. Cycle the outboard up and down 3-5 times to bleed out any air in the hydraulic lines.

4. Install and tighten fill plug.

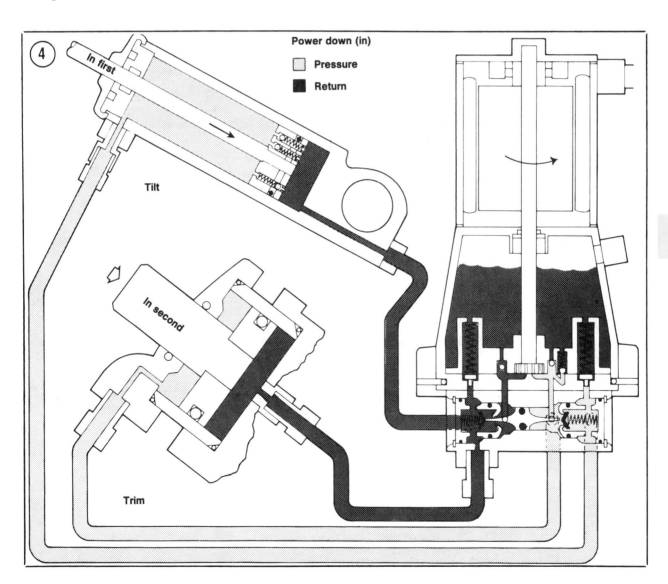

11

Troubleshooting

Whenever a problem develops in the power trim system, the initial step is to determine whether the problem is in the electrical or hydraulic system. If the problem appears to be in the hydraulic system, refer it to a dealer or qualified specialist for necessary service.

Use the following procedure to determine whether the problem is in the electric or hydraulic system.

1. Make sure the switch and pump cable plug-in connectors (**Figure 6**) are properly engaged.

2. Make sure the red power trim switch lead is connected to the positive battery terminal and the black pump lead to the negative battery terminal.

3. Check all terminals and wires for signs of corrosion or loose connections. Clean and tighten as required.

4. Make sure the battery is fully charged. Charge or replace as required.

5. Check the fluid level as described in this chapter. Top off if necessary.

Engine will not remain in trim or tilt position

1. Check the hydraulic pump oil level as described in this chapter. Top up as required.

2. Disconnect the lower hydraulic trim cylinder line at the pump. See **Figure 7**.

3. Connect a 0-2,000 psi pressure gauge to the trim cylinder fitting. Cap or plug the disconnected line. See **Figure 8**.

4. Operate the trim switch and lift the engine with the tilt cylinder. The pressure gauge should read over 1,000 psi:

   a. If the pressure remains over 1,000 psi, continue testing.

   b. If the pressure starts to decrease, there is a leak either in the tilt cylinder or the valve body/pump assembly.

5. Remove the pressure gauge and reconnect the trim cylinder line at the pump.

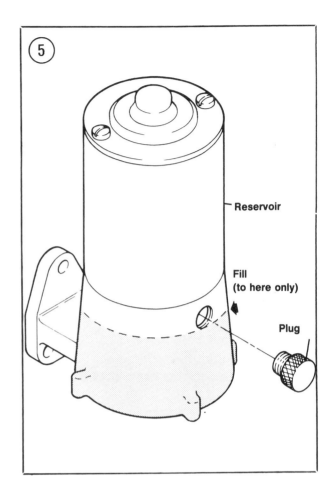

Reservoir

Fill
(to here only)

Plug

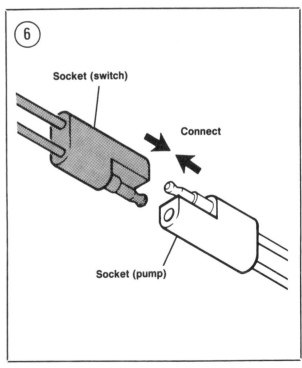

Socket (switch)

Connect

Socket (pump)

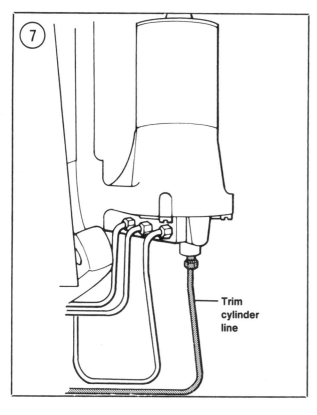

Trim
cylinder
line

6. If leakage is noted in Step 4, operate the tilt switch and raise the engine to the full tilt position. Let the motor stall out, then tap the switch downward once to relieve pump pressure. Remove the tilt cylinder for overhaul.

7. Recheck the pump reservoir oil level as described in this chapter. Top up if necessary.

8. Disconnect the lower hydraulic tilt cylinder line at the pump. See **Figure 9**.

9. Connect a 0-2,000 psi pressure gauge to the tilt cylinder fitting. Cap or plug the disconnected line. See **Figure 10**.

10. Operate the trim switch and lift the engine with the trim cylinder. The pressure gauge should indicate pressure:

    a. If the pressure gauge continues to indicate pressure, continue testing.

    b. If the pressure starts to decrease, there is a leak in the trim cylinder or the valve body/pump assembly.

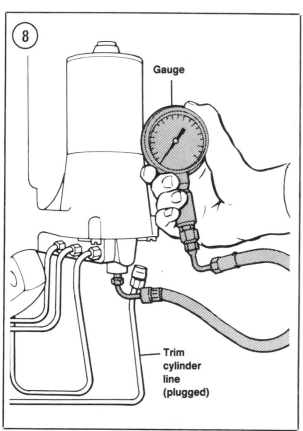

Gauge

Trim
cylinder
line
(plugged)

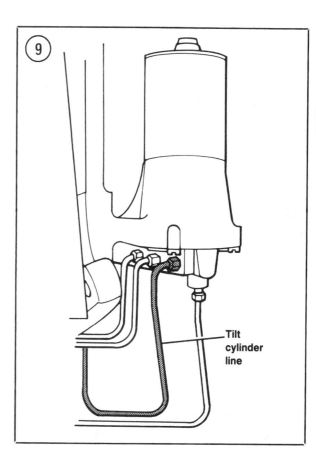

Tilt
cylinder
line

11

11. Remove the pressure gauge and reconnect the tilt cylinder line at the pump.

12. If leakage is noted in Step 10, operate the tilt switch and raise the engine to the full tilt position. Let the motor stall out, then tap the switch downward once to relieve pump pressure. Remove the trim cylinder for overhaul.

13. Repeat Step 7. If the unit still does not function properly, remove the pump assembly for overhaul.

Pump motor does not run

1. Check the battery. Recharge or replace as required.

2. Remove the trim switch and check for continuity with a self-powered test lamp. There should be continuity between the center and top terminals with the switch in the UP position and continuity between the center and bottom terminals with the switch in the DOWN position. Replace the switch as required.

3. Disconnect the wiring harness and check for continuity with a self-powered test lamp. Repair or replace as required.

4. Remove the pump assembly for further testing by a dealer or qualified marine specialist.

Engine tilts slowly or pump motor will not shut off when full tilt position is reached

1. Clean area around pump fill plug. See **Figure 5**. Remove plug from pump.

2. Cycle the outboard up and down 3-5 times to bleed out any air in the hydraulic lines.

3. Install and tighten fill plug.

Pump Removal/Installation

1. Disconnect the battery cable and black pump lead at the negative battery terminal.

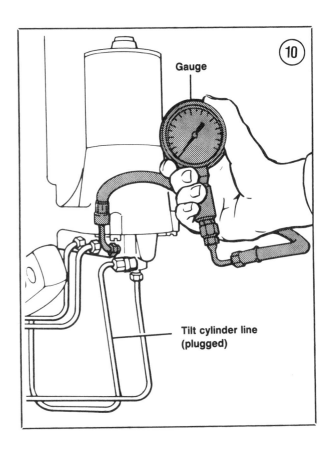

Gauge

Tilt cylinder line (plugged)

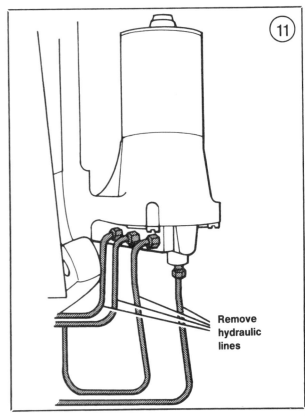

Remove hydraulic lines

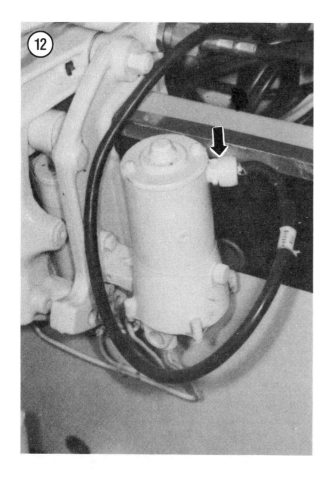

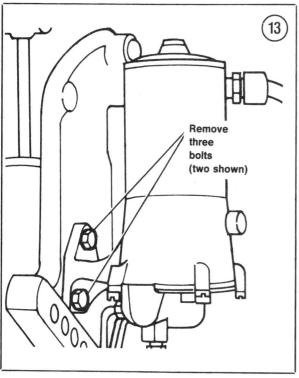

Remove
three
bolts
(two shown)

2. Disconnect the battery cable and red trim switch lead at the positive battery terminal.

3. Disconnect the trim switch and pump sockets in the wiring harness. See **Figure 6**.

4. Place shop rags underneath the pump hydraulic line fittings.

5. Disconnect all 4 hydraulic lines at the pump. See **Figure 11**.

6. Disconnect the hose fitting at the top of the pump. See **Figure 12**.

7. Remove the 3 bolts holding the pump to the stern bracket. See **Figure 13**. Remove the pump.

8. Installation is the reverse of removal. Fill the pump reservoir with non-detergent SAE 30 engine oil as described in this chapter.

**Trim and Tilt Cylinder
Housing Assembly
Removal/Installation**

1. Raise the outboard until the upper shaft clears the stern brackets. See **Figure 14** (pump shown removed).

2. Suspend a rope around the gearcase and attach it to a hoist or other solid object to provide support.

3. Hold the lower shaft stop nut on one side with a suitable wrench and loosen the stop nut on the other side of the shaft. Remove both stop nuts.

11

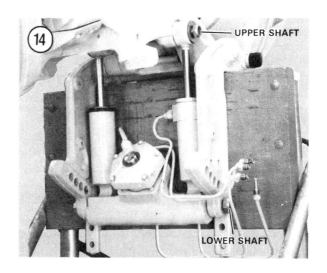

UPPER SHAFT

LOWER SHAFT

4. Repeat Step 3 to remove the upper shaft stop nuts and washers.

5. Place a jack under the cylinder housing assembly or have an assistant help support it while you tap the upper and lower shafts out.

6. Remove the cylinder housing assembly from the stern brackets.

> *NOTE*
> *The tilt cylinder and shock absorber can be removed from the housing assembly. The trim cylinder is an integral part of the housing base.*

7. Disconnect the hydraulic lines at the tilt cylinder. Remove the tilt cylinder and shock absorber from the cylinder housing assembly.

8. Installation is the reverse of removal.

Sender Switch Adjustment

Refer to **Figure 15** for this procedure.

1. Remove the sender unit from the stern bracket.

2. Raise and support the engine at the No. 4 lock bar position.

3. Adjust the sender potentiometer until the dash gauge indicates the No. 4 position.

4. Reinstall the sender unit on the stern bracket.

5. Check the sender unit arm for free operation.

REMOTE CONTROL

Remote control is standard on 3- and 4-cylinder engines and larger 2-cylinder models. It is optional on most other 2-cylinder engines of 6 hp and above.

Removal/Installation

1. Disconnect the negative battery cable.

2. Remove the 3 screws holding the remote control unit. Remove the unit.

3. From the back of the unit, remove the cotter pin holding the throttle cable terminal.

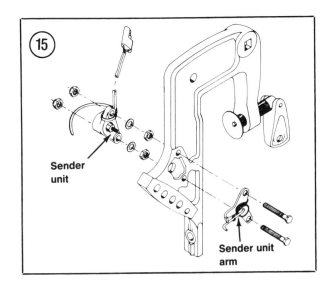

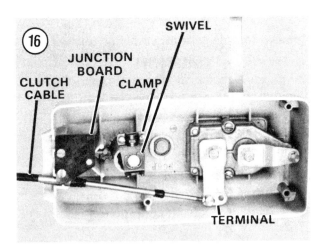

4. Remove the 2 screws holding the throttle cable clamp to the swivel. Disconnect the cable.

5. Remove the clutch cable terminal cotter pin.

6. Remove the 2 screws holding the junction board to the remote control unit housing. See **Figure 16**.

7. Disconnect the shift cable.

8. Installation is the reverse of removal.

Disassembly

Refer to **Figure 17** for this procedure.

1. Remove the hex bolt and lockwasher holding the throttle arm to the control unit housing.

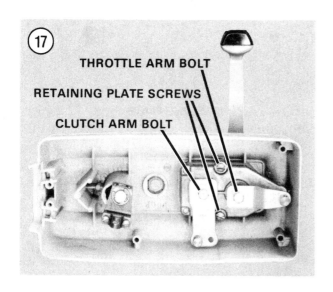

THROTTLE ARM BOLT

RETAINING PLATE SCREWS

CLUTCH ARM BOLT

2. Remove the hex bolt and lockwasher holding the clutch arm to the control unit housing.

3. Remove the 2 screws holding the retaining plate to the housing.

WARNING
Detent ball and spring are under tension. Cover the throttle gear assembly with shop cloths to prevent them from flying out unexpectedly and possibly causing serious personal injury.

4. Cover the throttle gear assembly with shop cloths and remove the setscrew holding the remote control handle to the throttle gear shaft.

WARNING
Detent ball and spring are under tension. Cover the clutch gear assembly with shop cloths to prevent them from flying out unexpectedly and possibly causing serious personal injury.

5. Remove the 2 bearings, bowed washer, throttle gear, detent ball and spring from the housing.

6. Pull clutch bearing, gear and shaft from housing, then remove the detent ball and spring.

7. If handle requires removal, remove the collar, felt washer and indicator tape. Carefully pry cap from top of handle.

8. Remove snap ring holding swivel to housing.

Cleaning and Inspection

1. Wash all metal parts in solvent and blow dry with compressed air.

2. Check clutch and throttle gears and shaft for wear or damage.

3. Check detent springs.

4. Check throttle and clutch arms for wear or damage.

5. Discard the felt washer.

6. Check plastic parts and housing for chips or cracks.

Assembly

1. Fit cap on handle, then place collar with indicator and a new felt washer on the handle.

2. Insert handle assembly in housing.

3. Drop one bearing in the housing throttle gear shaft bore.

4. Install the throttle gear, bowed washer and the second bearing in the housing. See **Figure 18**.

5. Install and tighten handle setscrew to hold handle to gear assembly.

6. Install clutch shaft, gear and bearing in housing. See **Figure 19**.

7. Install clutch detent ball and spring (smaller one) in slot (**Figure 19**).

8. Install throttle detent ball and spring in slot.

9. Install retaining plate over throttle and clutch gear assembly. See **Figure 20**.

10. Install swivel and fasten with snap ring.

11. Install clutch and throttle arms (**Figure 21**).

11

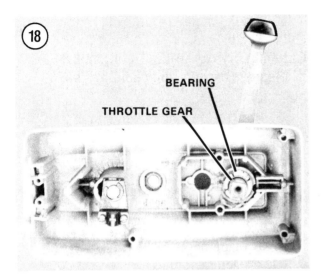

(18) BEARING
THROTTLE GEAR

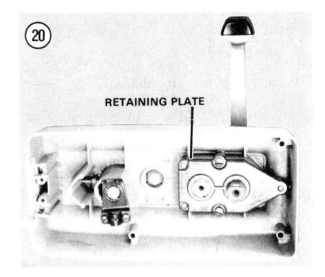

(20) RETAINING PLATE

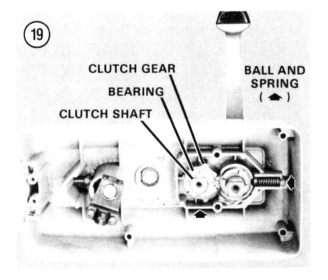

(19) CLUTCH GEAR
BEARING
CLUTCH SHAFT
BALL AND SPRING (➤)

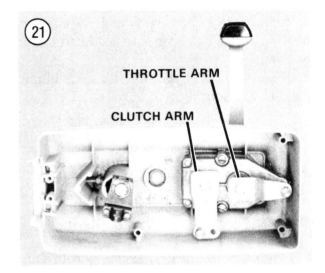

(21) THROTTLE ARM
CLUTCH ARM

Index

12

12

12

Wiring Diagrams

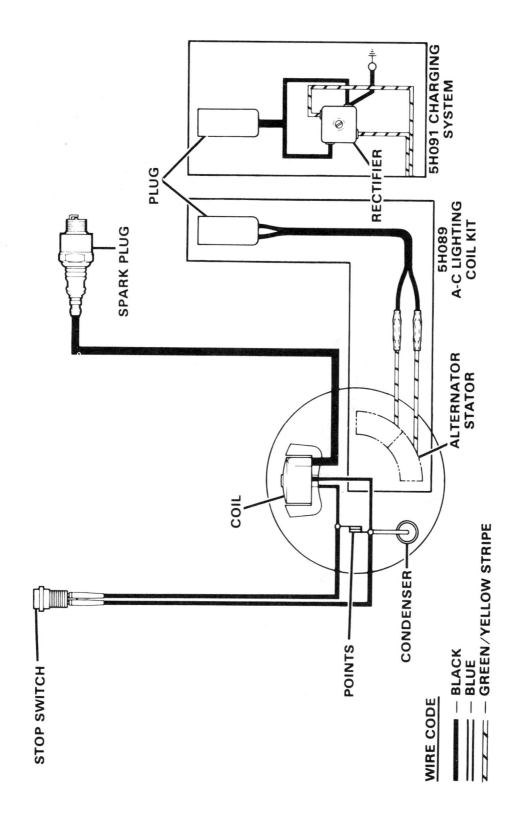

3.5-4 HP 1980-ON

5H091 CHARGING SYSTEM

RECTIFIER

5H089 A-C LIGHTING COIL KIT

PLUG

SPARK PLUG

ALTERNATOR STATOR

COIL

POINTS

CONDENSER

STOP SWITCH

WIRE CODE
— BLACK
— BLUE
— GREEN/YELLOW STRIPE

13

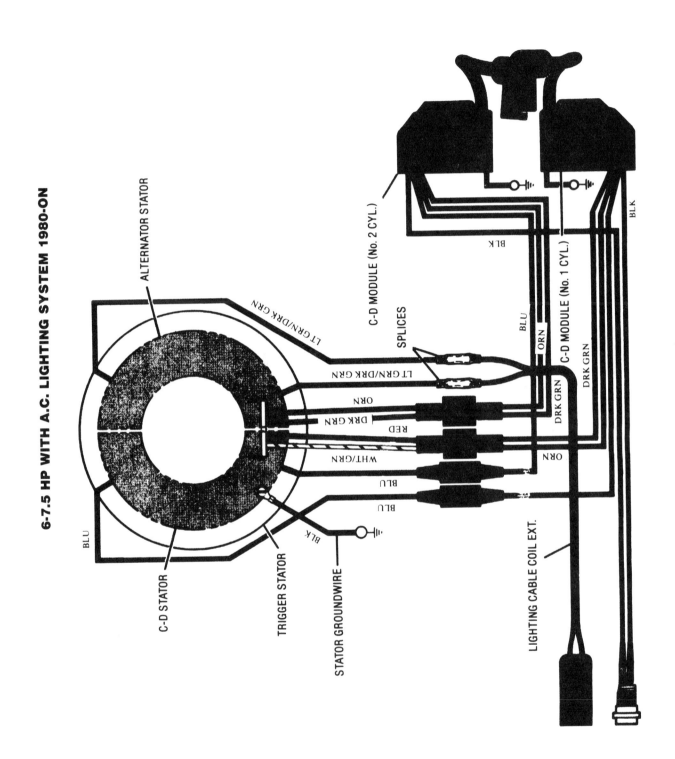

6-7.5 HP WITH A.C. LIGHTING SYSTEM 1980-ON

ALTERNATOR STATOR

LT GRN/DRK GRN

C-D MODULE (No. 2 CYL.)

SPLICES

C-D MODULE (No. 1 CYL.)

BLK

BLK

BLU

ORN

DRK GRN

DRK GRN

ORN

LT GRN/DRK GRN

ORN

DRK GRN

RED

WHT/GRN

BLU

BLU

BLU

BLK

C-D STATOR

TRIGGER STATOR

STATOR GROUNDWIRE

LIGHTING CABLE COIL EXT.

6-7.5 HP WITH D.C. BATTERY SYSTEM 1980-ON

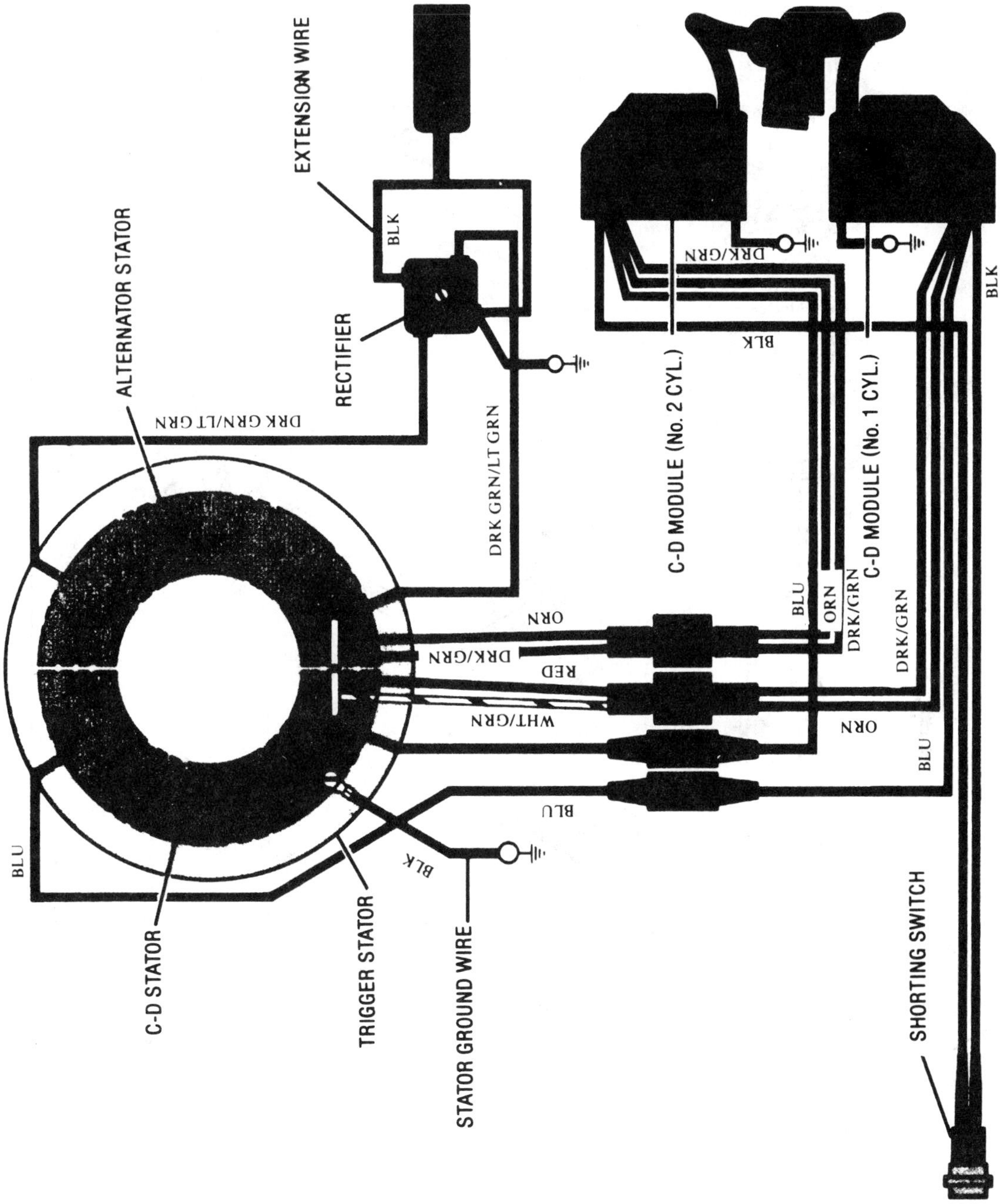

EXTENSION WIRE

ALTERNATOR STATOR

RECTIFIER

BLK

DRK GRN/LT GRN

DRK GRN/LT GRN

C-D MODULE (No. 2 CYL.)

C-D MODULE (No. 1 CYL.)

DRK/GRN

BLK

BLK

BLU

ORN

DRK/GRN

DRK/GRN

ORN

BLU

ORN

DRK/GRN

RED

WHT/GRN

BLU

BLU

C-D STATOR

TRIGGER STATOR

STATOR GROUND WIRE

BLK

SHORTING SWITCH

6-8 HP ALTERNATOR CD IGNITION

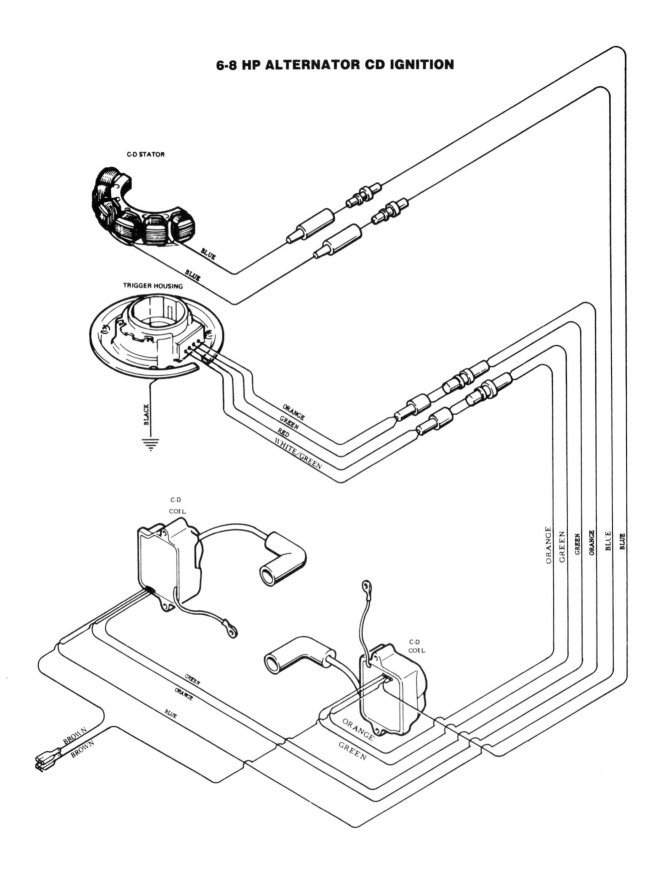

9.2 HP AUTOLECTRIC 1966-1967

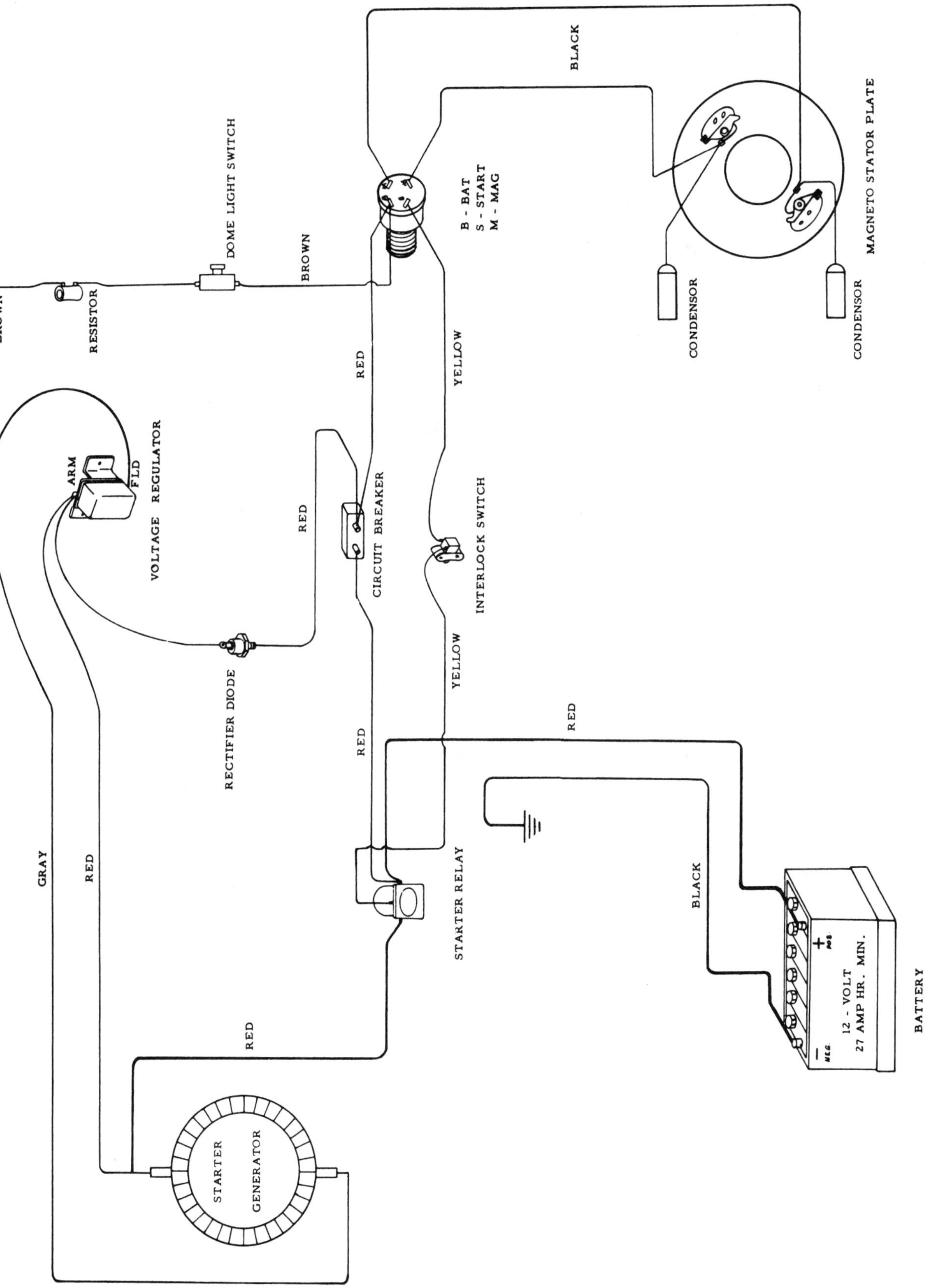

DOME LIGHT

BROWN

RESISTOR

DOME LIGHT SWITCH

BROWN

B - BAT
S - START
M - MAG

BLACK

MAGNETO STATOR PLATE

CONDENSOR

CONDENSOR

RED

YELLOW

ARM

FLD

VOLTAGE REGULATOR

RED

CIRCUIT BREAKER

INTERLOCK SWITCH

YELLOW

RECTIFIER DIODE

RED

RED

RED

STARTER RELAY

BLACK

GRAY

RED

RED

STARTER
GENERATOR

12 - VOLT
27 AMP HR. MIN.

+
MG

−
MEG

BATTERY

13

9.9 HP AUTOLECTRIC 1968

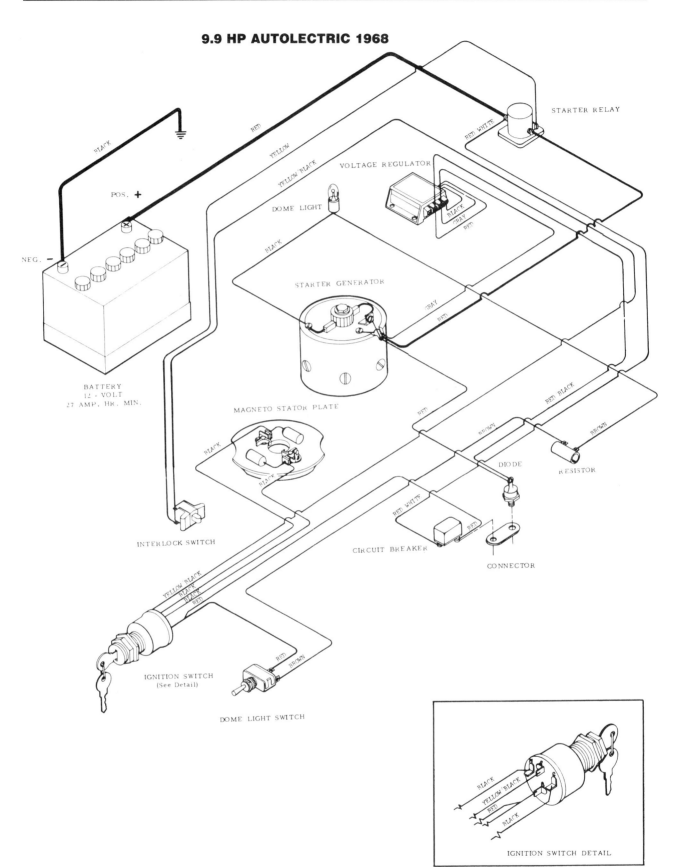

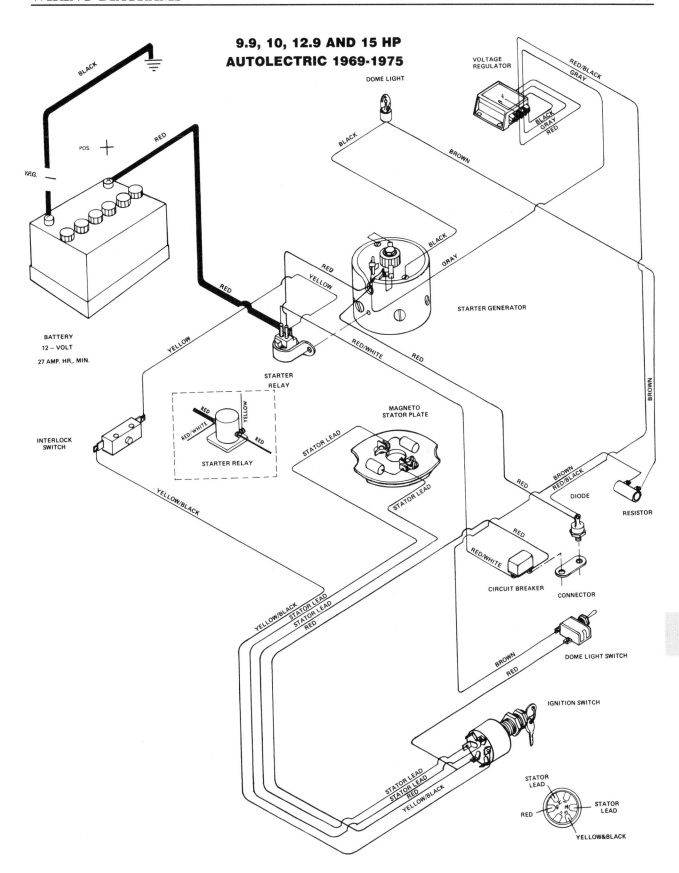

9.9, 10, 12.9 AND 15 HP AUTOLECTRIC 1969-1975

BLACK

DOME LIGHT

VOLTAGE REGULATOR

RED/BLACK

GRAY

BLACK
GRAY
RED

POS.

RED

BLACK

BROWN

X.F.G.

BLACK

GRAY

RED

YELLOW

STARTER GENERATOR

BATTERY
12 – VOLT
27 AMP. HR,. MIN.

YELLOW

RED/WHITE

RED

STARTER RELAY

BROWN

INTERLOCK SWITCH

RED

YELLOW

RED/WHITE

RED

MAGNETO STATOR PLATE

STARTER RELAY

STATOR LEAD

STATOR LEAD

RED

BROWN
RED/BLACK

DIODE

RESISTOR

YELLOW/BLACK

RED

RED/WHITE

CIRCUIT BREAKER

CONNECTOR

YELLOW/BLACK
STATOR LEAD
STATOR LEAD
RED

BROWN
RED

DOME LIGHT SWITCH

13

IGNITION SWITCH

STATOR LEAD
STATOR LEAD
RED
YELLOW/BLACK

STATOR LEAD

RED

STATOR LEAD

YELLOW&BLACK

9.9, 10, 12 AND 15 HP
MAGNAPOWER ALTERNATOR CD

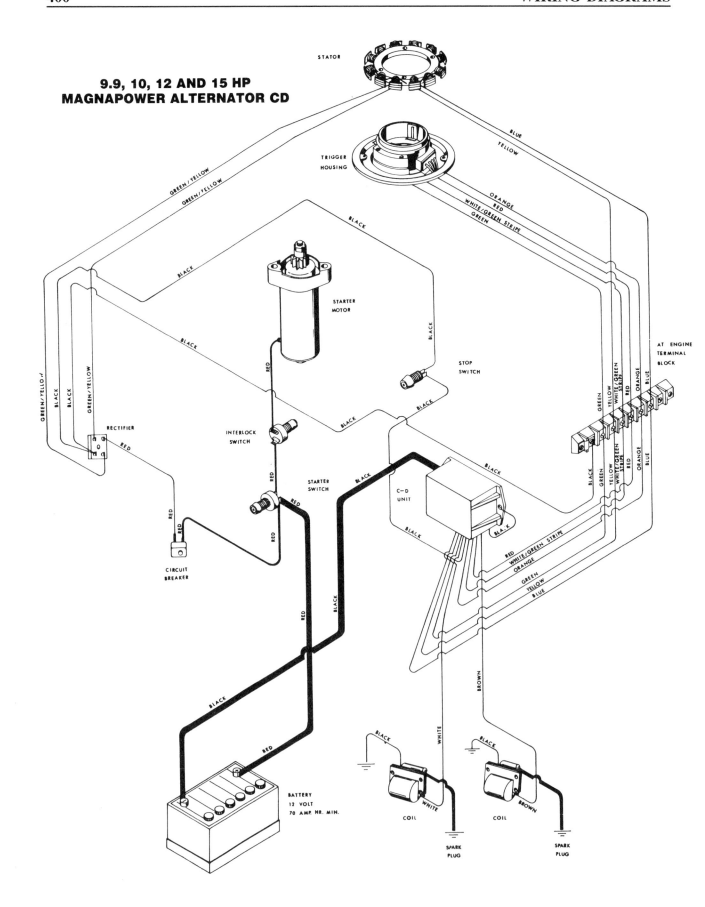

9.9-15 HP ALTERNATOR CD 1980-ON

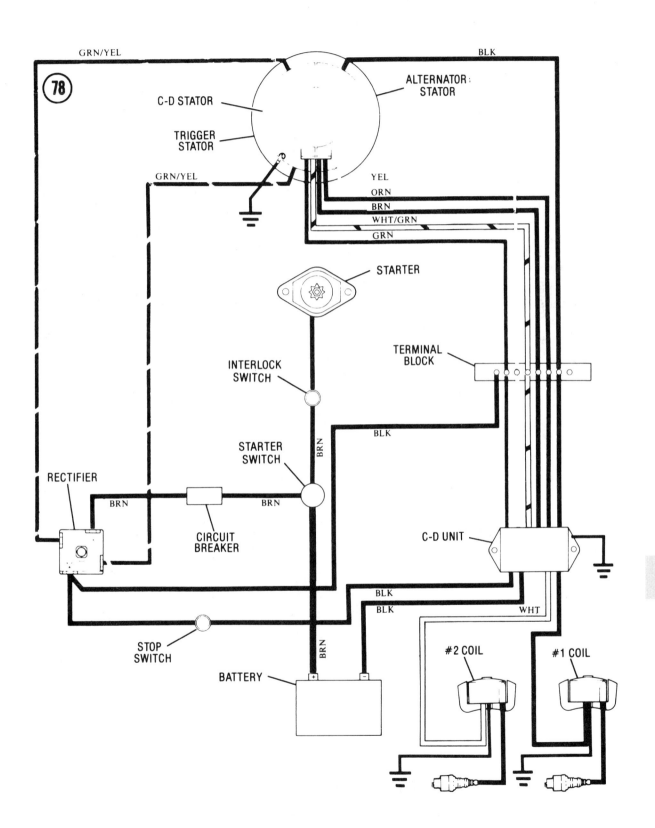

20 HP AUTOLECTRIC 1966-1967

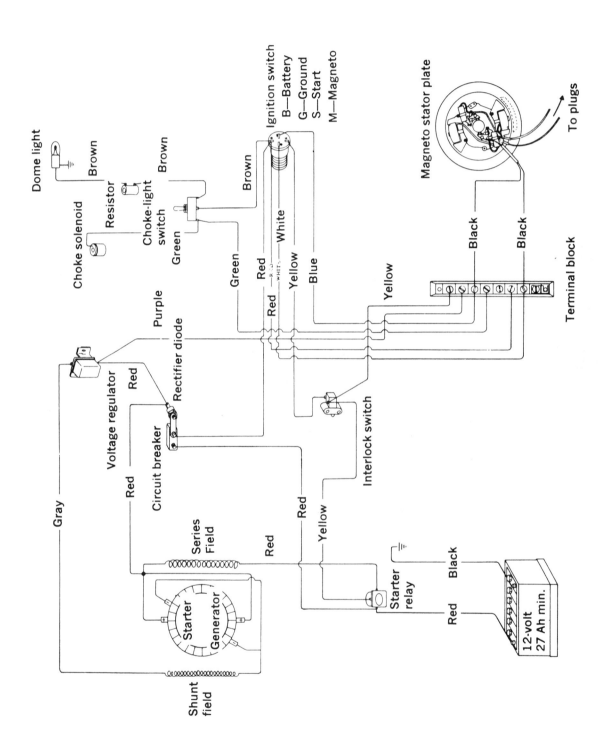

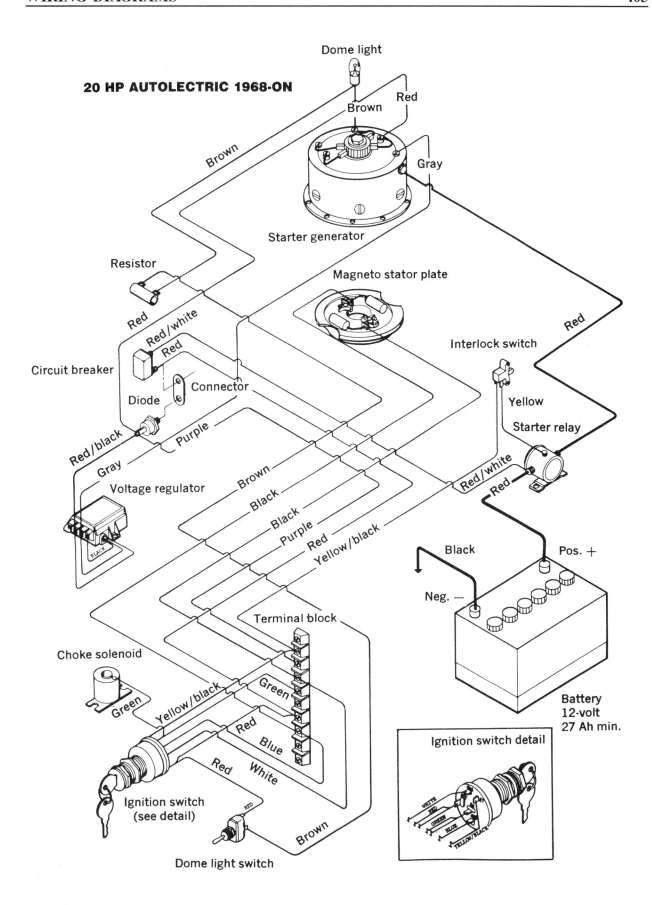

20 HP AUTOLECTRIC 1968-ON

Dome light

Red

Brown

Brown

Gray

Starter generator

Magneto stator plate

Resistor

Interlock switch

Red

Circuit breaker

Red

Red/white

Red

Connector

Diode

Yellow

Starter relay

Red/black

Purple

Gray

Red/white

Red

Voltage regulator

Brown

Black

Black

Purple

Red

Yellow/black

Black

Pos. +

Neg. —

Terminal block

Choke solenoid

Green

Yellow/black

Green

Green

Red

Blue

Red

White

Battery
12-volt
27 Ah min.

Ignition switch
(see detail)

Ignition switch detail

Brown

Dome light switch

13

20 HP AUTOLECTRIC WITH REMOTE IGNITION SWITCH 1968-ON

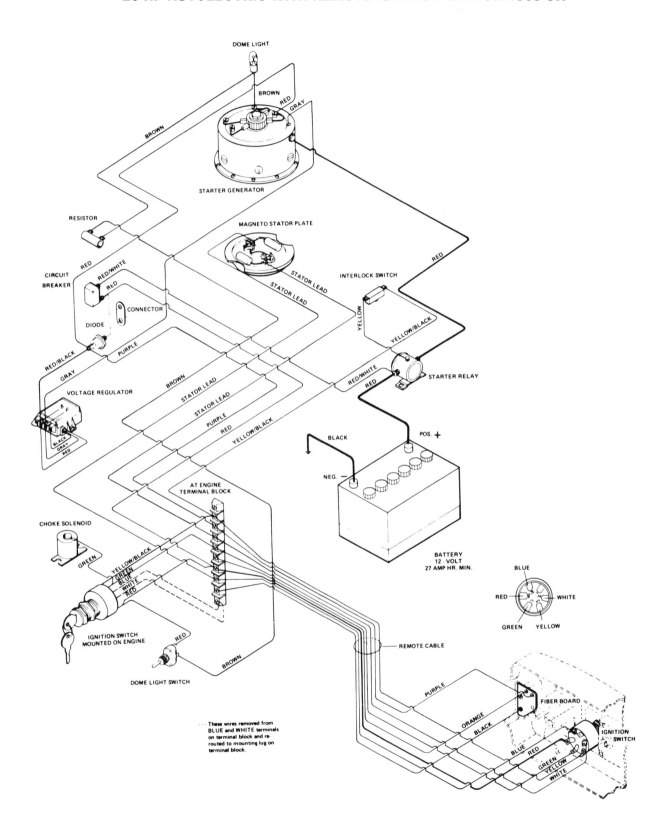

25-30 HP MAGNETO IGNITION

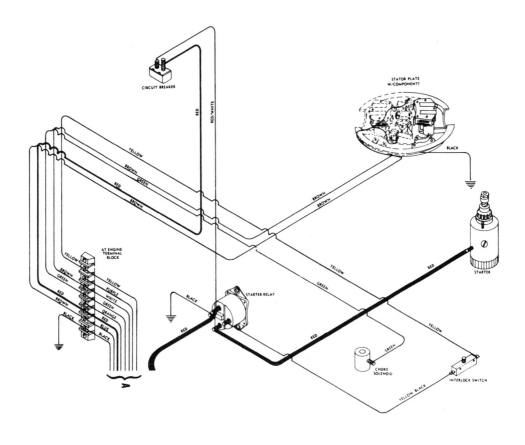

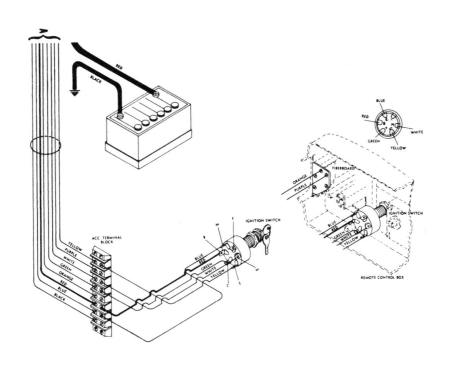

13

1980-ON 20 AND 30 HP, 1983-ON 35 HP (ELECTRIC START)

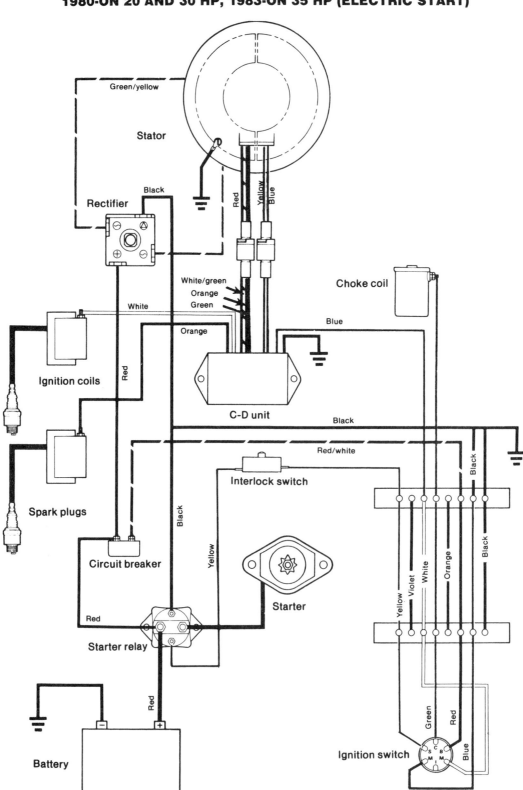

35, 45, 50 AND 55 HP MAGNETO IGNITION

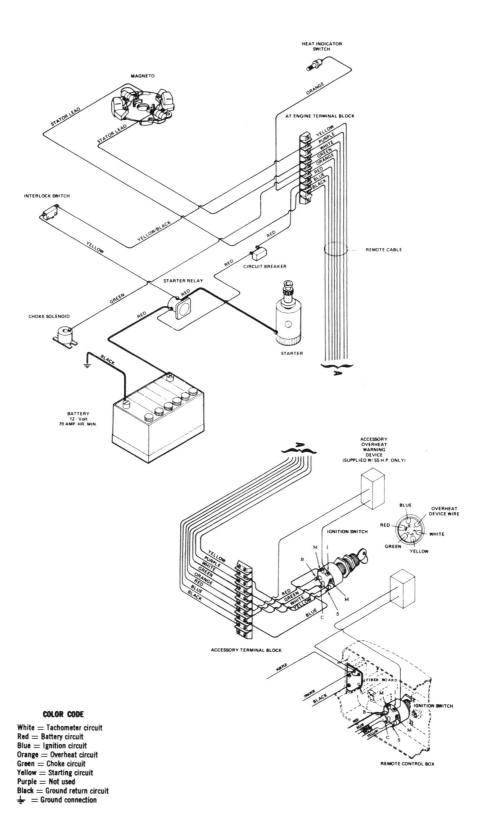

COLOR CODE

White = Tachometer circuit
Red = Battery circuit
Blue = Ignition circuit
Orange = Overheat circuit
Green = Choke circuit
Yellow = Starting circuit
Purple = Not used
Black = Ground return circuit
⏚ = Ground connection

13

35, 45, 50 AND 55 HP BATTERY IGNITION WITH ALTERNATOR

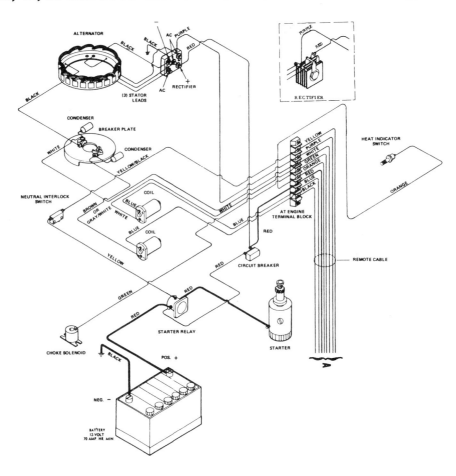

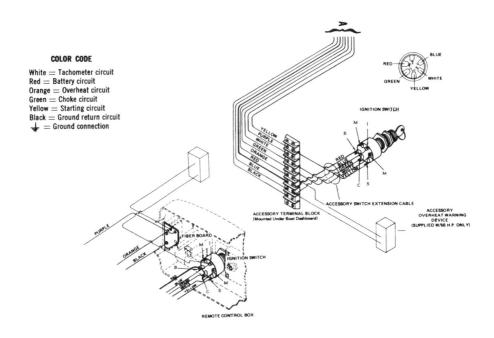

COLOR CODE

White = Tachometer circuit
Red = Battery circuit
Orange = Overheat circuit
Green = Choke circuit
Yellow = Starting circuit
Black = Ground return circuit
⏚ = Ground connection

1980-ON 35, 45 AND 50 HP

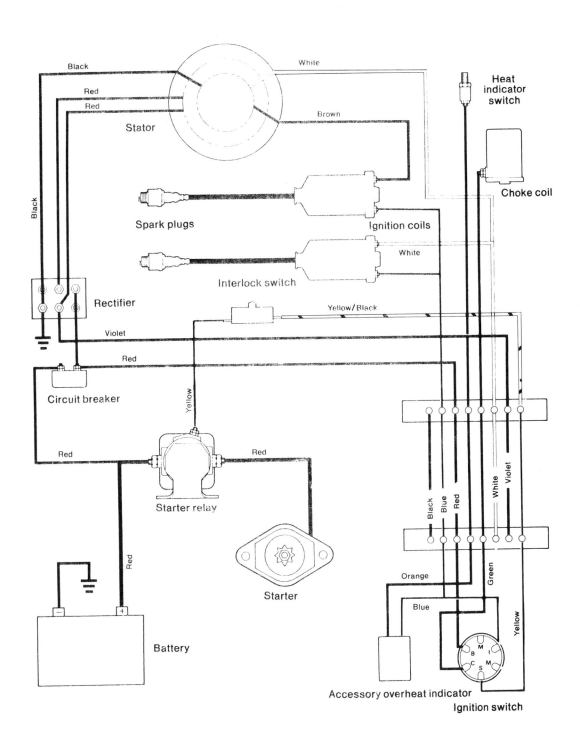

55-65 HP MAGNAPOWER CD IGNITION

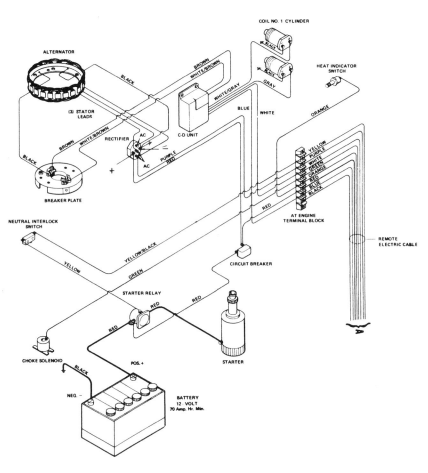

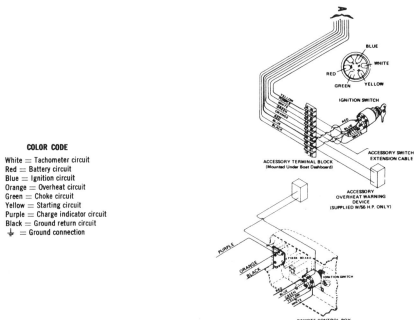

COLOR CODE

White = Tachometer circuit
Red = Battery circuit
Blue = Ignition circuit
Orange = Overheat circuit
Green = Choke circuit
Yellow = Starting circuit
Purple = Charge indicator circuit
Black = Ground return circuit
⏚ = Ground connection

1980-1981 55 HP

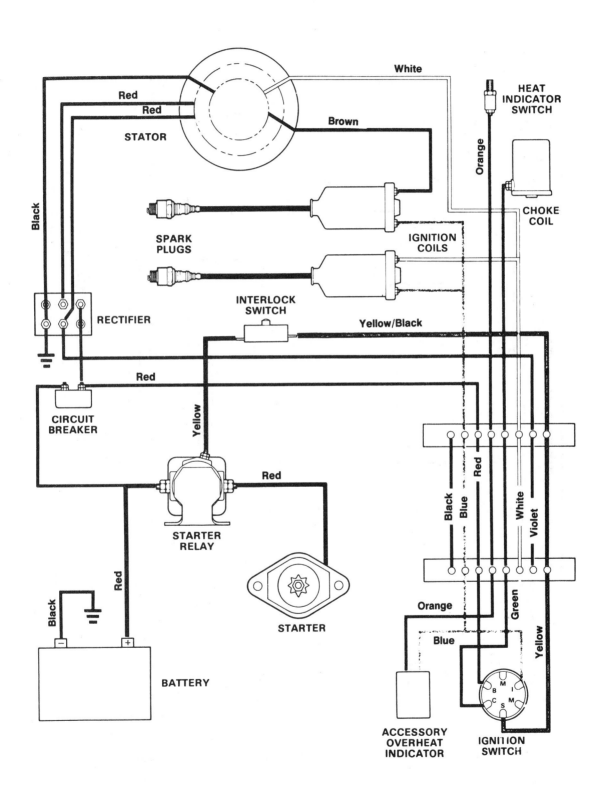

1982-ON 55 HP

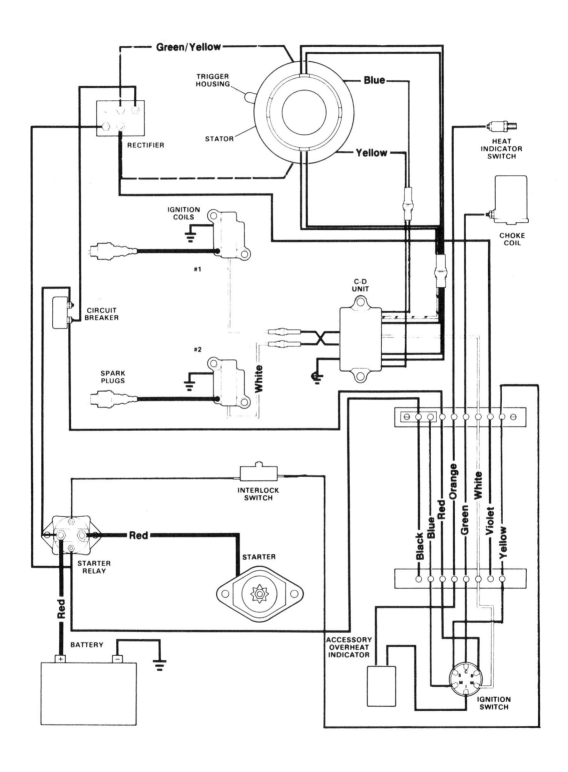

75-105 HP BATTERY IGNITION WITH ALTERNATOR

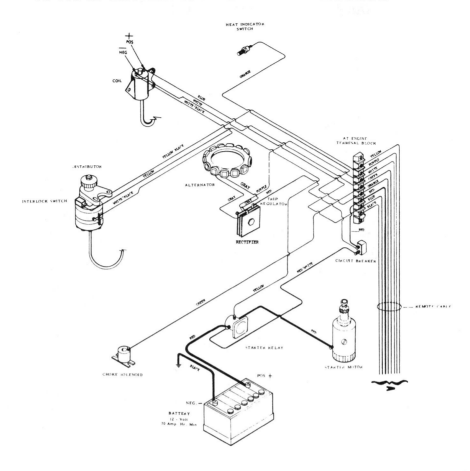

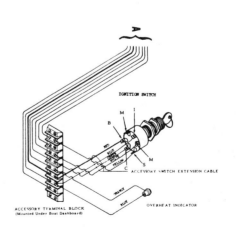

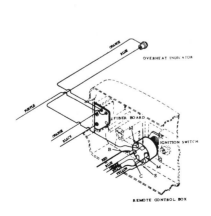

COLOR CODE

White = Tachometer circuit
Red = Battery circuit
Blue = Ignition circuit
Orange = Overheat circuit
Green = Choke circuit
Yellow = Starting circuit
Purple = Charge indicator circuit
Black = Ground return circuit
⏚ = Ground connection

13

70, 75, 85, 105, 120 AND 135 HP
MAGNAPOWER CD IGNITION (DELTA SYSTEM)

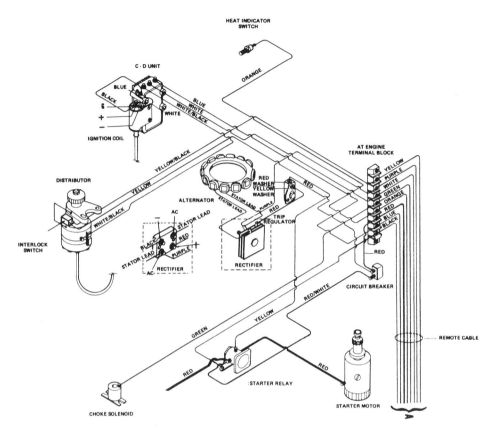

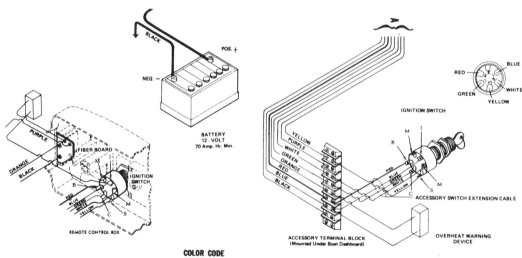

COLOR CODE

White = Tachometer circuit
Red = Battery circuit
Blue = Ignition circuit
Orange = Overheat circuit
Green = Choke circuit
Yellow = Starting circuit
Purple = Charge indicator circuit
Black = Ground return circuit
⏚ = Ground connection

70, 85, 105, 120 AND 135 HP
MAGNAPOWER CD IGNITION (MOTOROLA SYSTEM)

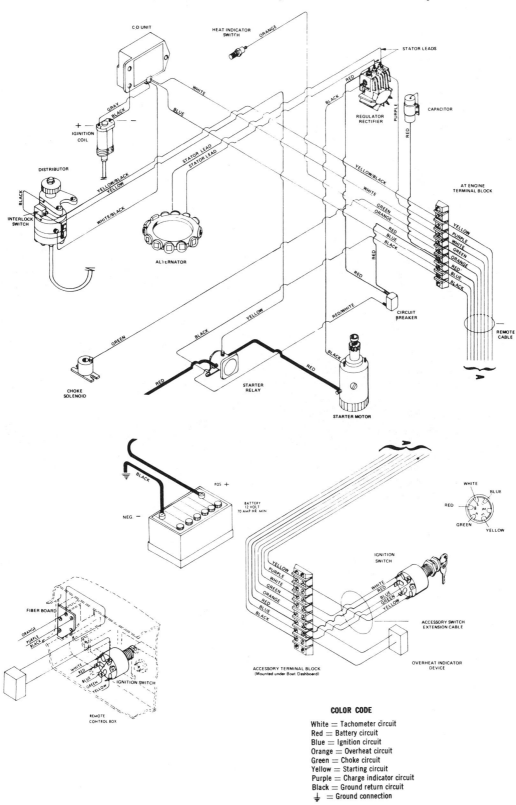

COLOR CODE

White = Tachometer circuit
Red = Battery circuit
Blue = Ignition circuit
Orange = Overheat circuit
Green = Choke circuit
Yellow = Starting circuit
Purple = Charge indicator circuit
Black = Ground return circuit
⏚ = Ground connection

13

70, 75, 85, 90, 105, 120, 130 AND 135 HP MAGNAPOWER CD IGNITION (MOTOROLA II SYSTEM)

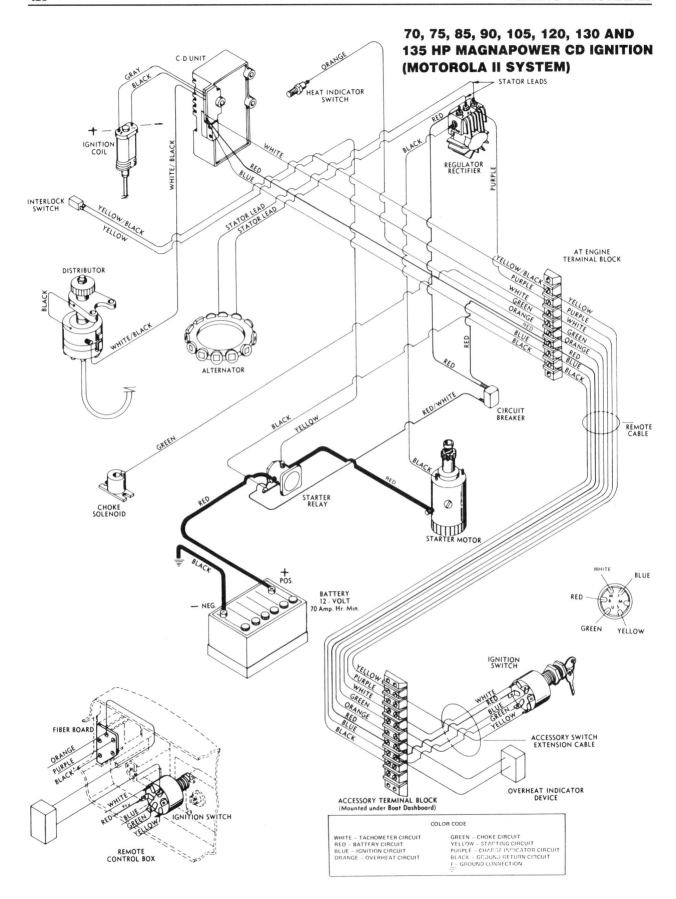

3- AND 4-CYLINDER MAGNAPOWER BREAKERLESS ELECTRONIC IGNITION

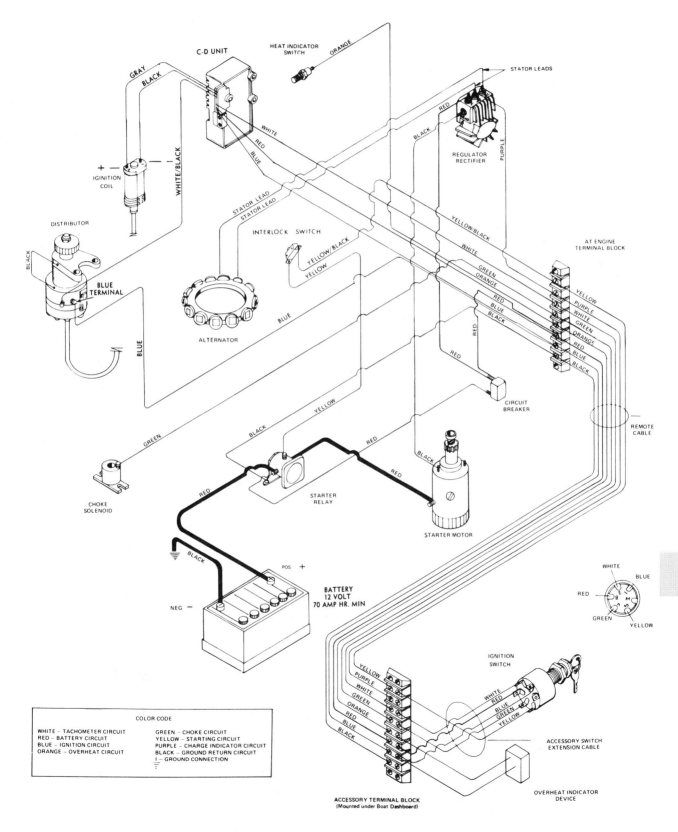

13

105, 120 AND 135 HP MAGNAPOWER II IGNITION

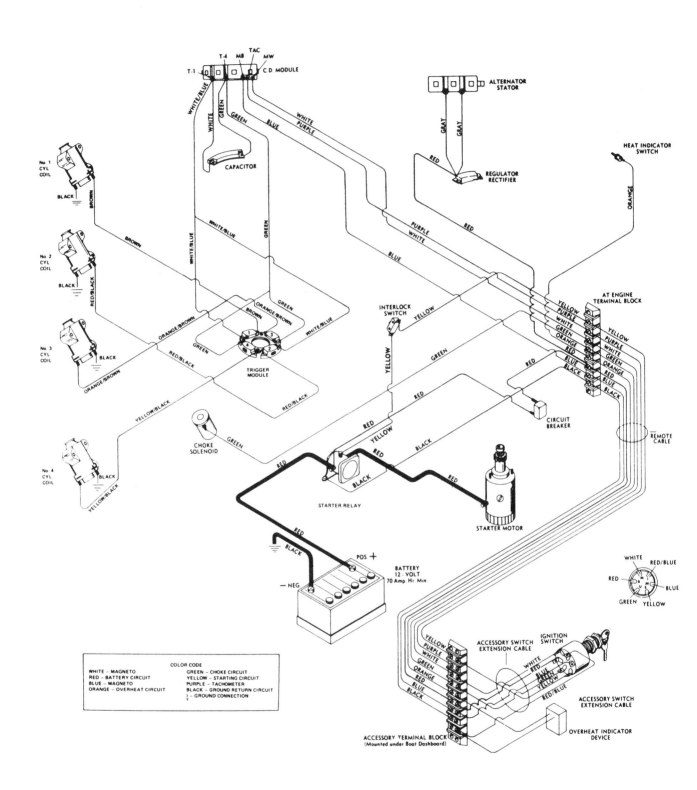

1980-ON 75 AND 85 HP (MOTOROLA)

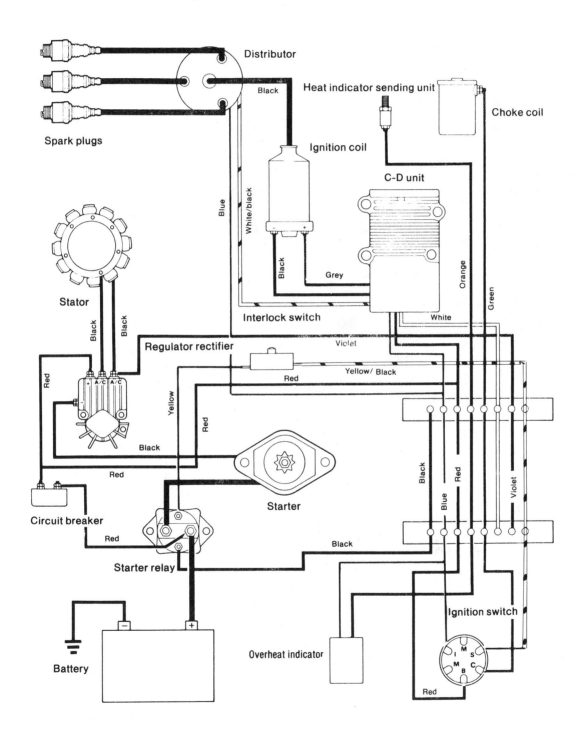

Distributor

Spark plugs

Black

Heat indicator sending unit

Choke coil

Ignition coil

C-D unit

Stator

Blue

White/black

Black

Grey

Orange

Green

White

Interlock switch

Violet

Regulator rectifier

Yellow/ Black

Red

Yellow

Red

Black

Red

Black

Blue

Red

Violet

Circuit breaker

Red

Starter

Black

Starter relay

Ignition switch

Battery

Overheat indicator

Red

I S
M B C

Red

13

1983-ON 75 AND 85 HP (PRESTOLITE)

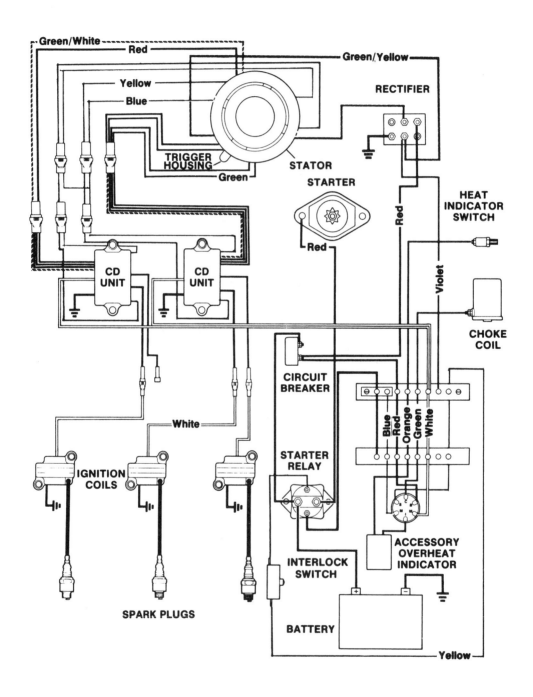

1980-ON 115 AND 140 HP (MOTOROLA)

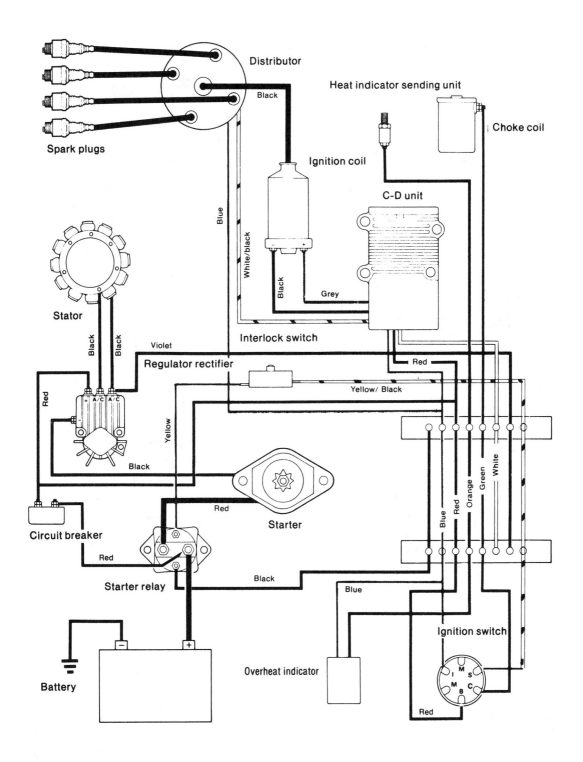

13

125 HP (PRESTOLITE)

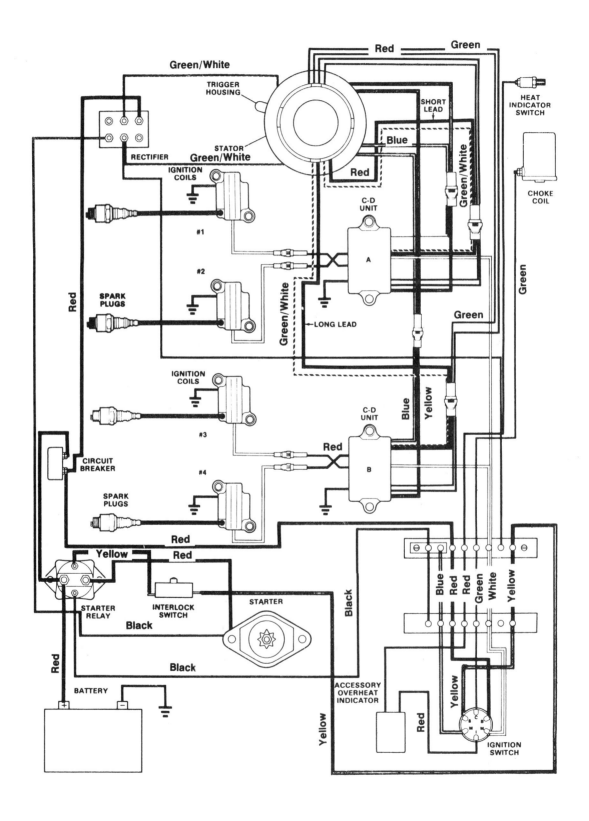

NOTES

MAINTENANCE LOG

Date	Maintenance performed	Engine hours